Electrical Engineering: Principles and Practice

Electrical Engineering: Principles and Practice

Edited by Jeremy Giamatti

CLANRYE
INTERNATIONAL
www.clanryeinternational.com

Clanrye International,
750 Third Avenue, 9ᵗʰ Floor,
New York, NY 10017, USA

ISBN: 978-1-63240-627-9

Cataloging-in-Publication Data

Electrical engineering : principles and practice / edited by Jeremy Giamatti.
    p. cm.
Includes bibliographical references and index.
ISBN 978-1-63240-627-9
1. Electrical engineering. 2. Electrical engineering--Equipment and supplies.
3. Electrical engineering--Materials. I. Giamatti, Jeremy.
TK145 .E44 2017
621.3--dc23

For information on all Clanrye International publications
visit our website at www.clanryeinternational.com

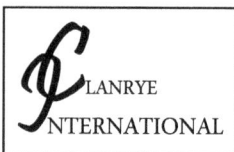

𝒞LANRYE
INTERNATIONAL

Printed in the United States of America.

Contents

Preface

The purpose of the book is to provide a glimpse into the dynamics and to present opinions and studies of some of the scientists engaged in the development of new ideas in the field from very different standpoints. This book will prove useful to students and researchers owing to its high content quality.

Electrical engineering involves the analysis and use of electromagnetism, electronics and electricity. The objective of this book is to explain the principles and practices utilized in this field. Developments in this field have been noticed since the later half of the 19th century, since the introduction of telegraph, telephone, etc. The aim of this book is to present researches that have transformed this discipline and aided its advancement. The ever growing need of advanced technology is the reason that has fueled the research in the field of electrical engineering in recent times. This book presents the fundamentals as well as modern approaches of electrical engineering. Coherent flow of topics, student-friendly language and extensive use of examples make this book a reliable resource guide.

At the end, I would like to appreciate all the efforts made by the authors in completing their chapters professionally. I express my deepest gratitude to all of them for contributing to this book by sharing their valuable works. A special thanks to my family and friends for their constant support in this journey.

Editor

A Common Signaling Mechanism for Coexistence between High-speed Power Line Communication

Hui-Myoung Oh[†], Sungsoo Choi*, Jimyung Kang* and Won-Tae Lee*

Abstract – In the field of high-speed power line communications, there are three major standards; ISO/IEC 12139-1, IEEE 1901-2010, and ITU-T G.99xx. However, they are not interoperable because of having their own physical and MAC layer specification. Actually, they cannot even avoid interfering with each other because they are using the same frequency band of 1 ~ 30 MHz. In this paper, a common signaling mechanism is suggested which uses multi-carrier partitioning and carrier sensing, so that all standards can be coexisted by sharing time resource. The suggested mechanism has been compared with the ISP protocol which is included in IEEE and ITU-T standard for coexistence.

Keywords: Power line communications, Standard, Coexistence, Signaling mechanism

1. Introduction

In the last 10 years, high-speed power line communications (PLC) technology has been developed with the development of information and communication technology. High-speed PLC uses the frequency band of about 1 ~ 50 MHz and has the data rate of more than 10 Mbps.

Currently in the field of high-speed PLC, there are three major standards; ISO/IEC 12139-1 [1], IEEE 1901-2010 [2], and ITU-T G.99xx [3]. However, they are not interoperable with each other because they have their own specification of physical and medium access control (MAC) layer. Actually, even one of them cannot even avoid interference from others because they are using the same frequency band of 1 ~ 30 MHz. In fact, three standards all could have a problem of coexistence.

Recognizing the issue of coexistence, IEEE had suggested the inter-system protocol (ISP) that is a coexistence mechanism between the heterogeneous systems [2]. However, the ISP currently covers only two standards of IEEE and ITU-T [3]. Consequently, it needs to be modified to include ISO/IEC standard. In this paper, regarding to modification of the ISP, a new common signaling mechanism is suggested which is based on multi-carrier partitioning and carrier sensing. The results of comparing between the mechanisms are also presented.

---

†   Corresponding Author: Korea Electrotechnology Research Institute, Korea (hmoh@keri.re.kr)

*   Korea Electrotechnology Research Institute, Korea ({sschoi, jmkang, wtlee}@keri.re.kr)

2. Overview of IEEE's ISP

The ISP allows for power line communications channel resources to be shared in the time domain (TDM) and the frequency domain (FDM) among systems that comply with IEEE and ITU-T standards. The basic mechanism of the ISP is based on the network status which is determined in accordance with the information on coexisting system presence, resource requirements, and resynchronization request. Each system uses the common signaling mechanism to exchange the network status [2].

Fig. 1 shows the ISP common signal format. One ISP signal consists of 16 orthogonal frequency division multiplexing (OFDM) symbols, and its first and last 2 symbols are multiplied by a window function. T_s, T_w, and T_{total} are 512, 1024, and 8192 samples, respectively. The duration of one sample is 0.01 usec (100 MHz sampling).

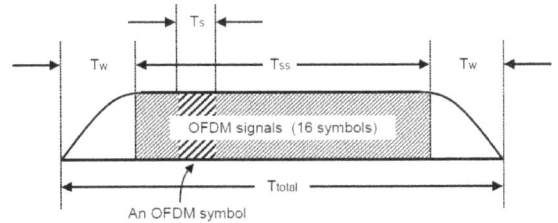

Fig. 1. The ISP common signal format [2].

The ISP assigns five different phases to the common signal using the start numbers as shown in Table 1. Each phase is selected as its usage. ACC means an access system, and IH-W, -O, and -G means an in-home system complying with IEEE Wavelet PHY, IEEE OFDM PHY, and ITU-T G.99xx, respectively. Actually, IEEE standard has two heterogeneous physical layer specifications.

Table 1. Phase vector start numbers [2]

	Start No.	Use
Phase 1	1	ACC
Phase 2	2	IH-W, resync
Phase 3	14	IH-O, resync
Phase 4	42	ACC FDM interference
Phase 5	58	IH-G, resync

The ISP also assigns time durations (called by ISP windows) synchronized with AC cycles to coexisting systems as shown in Fig. 2. However, the system complying with ISO/IEC standard cannot participate in coexistence because these ISP windows are fixed and there is no room for a new system.

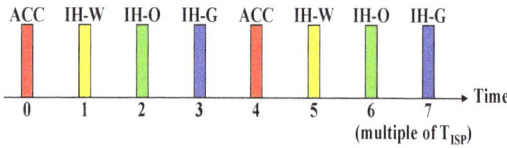

Fig. 2. Periodic allocation of ISP window [2].

Fig. 3 shows a general TDMA structure. The period between the starts of consecutive ISP windows is divided into three TDM units (TDMU), and each TDMU is subdivided into 8 TDM slots (TDMS). TDMSs are allocated by coexisting systems according to the network status as shown in Fig. 4.

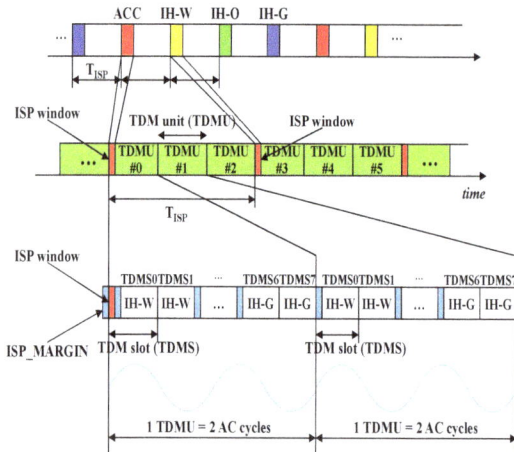

Fig. 3. General TDMA structure of the ISP [2].

Fig. 4. TDMS allocation [2].

Index	ISP Field ACC	IH-W	IH-O	IH-G	0	1	2	3	4	5	6	7
1	-	-	-	IH-G								
2	-	IH-W	-	-								
3	-	-	IH-O	-								
4	-	IH-W	-	IH-G								
5	-	IH-W	IH-O	-								
6	-	-	IH-O	IH-G								
7	-	IH-W	IH-O	IH-G								
8	ACC FB	-	-	-								
9	ACC PB	-	-	IH-G								
10	ACC PB	-	-	IH-G								
11	ACC PB	IH-W	-	-								
12	ACC FB	IH-W	-	-								
13	ACC PB	-	IH-O	-								
14	ACC FB	-	IH-O	-								
15	ACC PB	IH-W	-	IH-G								
16	ACC FB	IH-W	-	IH-G								
17	ACC PB	IH-W	IH-O	-								
18	ACC FB	IH-W	IH-O	-								
19	ACC PB	-	IH-O	IH-G								
20	ACC FB	-	IH-O	IH-G								
21	ACC PB	IH-W	IH-O	IH-G								
22	ACC FB	IH-W	IH-O	IH-G								

For the proper operation of the ISP, a coexisting system must monitor ISP windows of other systems and periodically transmit its own ISP signal. It needs to know IPS window timing through the synchronization and resynchronization mechanism between coexisting systems.

3. New Common Signaling Mechanism

Adopting the similar TDM mechanism with the ISP, we suggest more simple mechanism based on multi-carrier partitioning and carrier.

At first, we use a new common signal set which is partitioned as shown in Fig. 5 with Table 2. Of course, this mechanism includes ISO/IEC standard (IH-I). Even though this signal set is also based on the OFDM symbol specified by the same numerical parameters (such as frequency bandwidth, FFT size, sampling rate, signal length, and so on) as used in the ISP protocol, it can be differently used, that is, it allows each coexisting system to recognize the current network status through just sensing carriers of this signal set in the frequency domain.

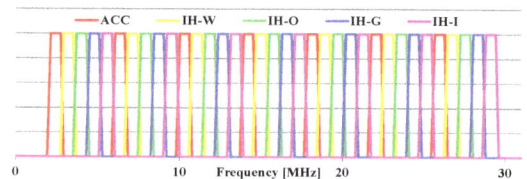

Fig. 5. Suggested common signal set.

Table 2. Multi-carrier partitioning

	Sub-carrier Number
ACC	$11+n+20m$: $(m=0, 1, \ldots, 6)$, $(n=0, 1, 2, 3)$
IH-W	$15+n+20m$: $(m=0, 1, \ldots, 6)$, $(n=0, 1, 2, 3)$
IH-O	$19+n+20m$: $(m=0, 1, \ldots, 6)$, $(n=0, 1, 2, 3)$
IH-G	$23+n+20m$: $(m=0, 1, \ldots, 6)$, $(n=0, 1, 2, 3)$
IH-I	$27+n+20m$: $(m=0, 1, \ldots, 6)$, $(n=0, 1, 2, 3)$

Another important fact is that each coexisting system can simultaneously transmit its own signal in an ISP window because all coexisting systems use the different sub-carriers which are determined and fixed in advance. The sub-carriers do not interfere in each other because they are originally orthogonal. Consequently, they can quickly update the network status with a period of one T_{ISP}. It is 4 times (if the ISP were covering ISO/IEC standard, then it was 5 times) faster than the original ISP. Fig. 6 shows an example of the network status monitoring.

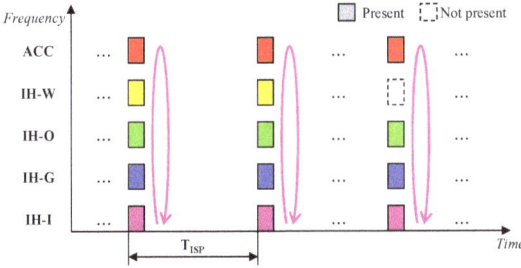

Fig. 6. Example of carrier sensing by IH-I.

Fig. 7 shows the modified TDMS allocation including three standards. As a result, the modified ISP can allow up to 1 access and 5 in-home systems to be coexisted. When resource allocation mechanism is considered, fairness is very important. Coexisting systems want to fairly share the given resource according to the network status. Another one of the important factors is the cost of complexity. This is related to system reliability. The modified TDMS allocation table has been designed so that coexisting system can occupy more than one TDMS slot(marked with W, O, G, I, and A) and maintain the occupancy as possible.

An example of coexistence situation according to the modified ISP is shown in Fig. 8. Assuming the proper operation, a system which wants to participate in coexistence should first monitor the current network status, and then it may transmit its own signal in the next ISP window, and it can finally be allowed to allocate some time slots. The existing system should continuously monitor the network status and accordingly change the TDMS allocation index. This process may take the multiple of T_{ISP}.

Fig. 7. Modified TDMS allocation.

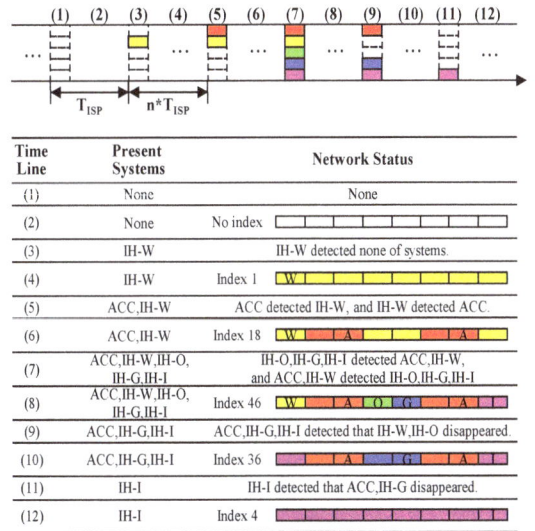

Index	ACC	IH-W	IH-O	IH-G	IH-I	0	1	2	3	4	5	6	7
1	-	IH-W	-	-	-	W							
2	-	-	IH-O	-	-			O					
3	-	-	-	IH-G	-				G				
4	-	-	-	-	IH-I								I
5	-	IH-W	IH-O	-	-	W		O					
6	-	IH-W	-	IH-G	-	W			G				
7	-	IH-W	-	-	IH-I	W							I
8	-	-	IH-O	IH-G	-			O	G				
9	-	-	IH-O	-	IH-I			O					I
10	-	-	-	IH-G	IH-I				G				I
11	-	IH-W	IH-O	IH-G	-	W		O	G				
12	-	IH-W	IH-O	-	IH-I	W		O					I
13	-	IH-W	-	IH-G	IH-I	W			G				I
14	-	-	IH-O	IH-G	IH-I			O	G				I
15	-	IH-W	IH-O	IH-G	IH-I	W		O	G				I
16	ACC FB	-	-	-	-			A				A	
17	ACC PB	IH-W	-	-	-	W		A				A	
18	ACC FB	IH-W	-	-	-	W		A				A	
19	ACC PB	-	IH-O	-	-			A	O			A	
20	ACC FB	-	IH-O	-	-			A	O			A	
21	ACC PB	-	-	IH-G	-			A		G		A	
22	ACC FB	-	-	IH-G	-			A		G		A	
23	ACC PB	-	-	-	IH-I			A				A	I
24	ACC FB	-	-	-	IH-I			A				A	I
25	ACC PB	IH-W	IH-O	-	-	W		A	O			A	
26	ACC FB	IH-W	IH-O	-	-	W		A	O			A	
27	ACC PB	IH-W	-	IH-G	-	W		A		G		A	
28	ACC FB	IH-W	-	IH-G	-	W		A		G		A	
29	ACC PB	IH-W	-	-	IH-I	W		A				A	I
30	ACC FB	IH-W	-	-	IH-I	W		A				A	I
31	ACC PB	-	IH-O	IH-G	-			A	O	G		A	
32	ACC FB	-	IH-O	IH-G	-			A	O	G		A	
33	ACC PB	-	IH-O	-	IH-I			A	O			A	I
34	ACC FB	-	IH-O	-	IH-I			A	O			A	I
35	ACC PB	-	-	IH-G	IH-I			A		G		A	I
36	ACC FB	-	-	IH-G	IH-I			A		G		A	I
37	ACC PB	IH-W	IH-O	IH-G	-	W		A	O	G		A	
38	ACC FB	IH-W	IH-O	IH-G	-	W		A	O	G		A	
39	ACC PB	IH-W	IH-O	-	IH-I	W		A	O			A	I
40	ACC FB	IH-W	IH-O	-	IH-I	W		A	O			A	I
41	ACC PB	IH-W	-	IH-G	IH-I	W		A		G		A	I
42	ACC FB	IH-W	-	IH-G	IH-I	W		A		G		A	I
43	ACC PB	-	IH-O	IH-G	IH-I			A	O	G		A	I
44	ACC FB	-	IH-O	IH-G	IH-I			A	O	G		A	I
45	ACC PB	IH-W	IH-O	IH-G	IH-I	W		A	O	G		A	I
46	ACC FB	IH-W	IH-O	IH-G	IH-I	W		A	O	G		A	I

Time Line	Present Systems	Network Status
(1)	None	None
(2)	None	No index
(3)	IH-W	IH-W detected none of systems.
(4)	IH-W	Index 1
(5)	ACC,IH-W	ACC detected IH-W, and IH-W detected ACC.
(6)	ACC,IH-W	Index 18
(7)	ACC,IH-W,IH-O, IH-G,IH-I	IH-O,IH-G,IH-I detected ACC,IH-W, and ACC,IH-W detected IH-O,IH-G,IH-I
(8)	ACC,IH-W,IH-O, IH-G,IH-I	Index 46
(9)	ACC,IH-G,IH-I	ACC,IH-G,IH-I detected that IH-W,IH-O disappeared.
(10)	ACC,IH-G,IH-I	Index 36
(11)	IH-I	IH-I detected that ACC,IH-G disappeared.
(12)	IH-I	Index 4

Fig. 8. Example of Coexistence.

The results of the comparison of the ISP and the modified ISP can be summarized as Table 3. Even though the fairness of the modified ISP is not good on the case of index 45 in Fig. 7, it is probably not a problem because the worst case that 5 systems (ACC, IH-W, IH-O, IH-G, and IH-I) are applied at the same time would be rare. The unfairness could be improved in such a way increasing the number of TDMS slots and adjusting the allocation pattern.

Table 3. Summary of results of comparison

	ISP	Modified ISP
Covering standards	IEEE, ITU-T	IEEE, ITU-T, ISO/IEC
Common signal multiplexing	Based on phase vectors	Based on subcarrier partitioning
System recognition method	Based on phase detection	Based on carrier sensing
Minimum duration of network status updating	$4*T_{ISP}$	T_{ISP}
TDMS allocation fairness	Good	Not bad
Resource occupancy	Good	Good

4. Conclusions

In this paper, after reviewing of IEEE's ISP, the new common signaling mechanism has been suggested, and the modified ISP with which has also been suggested. The modified ISP could cover all of three high-speed PLC standards and be simple and faster than the ISP. However, the problem of fairness and complexity is still a subject to be further studied.

References

[1] ISO/IEC 12139-1, Information Technology – Tele-communications and Information Exchange Between Systems – Powerline Communication (PLC) - High Speed PLC Medium Access Control (MAC) and Physical Layer (PHY) - Part 1 : General Requirements, July 2009.

[2] IEEE Std 1901-2010, IEEE Standard for Broadband over Power Line Networks : Medium Access Control and Physical Layer Specifications, Dec. 2010.

[3] ITU-T G.9960, G.9961, and G.9972, Series G : Transmission Systems and Media, Digital Systems and Networks, Access Networks – In Premises Networks, Unified High-Speed Wire-Line Based Home Networking Transceivers – System Architecture and Physical Layer Specification, Data Link Layer Specification, and Coexistence Mechanism for Wireline Home Networking Transceivers, June 2010.

Evaluation of Relationship between Demand Side Management and Future Generation Mix by Energy System Analysis

Taisuke Masuta[†], Akinobu Murata* and Eiichi Endo**

Abstract – Demand side management is an important part of the smart grid for alleviating the uncertainty in renewable energy generation including such as wind power generation and photovoltaic generation. The nuclear accident caused by the 2011 Tohoku Earthquake forced a re-evaluation of the future power generation system in Japan. As a result, demand side management has become critical for future Japanese power systems with its large integration of renewable energy sources, and the evaluation of the impacts of this management is needed for planning purposes. This paper proposes a new energy system model that expands the function of MARKAL (Market Allocation) and considers the characteristics of the total power system in more detail. The relationship between demand side management and the future power generation mix was evaluated using energy system analysis as part of the proposed model. The electric water heaters and heat pump water heaters are conceived as the controllable loads in the demand side management.

Keywords: Demand side management, Energy system analysis, Generation planning, MARKAL, Power system, Renewable energy sources, Smart grid.

1. Introduction

The nuclear accident caused by the 2011 Tohoku Earthquake had a major impact on energy policy in Japan. The accident prompted a re-evaluation of the country's future planning for power system generation. As a result, energy from nuclear power plants will likely be phased out and replaced with renewable energy sources, such as wind and photovoltaic power. Research has traditionally predicted the future power generation mix by selecting the optimal mix that minimizes the sum of the capital and operational costs, with a focus on the electric energy sector alone (which comprises only about 25% of all energy consumption in Japan). However, it is now essential that researchers consider all of the energy sectors as well as the balance between these sectors [1].

Several recent studies have considered wind and photovoltaic power in generation planning [1]. However, only the controllable plants (e.g., nuclear, thermal, and hydro power plants) have generally been considered generation plants in planning efforts [2]. Uncontrollable generation sources for, say, wind and photovoltaic power

†   Corresponding Author: National Institute of Advanced Industrial Science and Technology, Tsukuba, Japan (taisuke.masuta@aist.go.jp)

*   National Institute of Advanced Industrial Science and Technology, Tsukuba, Japan (aki.murata@aist.go.jp)

** National Institute of Advanced Industrial Science and Technology, Tsukuba, Japan (endo.e@aist.go.jp)

could have negative impacts on existing power systems (e.g., by causing an imbalance between supply and demand, voltage fluctuations, and frequency fluctuations.) Stabilization technologies, such as power output forecasts for renewable energy sources and energy storage in battery systems must be applied to the power system controls or operations. The impacts of such technologies on generation planning have not been fully evaluated in previous studies, where objectives were mainly to show the installation potential of renewable energy sources.

Demand side management is one approach for alleviating the uncertainty in renewable energy generation and has gained attention as a key technology in the smart grid. Although demand side control and operation methods have been studied previously [3], [4], the impacts of demand side management on power generation planning have not been evaluated as extensively.

This paper proposes a new energy system model by expanding the function of MARKAL (Market Allocation) [5] and by considering power system control technologies including the demand side management in more detail. This model allows for the consideration of all energy sectors in power system generation planning in Japan. This paper also focuses on the controllability of electric water heating appliances (e.g., electric water heaters and heat pump water heaters) in demand side management. The relationship between demand side management and the future power generation mix is evaluated using the energy system analysis based on the proposed model.

2. Energy System Model

2.1 MARKAL

MARKAL was developed in an international cooperative project lead by the International Energy Agency (IEA) [5]. It is an optimization model that considers all energy sectors and simulates the energy system as a network flow from the supply side to the demand side. MARKAL can describe the energy technologies in detail via their technical character-istics and can determine the multi-period composition of these technologies, including the final energy consumption and the electricity generation mix.

Power system control by both power plants and by the demand side must be considered in energy system analysis in order to consider the time change for the renewable energy output, the electricity demand, and the demand side management. This paper proposes a new energy system model which can analyze the relationship between demand side management and the future generation mix.

2.2 Indices and Parameters

The model contains parameters for analysis period index (T; 13 periods from 1990 to 2050 in 5 year intervals), climatic season (Z), and time period (Y). The parameter Z was comprised of a set of 9 patterns: weekdays, weekends, and special days (a week of light-loads) in the winter; weekdays, weekends, and special days (three days of heavy-loads) in the summer; and weekdays, weekends, and special days (a week of light-loads) in the spring/fall. The parameter Y included an hourly time series (i.e., 24 time periods in a day). The functions $QHR(Z, Y)$ and $FR(Z, Y, DM)$ also represented a set of time-series data. Here, $QHR(Z, Y)$ represented the proportion of time Y in season Z in a year, while $FR(Z, Y, DM)$ represented the proportion of the demand DM used during Y in Z in a year. $QHR(Z, Y)$ corresponded to the proportion of weekdays, weekends, and special days in each season. The parameterization of $FR(Z, Y, DM)$ is described in Section 3.

2.3 Modeling Power System Control

The constraint equations related to the power system considered in MARKAL included the maximum generation output, the balance between supply and demand, the capacity usage, the reserve capacity on peak load, and the base load operation. MARKAL was intrinsically designed as an optimization model for multiple time periods and seasons. Therefore, the constraint equations for maximum generation output, the balance between supply and demand, and the capacity usage were used without changing in the proposed model.

The reserve capacity constraint equation was determined in the proposed model such that the reserve capacity during the peak-load time on a weekday in winter or on a special day in summer was greater than 5% of the electricity demand.

The load-following capability of the power plants was approximated in the base-load operation constraint equation in MARKAL by suppressing the output of the base power plants to a certain level. The proposed model approximated the load-following capability in more detail by considering the economic-load dispatching control (EDC) and the load frequency control (LFC).

The constraint equation for the rate of change of the power plant output is given by equation (1):

$$\begin{cases} ACTE(T,Z,Y,ELA) - ACTE(T,Z,Y-1,ELA) \\ \quad \leq RCUP(ELA) \cdot QHR(Z,Y) \cdot CAP(T,ELA) \\ ACTE(T,Z,Y-1,ELA) - ACTE(T,Z,Y,ELA) \\ \quad \leq RCDOWN(ELA) \cdot QHR(Z,Y) \cdot CAP(T,ELA) \end{cases} \quad (1)$$

where $ACTE(T, Z, Y, ELA)$ is the electricity generation of power plant ELA during time Y in season Z in period T; $RCUP(ELA)$ is the upper limit of the increasing rate of ELA; $RCDOWN(ELA)$ is the upper limit of the decreasing rate; and $CAP(T, ELA)$ is the installed capacity of ELA in T.

The constraint equations for the upper limits of the LFC capacity are given by equations (2) and (3):

$$LFC(T,Z,Y,ELA) \leq \frac{CLFC(ELA)}{RPMIN(ELA)} \cdot ACTE(T,Z,Y,ELA) \quad (2)$$

$$\begin{aligned} LFC(T,Z,Y,ELA) \\ \leq CLFC(ELA) \cdot QHR(Z,Y) \cdot AF(ELA) \cdot CAP(T,ELA) \end{aligned} \quad (3)$$

where $LFC(T, Z, Y, ELA)$ is the LFC capacity of ELA during Y in Z in T; $CLFC(ELA)$ is the proportion of the LFC capacity of ELA to the rated capacity; $RPMIN(ELA)$ is the proportion of the minimum output of ELA to the rated capacity; and $AF(ELA)$ is the availability factor of ELA.

Equation (4) indicates that the total LFC capacity must be more than a certain proportion of the total electricity demand:

$$\sum_{ELA} LFC(T,Z,Y,ELA) \geq RD \cdot EDM(T,Z,Y) \qquad (4)$$

where *EDM (T, Z, Y)* is the total electricity demand during *Y* in *Z* in *T*; and *RD* is the proportion of the LFC capacity to the total electricity demand (set to 2% in this paper). The LFC capacity for the output fluctuation of renewable energy sources was not considered in this paper.

2.4 Electric Water Heating Appliances

Electric water heaters and heat pump water heaters were considered as electric water heating appliances in MARKAL. The proposed model assumed that these technologies have two operation patterns for heating water: a nighttime load pattern (operation in the nighttime), and a controllable load pattern (operation at any time). The heating period and the amount in the controllable load pattern were variables in the optimization problem that were determined by the problem's solution. The proposed model did not consider a pattern where operation was according to hot water demand on a moment-to-moment basis.

The total amount of hot water that would be demanded in a day was heated during the night in the nighttime load pattern. The electric power consumption and thermal output for heating were held constant for all time periods as shown in equation (5).

$$PNL(T,Z,Y',NL)$$
$$= \frac{QHR(Z,Y')}{\sum\limits_{Y'}^{Nighttime} QHR(Z,Y')} \cdot \sum_{Y}^{Allday} FR(Z,Y,'HW') \cdot \frac{CF(T,NL) \cdot CAP(T,NL)}{EFF(T,NL)} \qquad (5)$$

where, *PNL (T, Z, Y', NL)* is the appliance's electric power consumption in *NL* (i.e., the nighttime load pattern) during *Y'* (i.e., *Y'* represents *Y* in the nighttime.) in *Z* in *T*; *CF (T, NL)* is the capacity factor for *NL* in *T*; *EFF (T, NL)* is the efficiency of *NL* in *T*; and *HW* represents the hot water demand, which is a factor of the demand *DM*.

The heating period was changeable in the controllable load pattern as long as the total amount of heated water in a day was equal to the total hot water demand in a day. However, the maximum hot water heating amount in one hour was limited to 20% of the total amount in a day in this paper in order to prevent heating more water at a time than the rated output. Equation (6) is the constraint equation for the balance between the amount of heated water and the hot water demand:

$$\sum_{Y}^{Allday} PCL(T,Z,Y,CL) \qquad (6)$$
$$= \sum_{Y}^{Allday} FR(Z,Y,'HW') \cdot \frac{CF(T,CL) \cdot CAP(T,CL)}{EFF(T,CL)}$$

where, *PCL (T, Z, Y, CL)* is the electric power consumption for appliance *CL* (i.e., controllable load pattern) during *Y* in *Z* in *T*.

In addition, this paper assumed that the electric water heating appliances would be able to operate in the controllable load pattern after 2020 but would operate during nighttime even in the controllable load pattern before 2020.

3. Analysis Assumption

3.1 Objective Function

Japanese energy system was analyzed in this paper. Generally, MARKAL can determine the total system cost, the amount of CO_2 emissions, or a combination of cost and emissions in an objective function [5]. This option was also included in the proposed model, with the objective function given by equation (7):

$$ESLOPE = PRICE$$
$$+ \alpha \cdot \frac{1+r}{r} \cdot \left[1 - (1+r)^{-N}\right] \cdot \left[\sum_{T}^{All\ Periods} ENVCO2(T) \cdot (1+r)^{-N(T-1)}\right] \qquad (7)$$

where, ESLOPE [JPY] is the composite function, which represents the trade-off between PRICE and ENVCO2 (*T*); PRICE [JPY] is the total system cost; and ENVCO2 (*T*) [t-CO2] is the amount of CO_2 emission in *T*. The second term on the right-hand side of the equation indicates the penalty for CO_2 emissions (i.e., the cost of the emissions). The PRICE and the emissions penalty were discounted in the objective function. Furthermore, α [JPY/t-CO_2] is the weight coefficient (a constant); N is the number of years in each period (i.e., 5 years = 1 period); and r is the discount rate (5%; equivalent to the PRICE discount rate).

3.2 Data used in the Energy System Analysis

As in the MARKAL analysis, the data pertaining to demand (e.g., transportation, cooling and heating, hot water demand, etc.) were based on the effective energy demands supplied by the energy devices (e.g., passenger car, rail, gas heater, oil heater, gas water heater, electric water heater, etc.). These data were determined by solving the optimization problem in the proposed model. Table 1 shows the energy demand scenario obtained from [6]. The parameter *FR (Z, Y, DM)* (i.e., the proportion of demand *DM* in time *Y* in season *Z* in a year) was determined over 24 time periods in 9 patterns (i.e., the 3 daily patterns in 3 seasons) according to [7]. The duration of daytime and nighttime values are defined in Table 2. It was assumed that pumped storage power plants and battery energy storage systems generated/discharged during the day and pumped/charged at night.

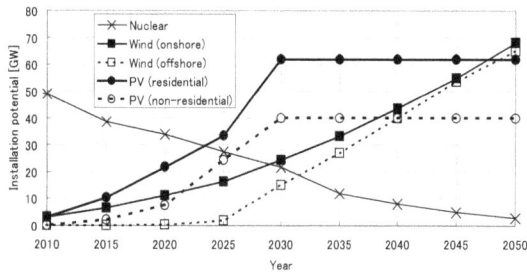

Fig. 1. Installation potential.

Fig. 1 shows the installation potential for nuclear power plants, wind power generation, and photovoltaic (PV) generation. It was assumed in the analysis that the capacity of nuclear power plants would be phased out in 40 years. The installation potential for wind and photovoltaic power was obtained from [8] and [9], respectively (Table 3). The construction and maintenance cost through 2030 were

obtained from [10], while those after 2030 were calculated based on the decreasing rate applied to pre-2030 values in [10]. It was assumed that the generated output from wind was constant at the rated power output multiplied by the capacity factor. The output generated from photovoltaic sources was assumed to change with time supposing a sunny day and the daily generated energy from this source was equivalent to the integral of the rated power multiplied by the capacity factor in a day. The technical characteristics for the other technologies considered in the analysis are based on data in [11].

The proposed model considered EDC and LFC from thermal and hydro power plants, which are described in Section 2. Table 4 shows the minimum output, the rate of change, and the LFC capacity for the thermal and hydro power plants as a percentage of the rated capacity [12].

4. Results

The impact of the penalty coefficient of CO_2 emission was evaluated. The electric water heating appliances operate in the nighttime load pattern here. Fig. 2 shows the quantity of CO_2 emissions and the saturation characteristics. Results showed that the penalty coefficient α is changed from 0 to 30,000 JPY/t-CO_2 and that CO_2 emissions were lower at larger penalty coefficients. Fig. 3 shows the installed capacity of electric power facilities in 5-year increments from 2010 to 2050. Results indicated that the power generation capacity was smaller for coal-fired power plants but larger for gas-fired power plants and renewable

Table 1. Energy demand scenario (increasing rate)

		Increasing rate[%]			
		2010-2020	2020-2030	2030-2040	2040-2050
Industory	Motor	+9.7	+7.7	+4.7	+3.9
	Boiler	+4.9	+2.8	+0.2	-0.4
	Furnace	+2.5	+1.8	-1.1	-1.8
	Iron reduction	-0.4	-0.4	-3.5	-4.6
	Cem ent kiln	+1.4	-5.5	-8.6	-9.6
	Chemical material	-0.8	-2.1	-5.1	-6.0
Customer	Light & app	+3.3	-1.0	-4.0	-5.6
	Space heating	+3.3	-1.0	-4.0	-5.6
	Water heating	+3.2	-1.1	-4.1	-5.7
	Air conditioning	+5.6	+0.7	-2.5	-3.7
Transportation	Ship	+1.4	-2.3	-5.4	-6.4
	Air	+4.1	-0.1	-3.4	-4.4
	Rail	+5.6	+0.8	-2.2	-3.5
	Automobile	+5.9	+1.1	-2.1	-3.3
	Bus & truck	+0.2	-3.3	-6.3	-7.4

Table 2. Duration of daytime and nighttime for each season

	Winter	Summer	Spring/fall
Daytime[o'clock]	7-16	6-19	6-18
Nighttime[o'clock]	17-6	20-5	19-5

Table 3. Cost data and capacity factors pertaining to wind power and photovoltaic generation in 10-year increments

		2010	2020	2030	2040	2050
Wind (onshore)	Construction cost [10^3JPY/kW]	275	188	177	173	170
	Maintenance cost [10^3JPY/kW]	15	11	11	11	11
	Capacity factor [%]	24	27	30	30	30
Wind (offshore)	Construction cost [10^3JPY/kW]	492	357	260	257	254
	Maintenance cost [10^3JPY/kW]	27	21	17	17	17
	Capacity factor [%]	31	36	40	40	40
PV (residential)	Construction cost [10^3JPY/kW]	515	230	189	156	128
	Maintenance cost [10^3JPY/kW]	8	4	3	2	2
	Capacity factor [%]	12	12	12	12	12
PV (non-residential)	Construction cost [10^3JPY/kW]	450	185	158	135	116
	Maintenance cost [10^3JPY/kW]	16	7	6	5	4
	Capacity factor [%]	16	16	16	16	16

Table 4. Data on thermal and hydro power plants

	Minimum outp ut (%)	Rate of change (%/h)		LFC capacity (%)
		Up	Down	
Coal	30	31	58	5.0
LNG-CC	30	82	75	5.0
Oil, Gas	30	100	100	5.0
Hydro	30	100	100	5.0

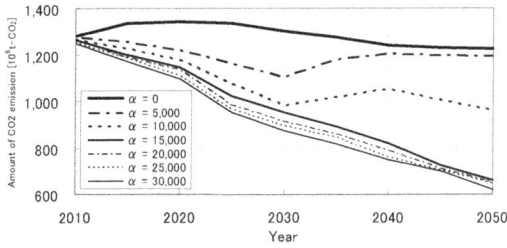

Fig. 2. Amount of CO_2 emission.

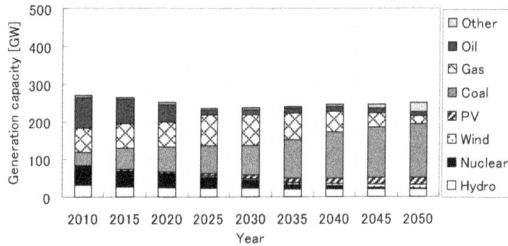

(a) $\alpha = 0$ [JPY/t-CO_2]

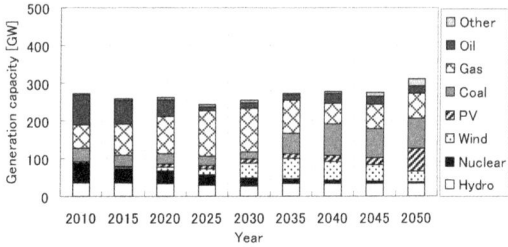

(b) $\alpha = 10,000$ [JPY/t-CO_2]

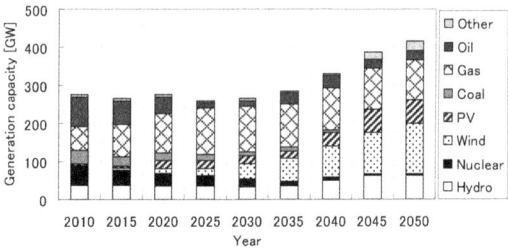

(c) $\alpha = 20,000$ [JPY/t-CO_2]

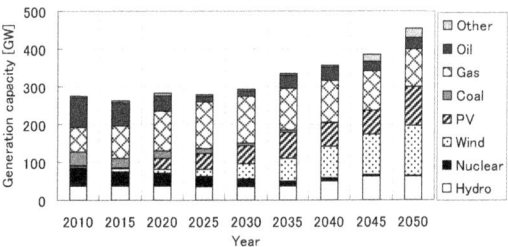

(d) $\alpha = 30,000$ [JPY/t-CO_2]

Fig. 3. Installed capacity of electric power facilities.

energy sources at larger penalty coefficients. The capacity of photovoltaic generation increased from 2040 to 2045 (Fig. 3d), but this capacity did not change in Fig. 3(c). The difference occurred because non-residential photovoltaic generation was installed in Fig. 3(d) (i.e., this power generation became cost competitive when α was set to 30,000 JPY/t-CO_2), but was not installed in Fig. 3(c).

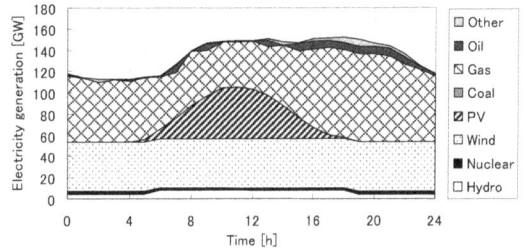

Fig. 4. Daily generation curve for weekends in the spring/fall of 2050 ($\alpha = 30,000$ JPY/t-CO_2).

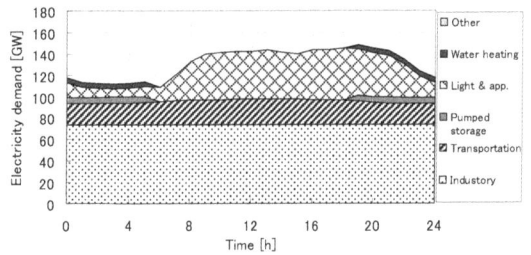

Fig. 5. Daily load curve for weekends in the spring/fall of 2050 ($\alpha = 30,000$ JPY/t-CO_2).

Figs. 4 and 5 show the daily generation and load curves as stacked bars for weekends in the spring/fall of 2050. Here, α was assumed to be 30,000 JPY/t-CO_2. The oil-fired power plants were operating during the peak-load time period in Fig. 4. The electric water heating appliances were operating during the night in Fig. 5. In this way, the proposed model made it possible to obtain the hourly time change for the power plant outputs and loads by type after solving the optimization problem.

The impact of demand side management by use of the electric water heating appliances was also evaluated. Fig. 6 shows the daily load curves for two cases during weekends in the spring/fall of 2020 and 2050. The electric water heating appliances operated in the nighttime load pattern in Case 1 and in the controllable load pattern in Case 2, with α equal to 30,000 JPY/t-CO_2 in both cases. Total electricity demand and power consumption for the electric water heating appliances are plotted in Fig. 6. Fig. 7 shows the daily power generation curve in 2050 for Case 2. In Case 2, the electric water heating appliances operated at night for load leveling in 2020, when the installed capacity of photovoltaic generation was small. However, these appliances operated during the day in 2050. By comparing Fig. 6(b) and Fig. 7, it was found that the electric water heating appliances operated while the output of photovoltaic generation was large in 2050 in Case 2. This pattern occurred because the efficient use of photovoltaic generation reduced the fuel cost of other power plants when the installed capacity of photovoltaic generation was large.

The relationship between load pattern and installed capacity of photovoltaic generation was considered. Fig. 8 shows the installed capacity of photovoltaic generation in 5-year increments from 2010 to 2050 (α = 20,000 or 30,000 JPY/t-CO$_2$). The installed capacity of photovoltaic generation based on the contribution of Case 2 was larger than that based on the Case 1 contributions from 2030 to 2045 in Fig. 8(a) and in 2030 and 2045 in Fig. 8(b). The results indicate that the installation of photovoltaic generation technologies was enhanced by the electricity demand increase that occurred during the day when CO$_2$ emission reductions were expected. The installed capacity did not change after 2045 in the both cases (Fig. 8a), because the installed capacity for the residential photovoltaic generation reached its upper limit, and non-residential photovoltaic

generation was not cost competitive. This also explained why the installed capacity in Case 1 was equivalent to that of Case 2 from 2035 to 2040 (Fig. 8b). The same capacity was installed under both cases in 2050 (Fig. 8b) because the installed capacity of both residential and non-residential photovoltaic generation reached their upper limits.

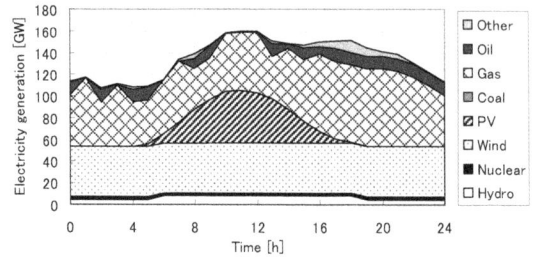

Fig. 7. Daily generation curve for weekends in the spring/fall of 2050 (α= 30,000 JPY/t-CO$_2$).

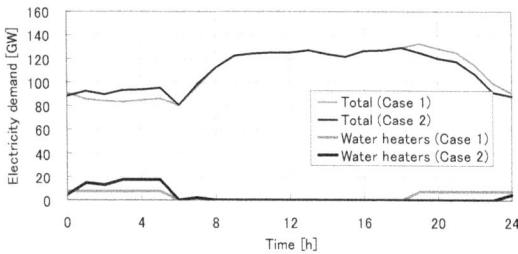

(a) α = 20,000 [JPY/t-CO$_2$]

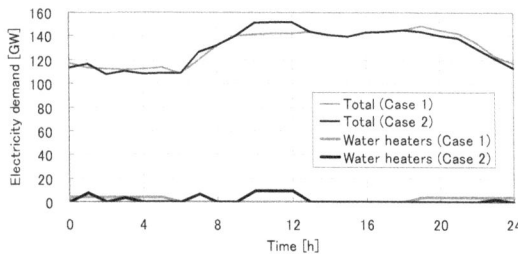

(b) α = 30,000 [JPY/t-CO$_2$]

Fig. 8. Installed capacity of photovoltaic generation.

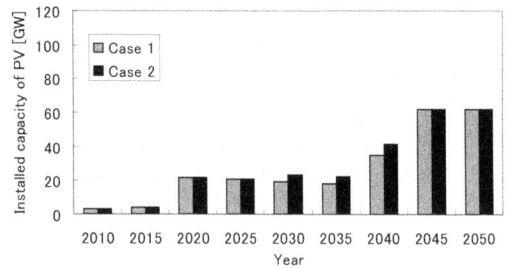

(a) 2020

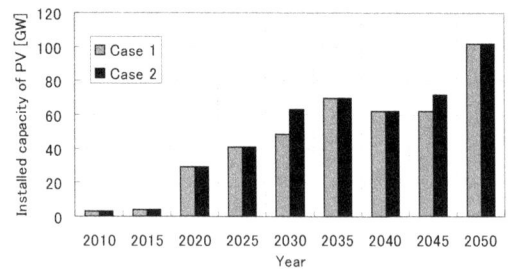

(b) 2050

Fig. 6. Daily load curve for weekends in the spring/fall of 2050 (α = 30,000 JPY/t-CO$_2$).

5. Conclusion

A new energy system model is proposed in this paper, which considers the power system controls that include EDC, LFC, and demand side management. The relationship between the electric water heating appliances' operational patterns and the future power generation mix was evaluated. Results showed that power system generation planning can be greatly impacted by demand side management. In future research, we will evaluate the impacts of the operational patterns of other energy devices, such as batteries, and cogeneration on the energy system.

Acknowledgements

This work was supported by JSPS KAKENHI Grant Number 24860074.

References

[1] N. Maïzi and E. Assoumou, "Electricity Generation and Renewables under Carbon Mitigation Policies", *in Proceedings of IEEE Energy 2030 Conference*, Atlanta, U.S., 2008.

[2] S. Jeong, J. Choi, J. Kim, and Y. Lee, "Flexible Best Generation Mix for Korea Power System Considering CO2 Constraint – Vision 2030", *in Proceedings of IEEE PES General Meeting*, Pittsburgh, U.S., 2008.

[3] T. Masuta and A. Yokoyama, "Supplementary Load Frequency Control by Use of a Number of Both Electric Vehicles and Heat Pump Water Heaters", *IEEE Trans. Smart Grid*, vol.3, issue 3, pp.1253-1262, 2012.

[4] M. D. Galus, S. Koch, and G. Andersson, "Provision of Load Frequency Control by PHEV, Controllable Loads, and a Cogeneration Unit", *IEEE Trans. Industrial Electronics* , vol.58, no.10, pp.4568-4582, 2011.

[5] IEA ETSAP, MARKAL. [Online]. Available: http://www.iea-etsap.org/web/Markal.asp.

[6] The Energy Data and Modeling Center (EDMC), "Handbook of Energy & Economic Statics in Japan", The Energy Conservation Center , 2012.

[7] Japan Industrial Technology Association (JITA), "Trend Survey on Residence in the Future", JITA, Tokyo, Japan, 1990 (in Japanese).

[8] Japan Wind Power Association (JWPA), "Wind Power Potential and Long Term Target along with the Road Map Ver. 1.1", JWPA, Tokyo, Japan, 2010.

[9] National Economic Development Office (NEDO), "PV Road Map toward 2030 (PV2030+)", NEDO, Kawasaki, Japan, 2009 (in Japanese).

[10] National Policy Unit, "Cost Verification Committee Report", Cost Verification Committee, Tokyo, Japan, 2011 (in Japanese).

[11] E. Endo, "Effects of the Low Nuclear Policy on Technology Competitiveness among Next Generation Vehicles in Japan", *2012 International Energy Workshop (IEA)*, Cape Town, South Africa, 2012.

[12] IEE Japan, "Demand and Supply Control Technology in Power System", *IEEJ Technical Report*, no.2-302, IEEJ, Tokyo, Japan, 1989 (in Japanese).

3

An Improved Modulation Scheme for Voltage Balancing in Modular Multilevel Converter

Shunke Sui[†], Rongfeng Yang* and Dianguo Xu*

Abstract – With the continuous development of power electronic technology toward the high voltage and high-power conversion, the Modular Multilevel Converter (MMC) has attracted research attention. Since a large number of capacitors, voltage balancing problem becomes one of the key technologies of the MMC. To resolve the voltage balancing problem, this paper presents a novel voltage balancing modulation scheme, which control the instantaneous power of each sub-module timely by adjusting carriers offset in PWM, and the respective capacitor voltage could be stabilized at a set value. Simulation results based on MATLAB / SIMULINK environment are provided to testify to the effectiveness of the proposed method.

Keywords: MMC, Voltage Balancing control, Itantaneous power, Ofset carrier

1. Introduction

The modular multilevel converter (MMC) provides an effective way for the power electronics technology deal with high voltage and high-power energy. The traditional multilevel converter requires a lot of clamping diodes and clamp capacitors with the voltage level increasing, and the power devices in the arm inside and outside work in imbalance situation which bring with complex control problems. MMC have characteristics of modular, high equivalent switching frequency, low switching stress, etc. However, MMC needs more sub-module capacitors, and this topology has circulating current problems [1], meanwhile the modular capacitor voltage needs maintain constant, and the relative control strategy is not yet mature.

MMC capacitor voltage balancing control researches have been focused on the capacitor voltage sorting control [2], the averaging and balancing control [3], the upper and lower leg balancing closed-loop control. The capacitor voltage sorting method is to sort all modules capacitor voltage in one phase, select the desired module capacitor to work in charging and discharging states according to the direction of current flow. However, this method has shortcomings that the switching frequency is not equivalent for each module, and the switching loss increases. Since the MMC has many modules, the method to balance the capacitor voltage with the PI controller is complex, and has

poor dynamic performance.

This paper proposed a novel capacitors voltage balance control method based on carrier offset adjusting in pulse width modulation (PWM). Firstly this paper decrypts the principle and structure of MMC briefly, and the equivalent mathematical model is derived. Then the question of the capacitor voltage balancing is particularly discussed, and the novel balance method by adjusting the carriers offset is proposed. Through adjusting the carrier offset, the duty cycle time of each module is regulated, thus the instantaneous power flow is changed and the capacitors voltage balancing is achieved. The method is simple to implement without extra switching losses and is suitable for a variety of the MMC topologies with carrier based modulation method. The validity of the proposed method is verified by simulation results.

2. Circuit Configuration of MMC

Fig. 1(a) is a three-phase MMC converter topology, and each phase consists of two stacks of multiple bidirectional cascaded modules and two none coupled connecting reactors. Each module works as a controllable voltage source that could generate 0 or u_C voltage, where the u_C is module capacitor voltage.

Fig. 1(b) illustrates sub-module structure. The capacitor voltage is set as u_C, and the states of S_1 and S_1 switch complementarily with the trigger pulse. The state of the two switches determines the state of the sub-module, thus the output voltage of sub-module changes between 0 and u_C.

Assuming all capacitors voltages are stable in normal

† Corresponding Author: Harbin Institute of Technology, China (suishunke@126.com)

* Harbin Institute of Technology, China (yrf@hit.edu.cn)

work condition, and n-modules are active in every arm, the MMC must meet the following conditions to work properly:

(1) The DC bus voltage is equal to n times of capacitors voltage at any time.

(2) In the same phase, if a module in the upper arm is plug-in, a module in the lower arm is plug-out complementarily.

The output voltage of each sub-module is controlled according to a certain manner. The upper and lower arms switching combinations alters, so the modules working status is constantly changing, and then the output AC voltage is $n+1$ level.

(a) Power Circuit

(b) Sub-module

Fig. 1. The circuit configuration of MMC.

When the system is running, according to KVL voltage law, the DC voltage, AC voltage, the sum of upper module voltage u_{psm} and the sum of lower module voltage u_{nsm} is given by

$$
\begin{cases}
u_{psmx} = \dfrac{U_{dc}}{2} - u_x - L\dfrac{di_{px}}{dt} \\
u_{nsmx} = \dfrac{U_{dc}}{2} + u_x + L\dfrac{di_{nx}}{dt}
\end{cases} (x = A,B,C) \quad (1)
$$

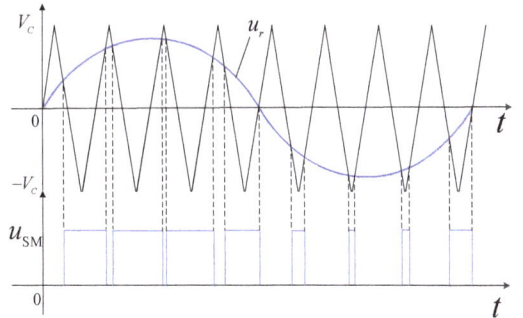

The upper and lower arms voltage, u_{px} and u_{nx}, can be expressed as

$$
\begin{cases}
u_{px} = u_{psmx} + L\dfrac{di_{px}}{dt} \\
u_{nx} = u_{nsmx} - L\dfrac{di_{nx}}{dt}
\end{cases} \quad (2)
$$

According to (1) and (2), the MMC can be characterized by the following equations:

$$
\begin{aligned}
u_x &= \frac{1}{2}(u_{nx} - u_{px}) - \frac{1}{2}\left(\frac{U_{dc}}{2} - \frac{U_{dc}}{2}\right) \\
&= \frac{1}{2}(u_{nx} - u_{px})
\end{aligned} \quad (3)
$$

3. Modulation Strategy of MMC with Voltage Balancing Control

3.1 Modulation scheme

When MMC is working properly, there are n working sub-modules. By changing the number of working sub-module in upper and lower arm, the multilevel output is achieved. The phase shift carrier pulse width modulation (PSC-PWM) is a conventional modulation method for cascaded converters [4] which has the nature of equivalent switching frequency and duty cycle time for different modules when generating multilevel voltages.

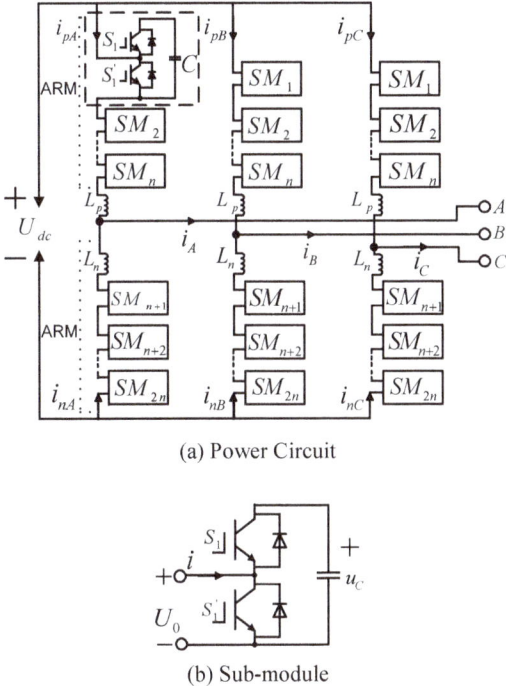

Fig. 2. The output voltage for a sub-module.

Fig. 2 shows the modulation method for a sub-module. u_r is the modulation wave, u_{SM} is the output voltage of the sub-module. According to equation (3), the upper and the lower arm can be looked as a two-level converter, so the two arms modulation waves' phase requires 180^0 difference. In the same arm, the neighbor carriers phase shifts with an angle of $2\pi / n$. In the other arm, the corresponding carrier phase shifted by an angle of π / n. In this case, it ensures that n sub-modules are plugged in at any moment. Note that when the phase-shift angle is different, the output levels are varied [5].

3.2 Voltage balancing control scheme

On the case that the sub-module is active, if the current flows into the capacitor the capacitor is charged and the voltage increases. On the other hand, if the current flows out the capacitor, the capacitor is discharged and the voltage decreases. But when the sub-module is bypassed (the output voltage is 0), the arm current has no influence on the capacitor voltage. Therefore, the scheme that change the capacitors charging and discharging time is proposed to control capacitor voltage fluctuations.

For example, when the capacitor voltage is higher than the set value, the capacitor needs to discharge, in order to reduce the voltage. Therefore, if the current flows into the Sub-module, it is needed to reduce the charging time. And if the current flow out the module, it is needed to increase the discharge time. Similarly, when the capacitor voltage is lower than the set value, it is needed to increase the charging time or reduce the discharge time, so as to increase the capacitor voltage.

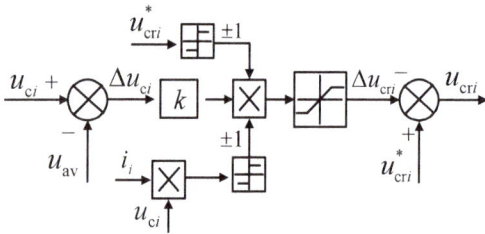

Fig. 3. The proposed system control block diagram.

The control block diagram is shown in Fig. 3. For the sub-module i in the leg of MMC, the capacitor voltage u_{ci} subtracts the capacitor average voltage u_{av}, to obtain a voltage deviation signal, and it is then multiplied by the factor k. When the instantaneous power of the sub-module $u_{ci} \cdot i_i$ is positive (the current of the sub-module is i_i), it means that the capacitor is charging. If the corresponding carrier is greater than zero, the carrier pluses the capacitor

voltage adjustments $k \Delta u_{ci}$, and if the carrier is less than zero, the carrier minuses the adjustment $k \Delta u_{ci}$. So it could shorten the charging time by reducing the duty ratio when the capacitor voltage is higher than average value. Similarly, when the instantaneous power is negative, it means that the capacitor is discharging and reverse operation should be employed.

The proposed control scheme changes the carrier amplitude offset according to the capacitor voltage, thus dynamically adjusts the capacitor voltage to follow the average value of the leg capacitor voltage. Fig. 4 illustrates the duty ratio variation of the sub-module through adjusting the carrier amplitude offset.

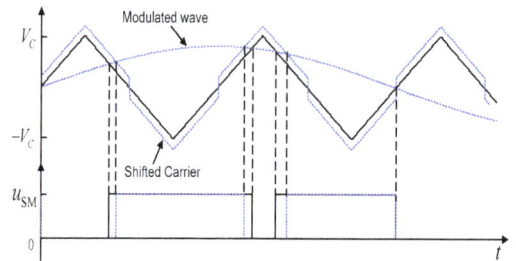

Fig. 4. The modified modulation and the output voltage of one sub-module.

After adjustment of the above process, it will increase the discharge time and shorten the charging time when the module has higher capacitor voltage. However, it will increase the charging time and shorten the discharge time of capacitor when the module has lower capacitor voltage. In the process of dynamically adjusting the carrier amplitude, it could achieve a good voltage balancing control.

4. Simulation results

verify the performance of the effectiveness of proposed modulation scheme, the simulation of MMC is conducted. The simulation parameters are shown in Table 1.

Fig. 5(a) is shown as a capacitor voltage fluctuations without balance control, It can be obviously seen that capacitor voltage fluctuations and result in a large deviation. After long-time running, the voltage will be reduced to zero or higher enough to cause the device malfunction. Fig. 5(b) shows the proposed control strategy performance after adjustment module instantaneous power. It could be seen that the voltage maintain stable and the voltage deviation is laminated under low level. It displayed that the proposed method has excellent voltage balance control effect.

Table 1. Simulation parameters

parameter	Value
DC bus voltage	600V
Number of sub-modules $2n$	6
capacitor	$2500\mu F$
Inductance	$4mH$
Load	10Ω
k	0.1
Carrier frequency	$1kHz$

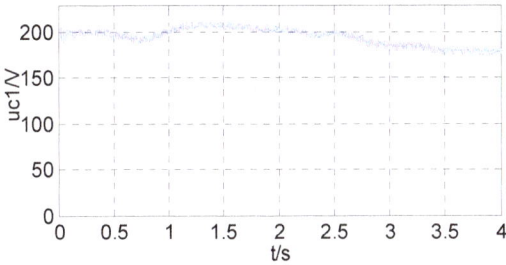

(a) Capacitor voltage without control.

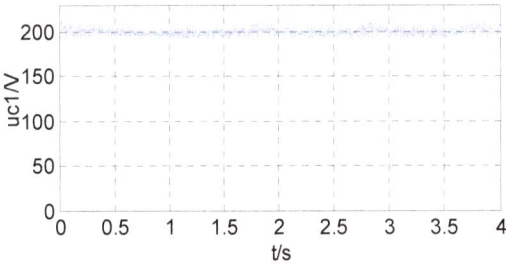

(b) Capacitor voltage control.

Fig. 5. Capacitor voltage fluctuations.

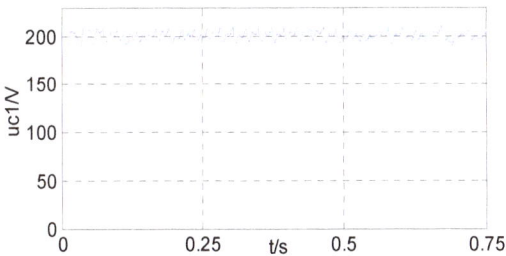

Fig. 6. The initial capacitor voltage value is 180V.

Fig. 6 shows that the module capacitor voltage initial value is not equal to the normal work setting, and this method is still able to adjust the voltage to setting value quickly and efficiently. The module capacitor voltage has initial value 180V, meanwhile the setting capacitor voltage value is 200V. It can be clearly seen that the capacitor voltage has been effective regulated to normal range after several cycles. It means that this method can resist perturbations and improve the MMC system stability.

5. Conclusion

In this paper, a novel voltage balancing modulation scheme is proposed to obtain the capacitor voltage balance. This method regulates the duty cycle time to change the capacitor charging or discharging energy through adjusting the carrier offset in PWM process, which has fast dynamic response and ability to resist perturbations. The simulation results have been presented to demonstrate the effectiveness and superiority of this method.

Acknowledgements

This work was financially supported by the Project 51237002 supported by National Natural Science Foundation of China and grants from the Power Electronics Science and Education Development Program of Delta Environmental & Educational Foundation.

References

[1] Hagiwara M, H. Akagi. PWM control and experiment of modular multilevel converters. Proceedings of IEEE Power Electronics Specialists Conference. Rhodes, Greece: IEEE, 2008 : 154-161.

[2] Zhao Xin, Zhao Chengyong, Li Guangkai, et al. Sub-module Capacitance Voltage Balancing of Modular Multilevel Converter Based on Carrier Phase Shifted SPWM Technique [J]. Proceedings of the CSEE, 2011, 21 (31) : 48-55.

[3] H. Akagi. Classification, Terminology, and Application of the Modular Multilevel Cascade Converter (MMCC). Power Electronics, IEEE Transactions on, 2011, 3119-3130.

[4] Konstantinou G S, Agelidis V G. Performance evalua-tion of half-bridge cascaded multilevel converters operated with multicarrier sinusoidal PWM techni-ques[C]. IEEE Conference on Industrial Electronics and Applications. Xi'an, China, 2009:3399-3404.

[5] A .Shojiaei, G. Joos. An improved modulation scheme

for Harmonic distortion reduction in modular multilevel converter. IEEE Power and Energy Society General Meeting. 2012 : 1-7.

[6] Wang Shanshan, Zhou xiaoxin, Tang Guangfu. Modeling of modular multilevel voltage source converter [J]. Proceeding of the CSEE, 2011, 31(24) : 1-8(in Chinese).

[7] Liu Zhongqi, Song Qiang, Liu Wenhua. VSC-HVDC system based on modular multilevel converters [J]. Automation of Electric Power Systems, 2010, 34(2) : 53-58(in Chinese).

[8] Li Xiaoqian,Song Qiang,Liu Wenhua,et al.Capacitor voltage balancing control by using carrier phase-shift modulation of modular multilevel converters[J]. Proceedings of the CSEE,2012, 32(9) : 49-55(in Chinese).

4

Comparison of Generator Performance of Small-Scale MHD Generators with Different Electrode Dispositions and Load Connection Systems

Toru Takahashi[†], Takayasu Fujino* and Motoo Ishikawa*

Abstract –The performances of experimental and small scale DCW, DIW, HCW, HIW and Faraday MHD generators driven by a scramjet engine are compared by three-dimensional numerical simulation. Numerical results show that the maximum power output is obtained in the DCW generator. The electrodes on the side walls in the DCW and the HCW generator suppress the loss of electrode voltage drop compared with the DIW and the HIW generator. Since the volume-averaged Hall parameter of plasma is less than unity for all the generators, the HCW and the HIW generator have less electric power output than other generators. The power output extracted by the Faraday generator is comparable to that extracted by the DCW generator. The Faraday generator, however, requires the load and the inverter to each number of electrode pairs, so that the complication of external circuits and the increase in cost may be apprehended. The authors conclude that the DCW generator is suitable for the experiment of scramjet engine driven MHD generators.

Keywords: Scramjet engine, MHD generator, Three-dimensional numerical simulation

Nomenclature

B	Magnetic flux density [T]
E	Strength of electric field [V/m]
E_0	Total energy [J/m^3]
f_{loss}	Thrust loss [N]
j	Electric current density [A/m^2]
p	Static pressure [Pa]
P_{out}	Electric power output [W]
T	Static temperature [K]
u	Gas velocity [m/s^2]
β	Hall parameter [-]
ΔV	Voltage drop against Faraday EMF [-]
$\Delta V_{y,U}$	Voltage drop near upper wall [V]
$\Delta V_{y,L}$	Voltage drop near lower wall [V]
η_e	Electrical efficiency [-]
κ	Thermal conductivity coefficient [J/(m·s·K)]
ρ	Mass density [kg/m^3]
σ	Electrical conductivity [S/m]
$\tau_{i,j}$	Viscous stress tensor [Pa]
ϕ	Electric potential [V]

†   Corresponding Author: National Institute of Advanced Industrial Science and Technology, Japan (t.takahashi23@gmail.com)

*   Dept. of Engineering Mechanics and Energy, University of Tsukuba, Japan (tfujino@kz.tsukuba.ac.jp)

1. Introduction

A scramjet driven magnetohydrodynamic(MHD) power generation has been proposed as an electric power source for advanced hypersonic aircrafts[1].

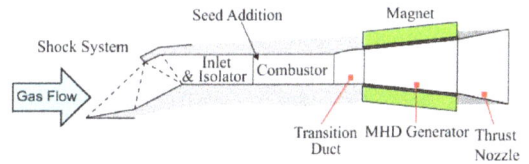

Fig. 1. Configuration of MHD generation system.

In 2001, Hypersonic Vehicle Electric Power System (HVEPS) program got started as a five-year research and development plan[1][2]. Fig. 1 illustrates the conceptual diagram of scramjet engine driven MHD power generation system in the HVEPS program. The MHD generator installed in the system utilizes the hypersonic airstream taken in from the atmosphere as an oxidizer. A near-stoichiometric scramjet combustion in the engine produces a high-temperature gas required to promote the MHD power generation. The addition of a seeding material(alkali metal) to the gas flow produces an electrically conducting plasma. The seeding material is injected into the hot air stream in the upstream of the pre-heater allow for its dissociation, vaporization and thorough mixing prior to the scramjet combustion. The exhaust passes into the MHD generator which is located in the downstream of combustor. Under the HVEPS program, the scramjet engine driven MHD power demonstration test was carried out by LyTec et al. in 2006[3]. The experiments were successfully concluded, and the results showed the feasibility of the scramjet engine driven MHD power generation. In the experiments, a Diagonal Conducting Wall(DCW) MHD generator was applied for the generation. In the experiment, the electric power and the electrode potential were measured.

In order to provide the suggestion for improving the performance of experimental generator, it is useful to understand the phenomena in the experimental generator by means of MHD numerical simulation. Hardiant, et al.[4] numerically examined that a combustion efficiency of the scramjet combustor operating point influences on the electrical power and the plasma behavior of the scramjet engine driven DCW MHD generator used in the HVEPS experiment. Harada, et al.[5] numerically showed the generator performance with considering the influence of non-uniformity of inlet temperature. It was also shown that even if the average value of inlet temperature is the same, the generator performance changes with the condition of non-uniform temperature distribution. These simulations, however, are carried out for only the DCW generator used in the experiment. There are different types of MHD generator, which have not been yet considered in the HVEPS experiment. By changing the type of MHD generator, the generator performance may improve.

The present study treats five MHD generators: DCW generator, DIW(Diagonal Insulating Wall) generator, HCW(Hall Conducting Wall) generator, HIW(Hall Insulating Wall) generator, and finite segmented Faraday generator. The DCW generator is one of the diagonal connected MHD generators, where window frame electrodes separated by insulators are placed on the walls. The DIW generator is another diagonal connected MHD generator, where several pairs of electrodes placed on upper and lower walls are electrically connected externally. The HCW generator and the HIW generator are Hall connected MHD generators. A finite segmented Faraday generator fundamentally differs in electrical connection from the diagonal connected generators and the Hall connected generators, and each separated electrode is individually connected to an external load.

The objectives of this study are to examine the difference of generator performance among the five MHD generators,

and to clarify the plasma behavior in each MHD generator by means of three-dimensional MHD numerical simulations.

2. Numerical Procedure

2.1 Analytical Domain and Operating Conditions

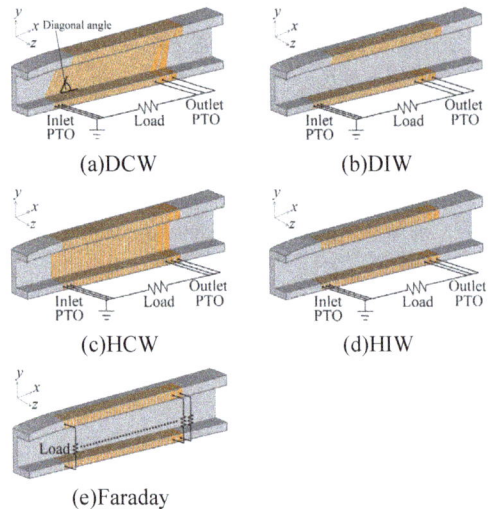

(a)DCW   (b)DIW

(c)HCW   (d)HIW

(e)Faraday

Fig. 2. Analytical domain and coordinate system.

Fig. 2 illustrates an analytical domain and the coordinate system. The analytical domain includes the transition duct, the generator section, and the thrust nozzle. The symmetric feature along the magnetic field with only the z component is taken into account and only the half region is treated in the z-direction.

Both the DCW and the DIW generator have 35 electrodes in the generator section. Insulators separate each electrode. The inlet power-take-off(PTO) electrodes from the 1st to the 3rd and the outlet PTO electrodes from the 33rd to the 35th are electrically connected to a load. The diagonal angle is varied from 60 to 90 degrees in the end of generator section. The DCW generator has window frame electrodes, and the DIW generator has separated electrodes on the upper and lower walls.

The Faraday generator has 35 pairs of electrodes in the generator section. Each electrode pair is placed vertically to the flow direction(the x-direction) on the upper and lower walls.

The electrode dispositions and the load connection systems of the HCW and the HIW generator correspond to those of the DCW and the DIW generator with the diagonal angle of 90 degrees. The Hall generator takes out not the Hall electromotive force(EMF) but the Faraday EMF. When the Hall parameter is more than unity, the Hall generator has the equivalent EMF obtained in the Faraday generator.

Table 1 shows the numerical conditions, which correspond to the operating conditions in the HVEPS experiment using the DCW MHD generator[3].

Table 1. Analytical conditions for experiment of MHD generator[3]

Combustion efficiency	89 [%]
Pre-heater H2 mass flow rate	0.031 [kg/s]
Air mass flow rate	0.595 [kg/s]
O₂ enrichment mass flow rate	0.659 [kg/s]
Seed (NaK) mass flow rate	0.038 [kg/s]
Fuel(ethylene) mass flow rate	0.160 [kg/s]
Total mass flow rate	1.483 [kg/s]
Inlet static pressure	0.283 [atm]
Inlet static temperature	2480 [K]
Channel length	1.1176 [m]
Inlet height	0.1329 [m]
Inlet width	0.1524 [m]
Outlet height	0.1670 [m]
Outlet width	0.1569 [m]
Diagonal angle	60 [degrees]

Fig. 3 shows the distribution of an externally applied magnetic field in the generator channel. We, here, assume that the magnetic flux density has the x- and z-components.

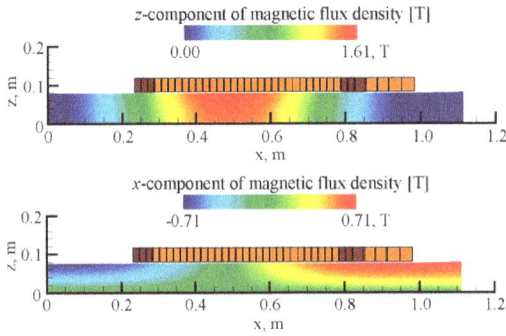

Fig. 3. Distributions of magnetic flux density on x-z plane with $y = 0$ m.

2.2 Basic Equations

The basic equations for gasdynamics are the mass conservation equation, the momentum conservation equations, and the energy conservation equation.

$$\frac{\partial \rho}{\partial t} + \nabla \cdot (\rho \boldsymbol{u}) = 0 \tag{1}$$

$$\frac{\partial (\rho \boldsymbol{u})}{\partial t} + \nabla \cdot (\rho \boldsymbol{uu}) = -\nabla p + \nabla \cdot \tau_{i,j} + \boldsymbol{j} \times \boldsymbol{B} \tag{2}$$

$$\frac{\partial E_0}{\partial t} + \nabla \cdot \{(E_0 + p)\boldsymbol{u}\} = \nabla \cdot (\tau_{i,j} \cdot \boldsymbol{u} + \kappa \nabla T) + \boldsymbol{j} \cdot \boldsymbol{E} \tag{3}$$

The basic equations of electrodynamics are the steady Maxwell equations and the generalized Ohm's law, where the induced magnetic field is neglected because the magnetic Reynolds number is much smaller than unity.

$$\nabla \times \boldsymbol{E} = \boldsymbol{0} \tag{4}$$

$$\nabla \cdot \boldsymbol{j} = 0 \tag{5}$$

$$\boldsymbol{j} = \sigma(\boldsymbol{E} + \boldsymbol{u} \times \boldsymbol{B}) - \frac{\beta}{|\boldsymbol{B}|}\boldsymbol{j} \times \boldsymbol{B} \tag{6}$$

The elliptic partial differential equation on the electric potential ϕ is obtained from these equations.

The basic equations for gasdynamics are discretized with the finite volume method. The Harten-Yee TVD scheme[6] is implemented for the numerical flux of the convection terms. The central difference scheme is used for the numerical flux of the diffusion terms. The algebraic turbulence model proposed by Cebeci and Smith and improved by Stock and Haase is adopted[7].

Values of static pressure, static temperature, and gas velocity in the x-direction of the core flow are fixed on the inlet cross-section. The fixed values are estimated by using the experimental data[3]. The boundary layer thickness at the inlet is assumed. The distributions of the temperature and the x-component of velocity in the boundary layer at the inlet are assumed to be given by the 1/5-th power law[8]. The distribution of pressure is assumed to be uniform over the inlet cross-section. The y- and z-components of velocity are fixed at zero over the inlet cross-section. The exit boundary condition is given by the first order extrapolation for all quantities, because the flow is kept supersonic. The no-slip condition is used as the wall boundary condition, and the wall temperature on the plasma side is fixed at 1000 K.

The second-order elliptic partial differential equation on the electric potential ϕ is solved by the Galerkin finite element method(FEM) with the first order tetrahedron elements[9]. The obtained linear equation is solved with BiCGStab2[10]. The electric potential of inlet PTO electrodes is given to zero, and that of outlet PTO electrodes is given to the equipotential value. The electric potentials on the electrodes between the inlet and the outlet PTO regions are calculated to satisfy a given load current.

The thermodynamic properties such as enthalpy and electric conductivity are calculated under the assumption of thermal equilibrium, where the experimentally inferred combustion efficiency is used. The scramjet combustion reactant mixtures contain the air with oxygen enrichment, hydrogen, seed (NaK), and scramjet combustor fuel (ethylene). The approximate functions of temperature and pressure for these properties are made from the calculation results. The minimum value of electrical conductivity is set to 0.1 S/m as an approximate arcing model.

3. Result and Discussion

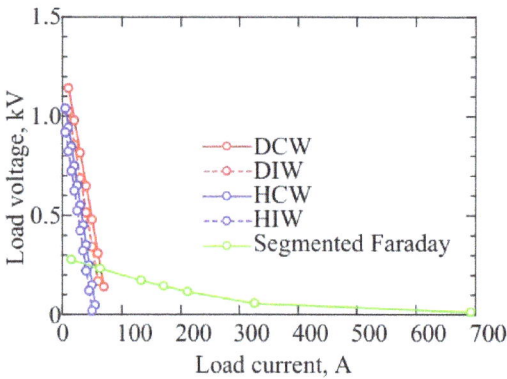

Fig. 4. V-I curves.

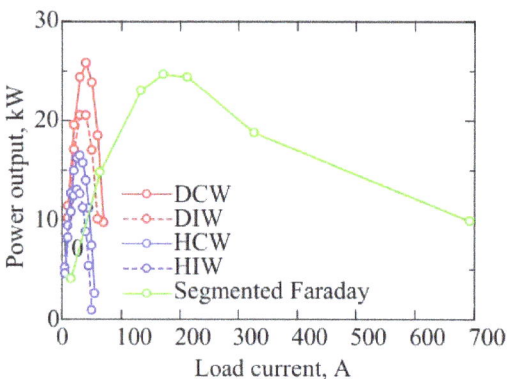

Fig. 5. Characteristics of electric power output.

Fig. 4 shows the V-I curves and Fig. 5 depicts the characteristics of electric power output P_{out}, which is calculated by the following equation:

$$P_{out} = -\iiint_{V_{all}} \boldsymbol{j} \cdot \boldsymbol{E} \, dV \tag{7}$$

The V-I curves of Diagonal(DCW and DIW) and Hall(HCW and HIW) generators are distributed over a wide range of load voltage. On the other hand, the V-I curve of Faraday generator is distributed over a wide range of load current. The DCW generator obtains the highest maximum electric power, followed by the Faraday generator and then the DIW generator. Since the volume-averaged Hall parameter of plasma is as small as about 0.95, the electric power outputs of the two Hall generators (DIW and DCW generators) show lower value compared with other generators. In the following discussion, the performances under the load condition where the maximum power is obtained in each generator are compared.

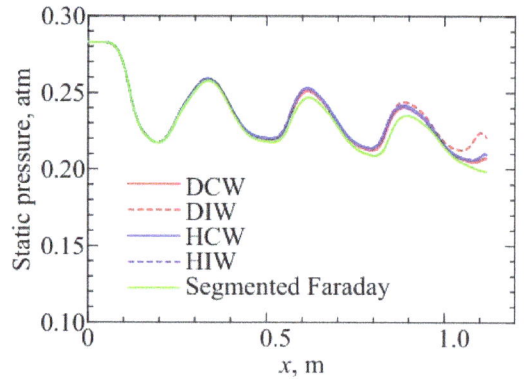

Fig. 6. Distributions of static pressure along center line.

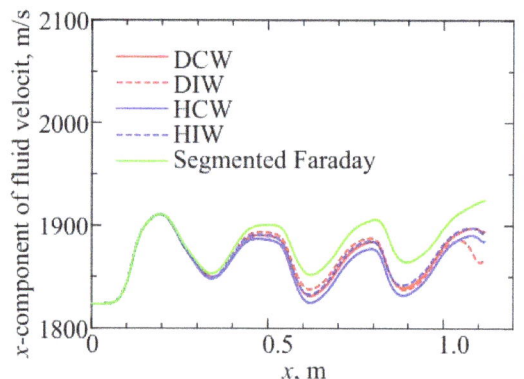

Fig. 7. Distributions of fluid velocity along center line.

Fig. 6 and 7 depict the distributions of static pressure and fluid velocity along the center line, respectively. Table 2 lists the values of MHD interaction parameter S_p, which is estimated with the following equation:

$$S_p = \frac{-\iiint_{V_{all}} j_y B_z \, dV}{\iint_{S_{in}} (p + \rho u_x^2) dS} \tag{8}$$

The difference of gasdynamic properties such as static pressure and fluid velocity among the five generators is rather small due to the low MHD interaction, as shown in Table 2.

Table 2. Value of MHD interaction parameter

Type	S_p
DCW	0.081
DIW	0.079
HCW	0.093
HIW	0.086
Faraday	0.052

Fig. 8 depicts the distributions of current density and vector trace on the y-z plane with x=0.5 m in the DCW and the DIW generator. It should be noticed that the lines in this figure illustrate a projection of the current on the y-z plane. The DCW generator induces the z-component of current density near the side wall due to the electrodes existing on the side walls, whereas the DIW and the Faraday generator have little current in the z-direction due to no electrodes on the side walls.

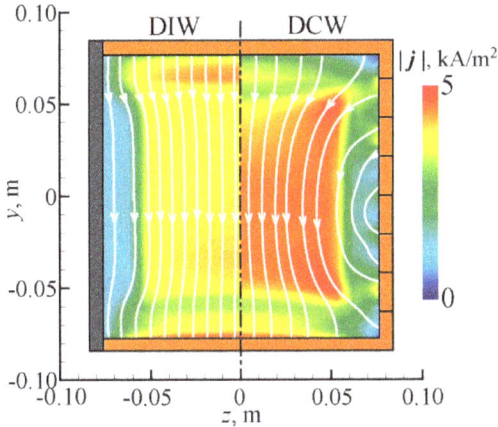

Fig. 8. Distributions of current density and vector trace on y-z plane with x=0.5 m.

Fig. 9 depicts the distributions of current stream line and the absolute value of electric current density on the x-y

plane with z=0 m for each generator. In the DCW and HCW generator, the electric current has a large value in the main flow rather than near the wall surfaces. In the DIW and the HIW generator, on the other hand, the electric current near the upper and lower walls is relatively larger than that in the main flow. This results in a large Joule heating near the upper and lower walls. The Faraday generator has the smallest electric current of the five generators.

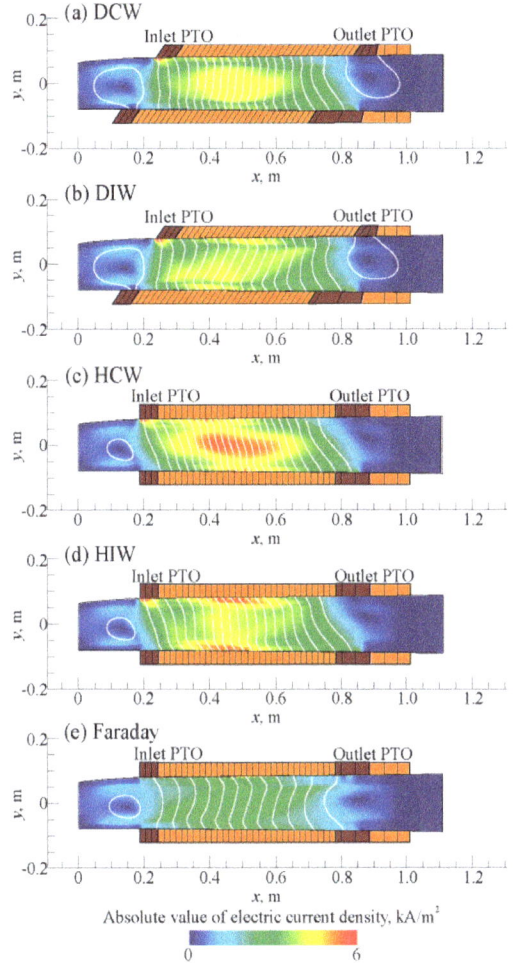

Fig. 9. Distributions of current stream line and absolute value of electric current density on x-y plane with z=0 m.

Fig. 10 depicts the distributions of electric potential on the y-z plane with x=0.5 m for the DIW and the DCW generator. For either generator, the voltage drop occurs near the upper and lower walls. In this paper, the voltage drop is defined as the subtraction of the electric potential drop on the wall from the estimated value linearly interpolated by the potential gradient with the main flow(y=0 m, z=0 m). Figure 11 depicts the electrode voltage drop ΔV against the Faraday electromotive force along the center line, which is estimated with the following equation:

$$\Delta V = \frac{\Delta V_{y,U} + \Delta V_{y,L}}{\int_{y_{min}}^{y_{max}} u_x B_z \, dy} \tag{9}$$

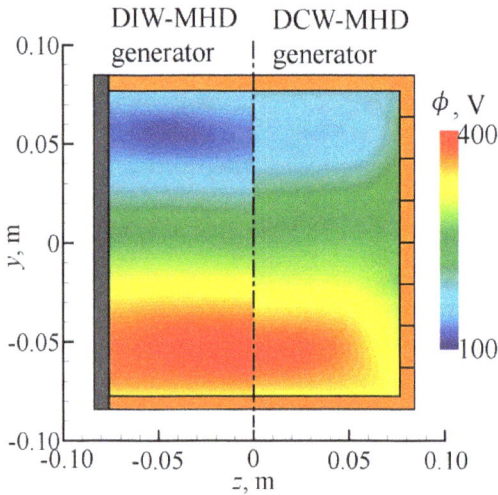

Fig. 10. Distributions of electric potential on y-z plane with x=0.5 m.

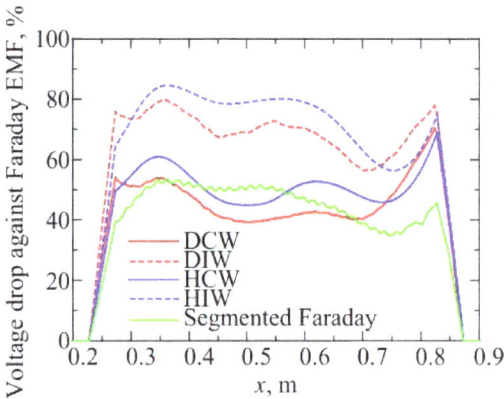

Fig. 11. Electrode voltage drop against Faraday EMF along center line.

The voltage drop of the DCW generator in the center region is smaller than that of the other generators. This indicates that a smaller electrode voltage drop yields a larger electric power output.

All electric currents flow out of the upper wall and flow into the lower wall in the perpendicular to the magnetic field in the cases of the DIW and the Faraday generator, whereas some currents flow out and into the side walls in parallel to the magnetic field in the case of the DCW generator. The voltage drop by the y-component of current density is several times larger than the voltage drop due to the z-component of current density. Thus, the DCW generator induces less voltage drop near the electrodes compared to the other generators, resulting in that the DCW generator produces the highest maximum power. In the case of the Faraday generator, on the other hand, the net currents in the x-direction are zero between neighboring electrodes. The Faraday generator, thus, induces less power loss than the DCW and the DIW generator.

Table 3. Electrical efficiency for each generator

Type	η_e
DCW	36.3 [%]
DIW	32.2 [%]
HCW	21.1 [%]
HIW	18.3 [%]
Faraday	50.7 [%]

Table 3 lists the values of electrical efficiency η_e, which is defined as:

$$\eta_e = \frac{- \iiint_{V_{all}} \boldsymbol{j} \cdot \boldsymbol{E} \, dV}{- \iiint_{V_{all}} \boldsymbol{u} \cdot (\boldsymbol{j} \times \boldsymbol{B}) \, dV} \tag{10}$$

The Faraday generator produces the highest electrical efficiency of the five generators. The existence of conducting sidewalls improves the electrical efficiency by 4 % for diagonal-type generator, and by 3 % for Hall-type generator, as can be seen from the comparisons of η_e between the DCW and the DIW generator, and also between the HCW and the HIW generator.

The Faraday generator has high electric power output and high electrical efficiency, as shown in Fig. 5 and Table 3. The Faraday generator, however, requires one load and one inverter for each of all electrode pairs. Therefore, the Faraday generator has many loads and inverters. This leads to the complication of external circuits and the increase in cost. On the other hand, the DCW generator needs only one load and one inverter. As shown in Fig. 5, DCW generator can produce the highest power output and a relatively high electrical efficiency. Therefore, the authors conclude that the DCW generator is suitable for the experiment of scramjet engine driven MHD generator.

4. Conclusions

The performances of the five MHD generators: DCW, DIW, HCW, HIW, Faraday generators under the HVEPS experimental conditions were compared by means of three-dimensional numerical simulation.

The numerical results showed that the maximum power output is obtained in the DCW generator. The electrodes on the side walls in the DCW and the HCW generator suppress the loss of electrode voltage drop compared with the DIW and the HIW generator. Since the volume-averaged Hall parameter of plasma is less than unity, the Hall-type generators: the HCW and the HIW generator have less electric power output than other generators. The power output obtained by the Faraday generator is comparable to that obtained by the DCW generator. The Faraday generator, however, requires one load and one inverter to each number of electrode pairs, so that the complication of external circuits and the increase in cost may be apprehended. Therefore, the DCW generator is suitable for the experiment of scramjet engine driven MHD generator.

References

[1]　J. T. Lineberry, L. L. Begg, J. H. Castro and R. J. Litchford: "Scramjet Driven MHD Power Demonstration - HVEPS Program", 37th AIAA Plasmadynamics and Lasers Conference, AIAA-2006-3080, San Francisco, USA(2006).

[2]　J. T. Lineberry, L. L. Begg, J. H. Castro and R. J. Litchford: "Scramjet Driven MHD Power Demonstration - HVEPS Program Overview", 14th AIAA/AHI International Space Planes and Hypersonic Systems and Technologies Conference, AIAA-2006-8010, Canberra, Commonwealth of Australia(2006).

[3]　J. T. Lineberry, L. Begg, J. H. Castro, R. J. Litchford and J. M. Donohue: "HVEPS Scramjet-Driven MHD Power Demonstration Test Results", 38th AIAA Plasmadynamics and Lasers Conference In conjunction with the 16th International Conference on MHD Energy Conversion, AIAA-2007-3881, Miami, USA (2007).

[4]　T. Hardianto, N. Sakamoto and N. Harada: "Computational study of a diagonal channel Magnetohydrodynamic power generation", International Journal of Energy Technology and Policy, Vol.6, No.1/2, pp.96-111 (2008).

[5]　N. Harada, T. Narikawa and T. Kikuchi: "Numerical Investigation on Non-Uniformity in Channel Cross Section of Scramjet Driven Magnetohydrodynamic Power Generation", 40th AIAA Plasmadynamics and Lasers Conference, AIAA-2009-3828, San Antonio, USA (2009).

[6]　H. C. Yee: "Upwind and Symmetric Shock-Capturing Schemes", NASA TM-89464 (1987).

[7]　H. W. Stock and W. Haase: "The Determination of Turbulent Length Scales in Algebraic Turbulence Models for Attached and Slightly Separated Flows Using Navier-Stokes Methods", 19th AIAA Fluid Dynamics, Plasma Dynamics & Lasers Conference, AIAA paper 87-1302, Honolulu, USA (1987).

[8]　H. Schlichting: "Boundary-layer Theory 6th edition", McGraw-Hill, New York, USA (1968).

[9]　M. Ishikawa, K. Itoh and K. Tateishi: "Comparison of Three-dimensional Constricted Electric Current near Anode and Cathode of Open-Cycle Faraday MHD Generator", 33rd AIAA Plasmadynamics and Lasers Conference, AIAA 2002-2237, USA (2002).

[10]　M. H. Gutknecht: "Variants of BICGSTAB for Matrices with Complex Spectrum", SIAM Journal on Scientific Computing, Vol. 14, No. 5, pp.1020-1033 (1993).

5

Design Method of Reliability Improvement of Voice Coil Type Linear Actuator for Hydraulic Valve

Baek-Ju Sung[†]

Abstract – The precise hydraulic valve is widely used in various industrial field like aircraft, automobile, and general machinery. Solenoid actuator is the most important device for driving the precise hydraulic valve. The reliable operation of solenoid actuator effects on the overall hydraulic system. The performance of solenoid actuator relies on frequency response and step response according to arbitrary input signal. In this paper, we performed the analysis for the components of solenoid actuator to satisfy the reliable operation and response characteristics through the reliability analysis, and also deducted the design equations to realize the reliable operation and fast response characteristics of solenoid actuator for hydraulic valve operation through the empirical knowledge of experts and electromagnetic theories. We suggested the design equations to determine the values of design parameters of solenoid actuator as like bobbin size, length of yoke and plunger and turn number of coil, and calculated the life test time of solenoid actuator for verification of reliability of the prototype. In addition, for reducing the life test time, the acceleration model of solenoid actuator is proposed and the acceleration factor is calculated considering the field operating conditions. And then, we verified the achieved design values through accelerated life test and performance tests using some prototypes of solenoid actuators adapted in servo valve.

Keywords: Solenoid actuator, Failure analysis, Design equation, Accelerated life test, Attraction force, Frequency response

1. Introduction

The solenoid actuator is a very economical motion converter due to its simple structure as electromagnetic energy converting to kinetic energy. And the solenoid actuator is used as key components in automobile and aircraft industry. For having higher response time and product reliability, two kinds of different techniques are needed. One is the optimal design method for solenoid actuator. A regarded point for design of solenoid actuator is flux density analysis, determination of plunger shape and mass, optimal bobbin design, selected magnetic analysis, determination of duty ratio, and calculation of coil turn number which is regarded temperature rising. For the optimal design of the solenoid actuator, theoretical and empirical knowledge are simultaneously needed. Theoretical knowledge governs the operational characteristics of the solenoid actuator, and empirical knowledge compensates for the theoretical limitation obtained from the designer's design and manufacturing experiences for various kinds of solenoid actuator[1]. They cannot be determined solely by

calculation or simulation because the empirical knowledge is more essential than theoretical knowledge for determination of the plunger shape and value of the space factor. When designer's accumulated experiences and expertise are added to these, the most proper shape and value of them can then be obtained.

In this study, we derived the design equations for design of solenoid actuator by a combination of electromagnetic knowledge and empirical knowledge. And also, to establish the reliability assessment technology of solenoid actuator, we performed the failure analysis considering the literature, failure data of manufacturing companies and the opinion of experts, and produced the reliability parameters for design and test. As the data of failure analysis, we draw out the Failure Modes Effect Analysis (FMEA), 2-Stage Quality Function Deployment (QFD), Fault Tree Analysis (FTA), and also analyzed failure cause and effects of solenoid actuator, and decided test items. And then the no-failure test time of solenoid actuator is calculated. In particular, for reducing of the no-failure test time, the acceleration model of solenoid actuator is proposed, and the acceleration factor is calculated with the reality. The validity of the proposed design method and deducted reliability parameters are proved by accelerated life test and performance test.

---
†   Corresponding Author: Dept. of System Reliability, Korea Institute of Machinery & Materials, Korea (sbj682@kimm.re.kr)

2. Solenoid Actuator

Fig. 1 represents the structure of a solenoid actuator for servo valve operation. It is composed of an excitation coil for generation of magnetic field, yoke for flux path, plunger for creation of mechanical stroke, stationary for attraction of the plunger, permanent magnet for assistance of attractive force and reduction of consumption power, and centering spring[2].

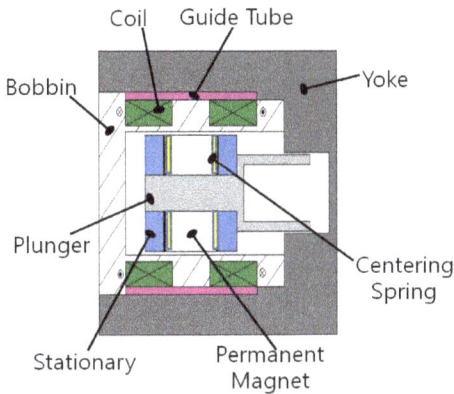

Fig. 1. Structure of solenoid actuator.

Fig. 2 presents the structure of a simple solenoid actuator to the exclusion of permanent magnet. Permanent magnet independently compensates the electromagnetic force of solenoid coil, and it contributes the reduction of consumption power and increasing of operational speed in comparison with the case only used solenoid coil for generation of same attraction power.

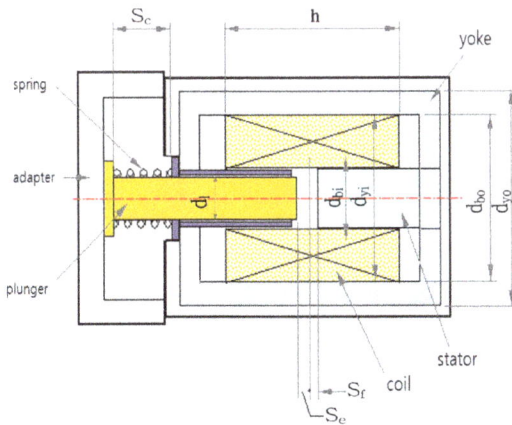

Fig. 2. Simple structure of solenoid actuator.

3. Failure Analysis Documents

3.1 Failure Modes and Effects Analysis

Failure mode and failure mechanism analysis represents the main failure mode of component parts, and the effect on total system and failure cause which was generated by analyzing failed component parts. Table 1 represents the result of failure mode and failure mechanism analysis[3-4]. Critical matrix analysis represents the severity and failure frequency as a matrix form according to the procedures of MIL-STD-882D such as table 2 in order to mark qualitatively distribution of severity at each failure modes, failure modes and critical analysis represents failure modes & failure mechanism analysis result and critical matrix analysis result which is performed previously with representing quantitative value such as table 3. Solenoid actuator used in automobile and aircraft industry is required to have long life time as well as its stable performance which is suitable for various usage situation, and that has recently been an active area of research[5-6].

Table 1. Failure modes & failure mechanism analysis results

#	Part	Function	Failure mode	Failure effect			Failure mechanism	
				Local effect	High-level effect		Mechanism	Load factor
1	Frame	Structure, Magnetic flux passage	Crack, Strain, Wear	Force Loss	Infelicitous motion		Corrosion, Fatigue	Humidity, Pollution, Vibration
2	Plunger	Motion	Jam, Constraint	Current Rise	Inoperable, Infelicitous motion		Scratch, Corrosion, Pollution	Temperature, Humidity, Pollution
3	Stationary	Structure, Magnetic flux passage	Strain, Wear	Noise Rise	Infelicitous motion		Wear, Scratch	Repeated motion
4	Sealing	Seal	Crack, Strain	Infelicitous motion	Leakage		Stress, Wear	Infelicitous assembly, Repeated motion
5	Spring	Return force generation	Fracture	Returning power loss	Infelicitous motion		Fatigue	Repeated motion
6	Coil	Force generation	Cut	Stoppage current	Inoperable		Fatigue	Temperature
7	Coil	Force generation	Short	Force loss	Inoperable		Dielectric breakdown	Temperature
8	Guide tube	Friction reduction	Pollution	Current rise	Infelicitous motion		Pollution	Dust, Temperature

Table 2. Critical matrix analysis results

	High	III	5,8	II	2	I	6,7
Severity	Medium	IV	4	III	1	II	
	Low	V	3	IV		III	
			Low		Medium		High
				Failure frequency			

Table 3. Failure modes, effects and severity analysis results of solenoid actuator

#	Part	Function	Failure mode	Failure effect		Failure mechanism		Criticality		
				Local effect	High-level effect	Mechanism	Load factor	Frequency	Severity	Criticality
1	Frame	Structure, Magnetic flux passage	Crack, Strain, Wear	Force Loss	Infelicitous motion	Corrosion, Fatigue	Humidity, Pollution, Vibration	Medium	Medium	5
2	Plunger	Motion	Jam, Constraint	Current Rise	Inoperable, Infelicitous motion	Scratch, Corrosion, Pollution	Temperature, Humidity, Pollution	Medium	High	7
3	Stationary	Structure, Magnetic flux passage	Strain, Wear	Noise Rise	Infelicitous motion	Wear, Scratch	Repeated motion	Low	Low	1
4	Sealing	Seal	Crack, Strain	Infelicitous motion	Leakage	Stress, Wear	Infelicitous assembly, Repeated motion	Low	Medium	3
5	Spring	Return force generation	Fracture	Returning power loss	Infelicitous motion	Fatigue	Repeated motion	Low	High	5
6	Coil	Force generation	Cut	Stoppage current	Inoperable	Fatigue	Temperature	High	High	9
7	Coil	Force generation	Short	Force loss	Inoperable	Dielectric breakdown	Temperature	High	High	9
8	Guide tube	Friction reduction	Pollution	Current rise	Infelicitous motion	Pollution	Dust, Temperature	Low	High	5

3.2 Fault Tree Analysis

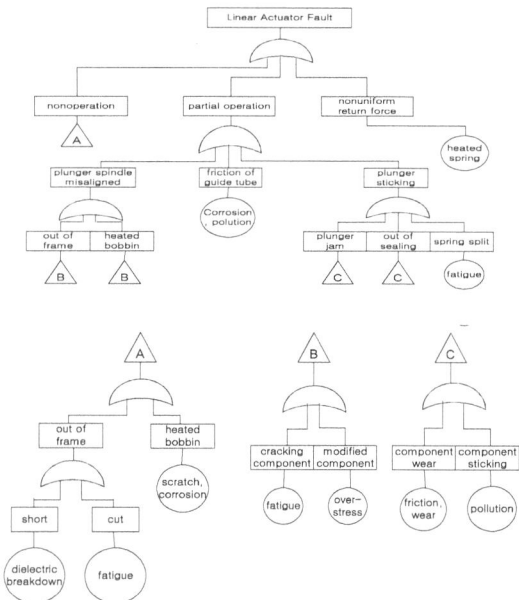

Fig. 3. Fault tree analysis of solenoid actuator.

In Fig. 3, contents in the circle are causes of the detailed failure and contents in the rectangular are phenomena of failure which are generated by causes of the detailed failure.

In case of there are many causes of detailed failure, the causes of failure are combined by OR gates and then linked to triangle having a connection character. By analysis of Fig. 3, we can know that linear actuator is unable to operate by causes of failure linked under △, and occur partial fault such as shaft bending of plunger, friction of guide tube, and plunger sticking by △ᴮ, △ᴄ, corrosion or pollution, fatigue, and also know that return force can be non-uniform by spring wearing.

3.3 Quality Function Deployment

Table 4 represents QFD(I) which expresses the failure modes corresponding to each required quality function. And the importance grade of each failure modes converted to scores, as it were, higher score means more important failure mode. Table 5 represents QFD(II). It is made by use of the analysis results of QFD(I), and determines the reliability test items.

Table 4. QFD level 1 of solenoid actuator

Failure mode Requirement	Frame crack & strain	Sealing wear	Plunger jam & Constraint	Stationary strain & Wear	Spring Strain & fracture	Coil cut	Coil short	Guide tube Pollution
Smoothing function of plunger	○	○	◎	○	◎	◎	◎	◎
Low consumption power	△			○			◎	○
High attraction force	△		○	◎	△	◎	◎	△
Low temperature rise of coil	○		◎	○	○	◎	◎	○
High insulation resistance						◎	◎	
Low pollution		◎	△					
Low noise					◎			△
Prevention of corrosion		◎	○		△		○	○
High durability	◎	△	○	△	◎	◎	◎	△
failure risk	25	26	36	27	38	45	59	33
priority	8	7	4	6	3	2	1	5

* Very important ◎:9, Important ○:5, Normal △:3

Table 5. QFD level 2 of solenoid actuator

failure mode/ risk test item	Frame crack & strain	Sealing wear	Plunger jam & Constraint	Stationary strain & Wear	Spring Strain & fracture	Coil cut	Coil short	Guide tube pollution	status	priority
	25	26	36	27	38	45	49	33		
Structure test	○							△	224	13
Noise test		△	△	◎					429	10
Consumption power Test			○			○	○		650	6
Maintain power test			○	○			○		560	9
Starting power test			○			○			405	11
Attraction force test	○		△	○	○		○	○	968	2
Temperature rising test			△			◎	○		758	4
Insulation Resistance test						◎	○		650	6
Voltage test						◎	○		650	6
High temperature test		○				○	◎		706	5
Low temperature test		○			△				265	12
Humidity test			△		△	△	○	◎	899	3
Lift test	○	◎	◎	○	◎	◎	◎	○	2171	1

* Very important ◎:9, Important ○:5, Normal △:3

4. Design Equations

For optimal design, theoretical and empirical knowledge are simultaneously needed. Theoretical knowledge governs the operational characteristics of the solenoid actuator, and empirical knowledge compensates for the theoretical limitation obtained from the designer's design and manufacturing experiences for various kinds of solenoid actuator. In this chapter, the derived design equations is presented table 6.

Table 6. Design equations

Items	Equations
Attraction Force	$F = \dfrac{B^2 \bullet S}{2\mu_0}[N]$
Design Coefficient	$K_f = \dfrac{\mu_0 \bullet S \bullet U_m^2}{2}$
Maximum Attraction Force	$F_{max} = \dfrac{K_f}{S_f^2}$
Minimum Attraction Force	$F_{min} = \dfrac{K_f}{d^2}$
Magnetic Flux Density	$B = 2 \bullet \dfrac{\sqrt{2 \bullet \mu_0 \bullet F_{min}}}{d_l \bullet \sqrt{\pi}}$
Magnetic Motive Force	$U = \dfrac{C_m \bullet B \bullet d}{\mu_0}$
Inner Diameter of Yoke	$d_{yi} = d_{bo} + C_g$
Outer Diameter of Yoke	$d_{yo} = \sqrt{d_{yi}^2 + C_p \bullet d_l^2}$
Rising Temperature	$T_f = \dfrac{q \bullet \rho}{d \bullet \lambda \bullet X_i \bullet w} \bullet \left(\dfrac{N \bullet W}{h \bullet V}\right)^2$
Coil Height	$h = \sqrt[3]{\dfrac{(q \bullet \beta \bullet \rho \bullet U^2)}{2 \bullet \lambda \bullet X_i \bullet T_f}}$
Coil Mean Length	$l_m = \dfrac{\pi(d_{bo} + d_{bi})}{2}$
Diameter of Bare Wire	$d_s = \sqrt{\left(\dfrac{2 \bullet \rho \bullet (d_{bo} + d_{bi}) \bullet U}{V}\right)}$
Total Turn Number	$N = n_c \bullet m_c$
Equivalent Resistance	$R_t = \dfrac{2 \bullet \rho \bullet (d_{bo} + d_{bi}) \bullet N}{\pi d_s^2}$
Coil Current	$I = \dfrac{V}{R_t}$
Consumption Power	$W = V \bullet I$

where,

F :	Attraction force	C_g :	Empirical constant
B :	Magnetic flux density	C_p :	Empirical constant
S :	Cross sectional area of plunger	T_f :	Rising temperature
μ_0 :	Permeability in the air	q :	Duty ratio
K_f :	Design coefficient	ρ :	Relative resistance
U_m :	Theoretical magnetic motive force	λ :	Dissipation coefficient
F_{max} :	Maximum attraction force	W :	Consumption power
F_{min} :	Minimum attraction force	N :	Total turn numbers

S_f :	Sum of fixed air gap	h :	Coil height
d :	Maximum distance between plunger and stationary	V :	Supply voltage
d_l :	Radius of plunger	l_m :	Coil mean length
U :	Actual magnetic motive force	m_c :	Total layer number of coil in the radial direction
C_m :	Empirical compensation coefficient	n_c :	Total turn number in shaft direction
d_{yi} :	Inner diameter of yoke	d_s :	Diameter of bare wire
d_{yo} :	Outer diameter of yoke	R_t :	Equivalent resistance
d_{bo} :	Outer diameter of bobbin	I :	Coil current

[1,7-9]

We obtained the results of design in table 7 through the design equations and input parameters. The prototype solenoid actuator is represented Fig. 4. And also, Fig. 4 is represented the servo valve which is adopted the solenoid actuator prototype.

Table 7. Results of design

Parameters	Values
Attraction force	150 (N)
Magnetic flux density	2.3 ~ 2.4 (T)
Permanent magnet width	8 (mm)
Coil turn number	330 (No.)

Fig. 4. Prototype of solenoid actuator & servo valve assembly.

5. Acceleration Model of Solenoid Actuator

Domestic industries surveyed integral servo valve operating conditions the lifetime of the field by considering the 90% confidence level B_{10} life of $1.0*10^7$ cycles that were guaranteed. According to the survey of the literature, shape parameter of 1.1 Weibull distributions follows. Reliability standards for the evaluation of servo valve in the prescribed lifetime of $1.0*10^7$ cycles (B10 life) means to guarantee the following.

-Lifetime distribution : Shape parameter(β) 1.1 Weibull distribution[10]

- Insurance life : $1.0*10^7$ cycles (B_{10} Lifetime)

- Confidence level : 90 %

- Prototype : 3ea

At this point, no-failure test time was calculated (1) using, the result is $6.1*10^7$ cycles.

$$t_n = B_{100p} \bullet \left[\frac{\ln(1-CL)}{n \bullet \ln(1-p)} \right]^{\frac{1}{\beta}} \tag{1}$$

$$t_n = 1.0 \times 10^7 \bullet \left[\frac{\ln(1-0.9)}{3 \bullet \ln(1-0.1)} \right]^{\frac{1}{1.1}} \cong 1.0 \times 10^7 \, cycles$$

Where,

t_n : No failure test time

B_{100p} : Assurance life

CL : Confidence level

n : Number of prototype

p : Unreliability (if B_{10}, p =0.1)

β : Shape parameter

However, because no-failure test time is too long to accelerate the model chosen, and accelerated life test of time should be calculated. Failure modes related to the pressure and flow of the servo valve. Pressure and flow are chosen to acceleration stress. Considering the pressure and flow General Log-Linear acceleration model applied to the test conditions. So the acceleration factor is calculated acceleration time fault-tolerance test. 7.0 MPa, 50 L/min and acceleration, conditions 25.2 MPa, 88 L/min was chosen as the acceleration factor calculation Thus, equation (2) 22.8096.

$$AF = \left(\frac{P_{test}}{P_{field}} \right)^m \times \left(\frac{F_{test}}{F_{field}} \right)^l \tag{2}$$

$$= \left(\frac{25.2}{7.0} \right)^2 \times \left(\frac{88}{50} \right)^1 = 22.8096$$

Where,

AF : Acceleration Factor

P_{test} P_{field} : Acceleration & field pressure (MPa)

ω_{test} ω_{field} : Acceleration & field flow (L/min)

m, l : Acceleration index (m =2, l =1)

Calculated acceleration factor equation (3) by substituting the acceleration test, time (t_{na}) is produced.

$$t_{na} = \frac{t_n}{AF} = \frac{61,000,000}{22.8096} \cong 2.7 \times 10^6 \, cycles \qquad (3)$$

6. Life Test and Performance Test

6.1 Accelerated Life Test

We have proven the propriety of the design equations through the accelerated life test. We used 3 unit the valve assembly adopted the solenoid actuator. Fig. 5 shows the accelerated life test.

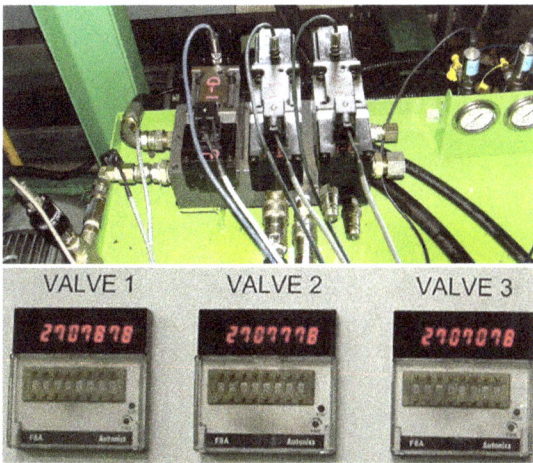

Fig. 5. Accelerated life test.

6.2 Attraction Force and Linearity Test

For attraction test, firstly, the prototype solenoid actuator is to be fastened on the attraction force test equipment, and it is connected to load cell by mechanical coupling. When the work of coupling connection is done, the plunger must be surely tightened not to arise the eccentric phenomenon, and also it must be checked that the value of load cell is 0. The attraction force should be measured changing the value of current form 0 to +3 A and form 0 to -3 A. Fig. 6 represents the measuring result of the attraction force. From

Fig. 8, we can know that the attraction force is about 200 N at maximum current ±3 A and about 153 N at rated current ±2.2 A. And, the linearity is almost approaching to the first order function, f(x)=3.5x. At this time, for overall region, the error rate of linearity is 1.90 %, and the error rate of symmetrical characteristic is 3.05~-2.00 %. These mean that the test results for attraction force and linearity are generally satisfactory.

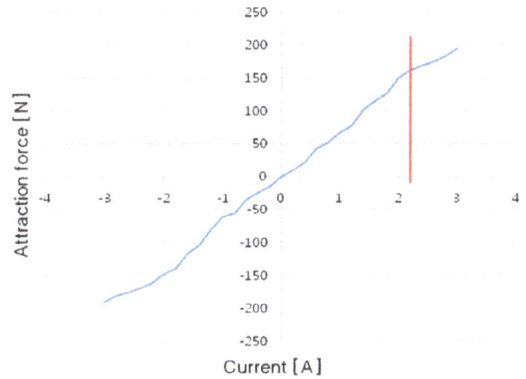

Fig. 6. Results of attraction force test.

6.3 Test of Step Response

This test is to measure the time difference between supplying time of input step signal and reaction time of plunger. At here, the 80 % control signal(8 V) to controller is used as input step signal, and reaction of plunger is detected by output signal of LVDT. Referring to Fig. 7, the control signal 8 V is applied to controller at point of time 30 ms, and the reaction signal 5 V of LVDT is detected at point of time 35 ms. So, we can know that the step response time is 5 ms.

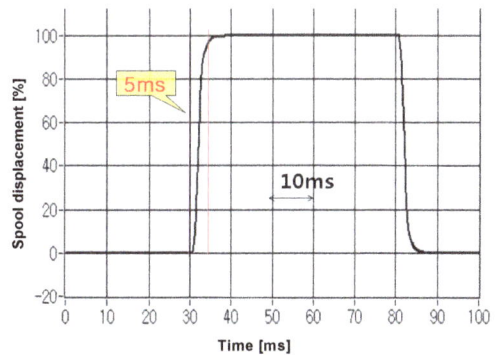

Fig. 7. Results of step response test.

6.4 Test of Frequency Response

This test is similar to the test of step response. The input is control signal of controller and output is reaction signal of LVDT. This test performed at 25 % and 100 % magnitude of input signal with 0.01 Hz~200 Hz carrier frequency region. Figure 8 is the test results for 25 % control signal of controller. It shows that the -3dB frequency is about 189 Hz.

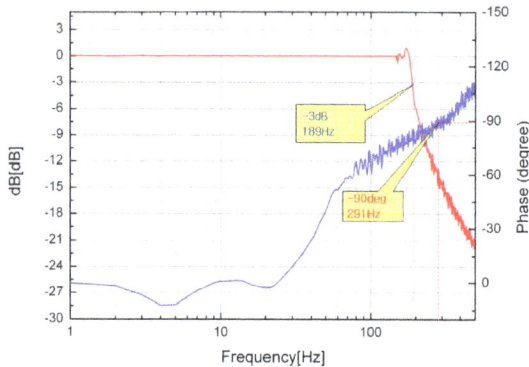

Fig. 8. Results of frequency response test.

7. Conclusion

In this paper, as a reliability assessment study of solenoid actuator for valve operation, we performed detailed analysis about potential failure modes of linear actuator and the effects and converting the importance grade of each failure modes into score, and determine test items for prior check the failures observed through failure analysis. And also, design equations of solenoid actuator for valve operation was composed.

1) The high speed solenoid actuator reliability assessment techniques were established. In addition, the failure analysis data such as FMEA, CMA, FTA, 2-Stage QFD was written.

2) The design equations are induced for design using between electromagnetic theories and empirical knowledge.

3) The major test items of servo valve takes a long time for the accelerated life test. 3ea samples were calculated for the accelerated life test of time.

As results of experiments, we can know that the step response of prototype solenoid actuator approaches 5ms in Fig. 7, the operating frequency   is about 189 Hz in -3dB gain and 291 Hz in -90 degree at   25 % input signal in Fig. 8.

Even though a conclusion may review the main results or contributions of the paper, do not duplicate the abstract or the introduction. For a conclusion, you might elaborate on the importance of the work or suggest the potential applications and extensions.

References

[1]   C. Roters, "Electro Magnetic Device", *John Wiley & Sons*, Inc, 1970.

[2]   B.J.Sung, E.W,Lee, H.E.Kim, "Development of Design Program for On and Off Type Solenoid Actuator", *Proceedings of the KIEE Summer Annual Conference 2002(B)*, pp929~931, 2002.7.10.

[3]   C. Lie et al., "Micromachined Magnetic Actuators using Electroplated Permalloy", *IEEE transactions on magnetic*, 5(3), pp.1976~1985, 1999.

[4]   A. Steck, "Modeling the Magnetic Propertics and Dynamic Behaviour of MRF-Valves in Flow Mode, Actuator 2002", No. B5.3 pp.347, 2002.

[5]   Ministry of Knowledge Economy, 2002, "Reliability Assessment Technology of high-speed and intelligent system", *Ministry of Knowledge Economy*, 2002. 8.

[6]   Takasi Arakawa, Shigeki Niimi, "Optimization Technology of Magnetic Circuit for Linear Solenoid", *SAE Technical Paper Series*, 2002-01-0565.

[7]   Hydraulic and Pneumatic Lap. of KIMM, "Development of low Consumption Power Type Solenoid Valves" *KIMM-CSI annual report*, 2001.

[8]   Baek ju Sung, Eun woon Lee and Hyoung eui Kim, "Characteristics of Non-magnetic Ring for High-Speed Solenoid Actuator", *The Eleventh Biennial IEEE Conference on Electromagnetic Field Computation*, Korea, pp.342, June 2004.

[9]   Kanda Kunio, "Design Concept for DC Solenoid of Pneumatic Valve" *KIMM research reporter*, 1997.

[10] Barringer Associates, Inc., "Weibull database".

6

A New RBC handover scheme for LTE-R system

Jong-Hyun Kim[†], Soon-Ho Kim* and Kyu-Hyoung Choi**

Abstract – RBC(Radio Block Center) handover of high-speed railway impacts greatly on the safety and efficiency of train operation. However, there are still many problems to be solved in the high mobility applications of GSM-R, especially the higher handover failure probability, which seriously degrades the reliability of railway communication. This paper proposes an optimized handover scheme, in which the coordinated multiple point transmission technology is applied to improve the traditional hard handover performance of GSM-R system.

Keywords: Radio Block Center(RBC), Coordinate multiple point transmission(CoMP), Soft handover, Outage probability

1. Introduction

While high-speed train enters the section of RBC handover, the continuous, reliable data transmission will become one of the prominent problems, which directly affects the safety and efficiency of train operation.

Nowadays, the development of high-speed railway makes people's lives more and more convenient. Meanwhile, it puts forward higher requirements on high-speed railway communication services.

The existing GSM for Railway (GSM-R) network is mainly based on the second-generation Global System for Mobile Communications (GSM), and its data rate is too low to meet the broadband mobile communication access and other value-added service demands of passengers.

In order to provide broadband services and applications for users not only at home but also on trip, long-term evolution (LTE) has been chosen as the next generation's evolution of railway mobile communication system by International Union of Railways (UIC), which supports significant higher data rates and lower system latency. This paper focuses on the applications of LTE technology in railway broadband wireless communication network.

Normally, the main problems caused by user's high-speed movement in cellular wireless communication system are frequent handover, Doppler shift and large penetration loss, among which over frequent handover needs to be paid special attention as it seriously affects the communication

quality of service (QoS) and traffic reliability.

Currently only traditional hard handover scheme is supported in LTE, which encounters two challenges under high speed movement circumstance.

On the one hand, the handover delay caused by hard handover is relatively large. The high-speed train passes through the overlapping areas so fast that the handover procedure can't be accomplished timely. On the other hand, the speed of MRS is so fast that it would miss the optimal handover position, which degrades the handover success probability. In order to overcome the challenges mentioned above, the existing handover scheme of GSM-R should be optimized to improve the handover success probability in high-speed movement circumstance.

Currently, more and more researches focus on the Broad band communication access issues in railway communication.

Coordinated multiple point transmission (CoMP) transmission and reception allows geographically separated base stations to joint sending data to one terminal and joint receiving data from one terminal, by which the inter-cell interference could be reduced and the system frequency spectral efficiency would be improved.

From the analysis above we can see that, there are few researches on the application of base stations interaction and multiple vehicle stations cooperation. In order to take full advantage of the multiple base stations cooperation feature of CoMP systems under high speed scenarios, this paper proposes a seamless soft handover scheme based on CoMP, which allows the train to receive signals from both adjacent base stations when the train travels through the overlapping areas. Thus, the handover failure rate is degraded and the reliability of train to ground communication is guaranteed. The rest parts of the article are

---

†   Corresponding Author: Hunter technology Co.,LDT, Korea (kjh85@htt.co.kr)

*   Hunter technology Co.,LTD, Korea (snow@htt.co.kr)

**   Graduate School of Railroad, Seoul National University of Science and Technology, Korea (khchoi@seoultech.ac.kr)

arranged as follow: Section 2 introduces the current handover scheme in GSM-R systems. Section 3 describes the proposed handover scheme in detail. Section 4 analyzes the system performance.

Section 5 illustrates the simulation results and the improvement achieved by the proposal. And finally Section 6 concludes the whole paper.

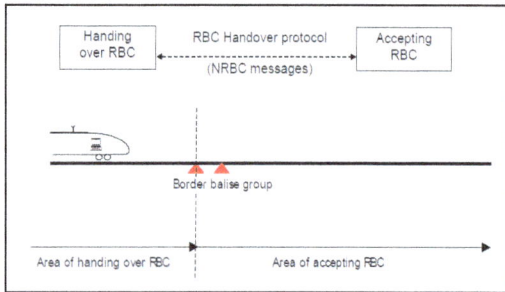

Fig. 1. RBC/RBC handover.

2. Current hard handover scheme in GSM-R system.

The current hard handover scheme in GSM-R systems is shown in Fig. 2. The eNodeB is with the coverage radius R. The width of overlapping area is L. The vertical distance between an eNodeB and track is d_{min}. The whole railway mobile communication network has a feature of linear coverage topology. The overall handover procedure could be described as follow.

As the train moves into the overlapping area from cell i, eNodeB i decides whether to handover or not according to the reported received signal strength indication (RSSI), reference signal received power (RSRP) or reference signal received quality (RSRQ) measurement information by the train and the radio resource management (RRM) information, as shown in Fig. 2. Once the handover is triggered, the train disconnects with eNodeB i and tries to synchronize with the target eNodeB j. Then the Mobility Management Entity (MME) switches communication route and the previous station eNodeB i releases both user plane resources and control plane resources when the handover procedure is completed, as shown in Fig. 2.

As analyzed above, the current handover scheme in GSM-R systems is a Break-Before-Make approach. The scheme allows the train to receive signals from only one base station at one time. It will cause a larger outage probability and interrupt latency, which severely affect the

reliability of train to ground communication.[1]

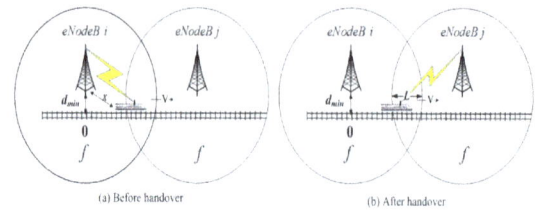

Fig. 2. Current hard handover procedure in GSM-R system.

3. A CoMP based soft handover scheme in LTE system

This paper proposes a seamless soft handover scheme utilizing CoMP joint processing and transmission technology, which can significantly improve the handover performance when the train moves through the overlapping areas.

As shown in Fig. 3, as the front on-vehicle station enters into the overlapping area, the source eNodeB i activates the cooperative transmission set (CTS) composed of eNodeB i and eNodeB j. The two or more eNodeBs communicating with MRSs simultaneously are called CTS. The CTS activation is based on the measurement information reported by the moving train and the position information supplied by the communication based train control system (CBTC). Once the CTS are activated, the source eNodeB i shares the entire user plane data of users inside the train to the target eNodeB j by the high-speed backhaul of LTE network. The two adjacent eNodeBs both use the same frequency resource to communicate with the train. Signals from the eNodeBs in the cooperative set are in-phase superposed by pre-coding, which provides a diversity gain and power gain.[2]

It should be noted that CoMP CTS always contains two eNodeBs in the linear coverage topology of high-speed railway. As shown in Fig. 3, eNodeB i and eNodeB j keep the cooperative relation and communicate with the train simultaneously when the train body entirely enters into the overlapping area. Both the front and the rear on vehicle stations can receive signals from the two cooperative eNodeBs, but the measurement information is only forwarded to the source eNodeB i. The source eNodeB i decides whether to handover or not according to the measurement information reported by the moving train and the RRM information of the target eNodeB j, without interrupting the data transmission in user plane. The signal from the source eNodeB i drop gradually as the train moves farther away from it. Once the signal strength of target eNodeB j is larger than that of source eNodeB i for hdB, the handover is triggered. The signaling switch in control plane is a 'break before make' process, which is similar to current hard handover scheme. Meanwhile, both eNodeB i and eNodeB j transmit the user plane data to users inside the train with the same frequency resource by Physical Downlink Shared Channel (PDSCH) without interrupting communications, which avoids the interrupt latency caused by current hard handover in GSM-R systems. After the handover procedure having been done, the new source eNodeB j kicks the former source eNodeB i off the cooperative set, when the signal strength of eNodeB i drops to a certain threshold, as shown in Fig. 3.[3]

Summarily, the optimized handover scheme based on CoMP can significantly degrade the outage probability and improve the handover performance, by which the train can communicate with the two adjacent eNodeBs in the overlapping area. The scheme achieves a seamless handover performance as soft handover.

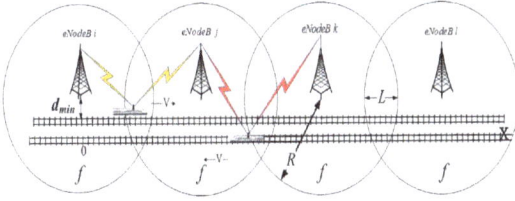

Fig. 3. CoMP based soft handover in LTE-R system.

4. Performance analysis

Without loss of generality, let the train be located at x away from the eNodeB i, the received power in dBm in decibels can be expressed as

$$P_r(i) = P_t(i) - PL(i,x) - A(i,x,\sigma) \quad (1)$$

Where P_t is the transmit power in dBm, PL(i,x) is the path loss between eNodeB i and location x, and $A(i,x,\sigma)$ is the shadow component at the location x, generally modeled as a Gaussian random variable with mean zero and standard deviation. Typically, σ are 6 or 8 in dB.

As the railway communication network is the linear coverage topology, we consider a two-cell system. The shadow can be expressed as

$$A_i = a\xi_0 + b\xi_i \quad (2)$$
$$A_j = a\xi_0 + b\xi_j$$

Where $a^2 + b^2 = 1$ Note that i is the link number. ζ_0 is common to both A_1 and A_2, ζ_i represents the independent, both ζ_i and ζ_j are independent Gaussian random variables with mean zero and standard deviation σ. Meanwhile, because the tail of Gaussian distribution extends to infinity, a fade margin F dB is added to the transmit power.

4.1 Outage in the CoMP handover scheme

In our proposal, the adjacent eNodeBs communicate with the train simultaneously in the overlapping area. Outage will happen if and only if both signals are of unacceptable quality. So the outage probability using the proposed handover scheme is

$$P_{out} = P_r\{Min[10m\log(r_1) + \zeta_1, 10m\log(r_2) + \zeta_2]\}$$
$$> F$$

Where, $A(i,x,\sigma)$ and $A(i,x,\sigma)$ represent the shadow fading losses of eNodeB i and eNodeB j respectively. F is fade margin and $F = P_t - PL - R_s$, PL is the path loss and R_s is the receiver sensitivity. By straightforward manipulation, the outage probability can be shown as [7]

$$a\zeta_0 + b\zeta_i = F$$
$$a\zeta_0 + b\zeta_j = F$$
$$\zeta_i = \frac{F - a\zeta_0}{b} = \frac{\zeta_i}{\sigma} = \frac{F - a\zeta_0}{b\sigma}$$
$$\zeta_j = \frac{F - a\zeta_0}{b} = \frac{\zeta_j}{\sigma} = \frac{F - a\zeta_0}{b\sigma}$$
$$P_{out} = \frac{1}{(2\pi\sigma)^{3/2}} \int_{-\infty}^{\infty} e^{-\frac{t^2}{2}} \left(\int_{\frac{F-at}{b}}^{\infty} e^{-\frac{\zeta_i^2}{2\sigma^2}} d\zeta_i \int_{\frac{F-at}{b}}^{\infty} e^{-\frac{\zeta_j^2}{2\sigma^2}} d\zeta_j \right) dt$$
$$P_{out} = \frac{1}{\sqrt{2\pi}} \int_{-\infty}^{\infty} e^{-\frac{x^2}{2}} \left[\Phi(\frac{F - a\sigma x}{b\sigma}) \right]^2 dx$$

For $a^2 + b^2 = 1$ and $\sigma = 6$, the above equation is simplified with the help of MATLAB software and the margin of power by varying the cell distance, and variation in P_{out} with margin of power are calculated.

Calculations for P_{out} by varying margin of power on simplification the above equation can be written as

$$P_{out} = \frac{1}{\sqrt{2\pi}} \int_{-\infty}^{\infty} e^{-\frac{x^2}{2}} \left[\frac{1}{\sqrt{2\pi}} \int_{\frac{F-3\sqrt{2}x}{3\sqrt{2}}}^{\infty} e^{-\frac{x^2}{2}} dx \right]^2 dx$$

The integral cab be simplified with the MATLAP software and can be written as

$$\Phi(\frac{F-a\alpha x}{b\sigma}) = \frac{1}{2}\frac{\sqrt{2}}{\sqrt{\pi}}\left[\frac{1}{2}\sqrt{2}\sqrt{\pi} - \frac{1}{2}erf\left(\frac{F}{6} - \frac{\sqrt{2}x}{2} \right) \right]\sqrt{2}\sqrt{\pi}$$

$$\Phi(\frac{F-a\alpha x}{b\sigma}) = [0.5 - 0.5erf(0.166F - 0.707x)]$$

Put this value of Φ in equation(x), we have

$$P_{out} = \frac{1}{\sqrt{2\pi}} \int_{-\infty}^{\infty} e^{-\frac{x^2}{2}} [0.5 - 0.5erf(0.166F - 0.707x)]^2 dx.$$

4.2 Outage in current hard handover scheme

In the current handover scheme, the MRS can connect to only one eNodeB at one time. For ideal case, the MRS is always switched to the eNodeB with the best signal quality. However, this may lead to the well-known 'ping pong' effect around the cell boundary. In practical systems, handover will be triggered on the condition that the received power of the source eNodeB is lower than that of the target eNodeB by hysteresis level h.

In most of the existing researches, the outage probability in hard handover system is obtained with the hysteresis level h being assumed infinite so that it is impossible to handover to the neighboring cell. Let shadow fading loss $A = \xi$, where ξ is Gaussian random variable with mean zero and standard deviation σ. An outage occurs when the fading component is larger than fade margin, which is expressed as[7]

$$P_{out} = P_r[10m\log(r) + \zeta] > F$$

$$P_r(\zeta > F) = \frac{1}{\sqrt{2\pi}\sigma} \int_{F}^{\infty} e^{-\frac{\zeta^2}{2\sigma^2}} d\zeta$$

$$P_{out}(\frac{F}{\sigma} = t) = \frac{1}{\sqrt{2\pi}} \int_{F}^{\infty} e^{-\frac{t^2}{2}} dt$$

$$= \Phi(\frac{F}{\sigma}) = \frac{1}{2}erfc(\frac{F}{\sqrt{2}\sigma})$$

Where

$$\Phi(x) = \int_{x}^{\infty} \frac{1}{\sqrt{2\pi}} e^{-\frac{t^2}{2}} dt, \ erfc(x) = \frac{2}{\sqrt{\pi}} \int_{x}^{\infty} e^{-t^2} dt$$

5. Simulation and analysis

To make a comparison between current scheme and our proposal, the MATLAB simulation results and analysis are presented. Detail simulation parameters are shown in Table 1.

Table 1. Simulation parameters

Parameters	Value
Channel bandwidth	10MHz
Antenna pattern	Omni
Carrier frequency	2.5Ghz
P_t	46dBm
h_{at}	1.5m
h_{eNodeB}	35m
d_{min}	30m
N_o	-174dBm/Hz
Cell radios(R)	3km
Site-to-site distance	4.8km
Path loss model	ITU-R M.1225

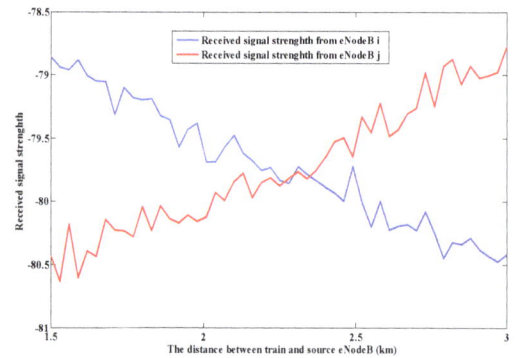

Fig. 4. Received signal strength according to train location.

The train moves through the overlapping area, the received signal strength of the two adjacent eNodeBs is shown in Fig. 4. The red and blue curves represent the received signal strength of eNodeB i and that of eNodeB j respectively. Due to the influence of path loss and shadow fading loss, both signals strength of eNodeB i and eNode j are extremely lower at cell boundary than those in the interior of a cell, which would cause a higher handover failure probability.

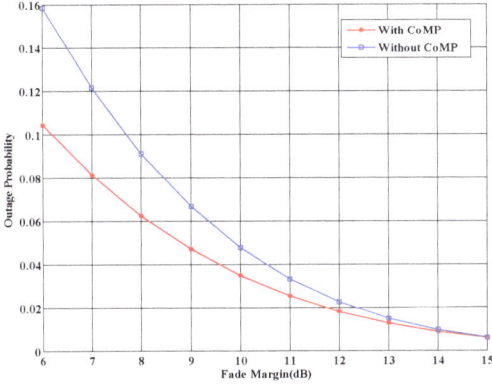

Fig. 5. Outage comparison When σ=6 dB.

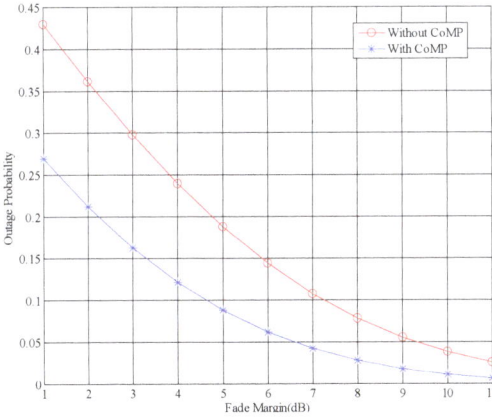

Fig. 6 Outage comparison When σ=8 dB

Table 2. Outage comparison When σ=6 dB

No	Fade margin	With CoMP	Without CoMP
1	6	0.1044	0.1587
2	7	0.0814	0.1217
3	8	0.0625	0.0912
4	9	0.0427	0.0668
5	10	0.0350	0.0478
6	11	0.0256	0.0334

Table 3. Outage comparison When σ=8 dB

No	Fade margin	With CoMP	Without CoMP
1	6	0.0622	0.2266
2	7	0.0426	0.1908
3	8	0.0282	0.1587
4	9	0.0182	0.1303
5	10	0.0113	0.1056
6	11	0.0068	0.0846

6. Conclusion

The probability of outage is an important parameter for performance evaluation of the railway communication system. This paper proposes a CoMP based soft handover scheme for LTE-R in high speed railway. The proposal allows the train to receive signals of both adjacent eNodeBs, which significantly improves the handover performance, degrades the outage probability and guarantees the reliability of train to ground communication, as the simulation results show.

It is clear that the required margin of power for Soft handover (proposed LTE-R) is about 6 dB less than for hard handover (GSM-R).

The reduction in margin of power may increase the life of battery and the cell area for soft handover or number of cell decrease. The lesser number of cell required less number of base station, it is because of the dependence of number of user on the cell size. Thus for better results in LTE-R communication system the value of standard deviation σ must be minimum as possible.

References

[1] S.A.Mawjoud, S.H.Fasola, Performance Evalustion of Soft Handover in WCDMA System. 2011

[2] Kiran M. Rege, Fade margins for soft and hard handoffs. Wireless Networks 2 (1996) 277-288).

[3] Un-ho Jeong and Dong-Hoi Kim, Hard Handover by the Adaptive Time-to-trigger Scheme based on Adaptive Hysteresis considering the Load Difference between Cells in 3GPP LTE System. 2010.

[4] Wantuan Luo, Ruiqiang Zhang and Xuming Fang, a CoMP soft handover scheme for LTE Systems in high speed railway. EURASIP Journal on Wireless Communications and Networking 2012.

[5] Fading Models, http://www.comlab.hut.fi/opteus/333/ 2004 2005 slides/Fading models. Pdf.

[6] Amit Dixit, S.C.Sharma, R.P.Vats, Vishal Gupta, A Comparative Study of Call Outage for Hard and Soft

Handoff in CDMA Communication Networks. 2008 IEEE.

[7] Amit Dixit, S.C.Sharma, R.P.Vats, Vishal Gupta, A Comparative Study of Call Outage for Hard and Soft Handoff in CDMA Communication Networks. 2008 IEEE.

[8] Andrew J. Viterbi, Audrey M. Viterbi, Soft Handoff Extends CDMA Cell Coverage and Increases Reverse Link Capacity. IEEE Journal on selected areas in communications, Vol. 12. No.8. OCTOBER 1994.

[9] H Jijing. M Jun, ZH Zhangdui, Reserch on handover of GSM-R network under high-speed scenarios. Railway Communication Signals. 42, 2006.

7

Experimental Study and Analysis of Insulator Breakdown Characteristics with Short-tail Lightning Impulse

Zhao Yuan[†], Li Yu[*], Deng Chun[*], Yuan Yi-chao[*], He Jin-liang[**] and Wang Xi[**]

Abstract – The voltage waveform on the insulator by a direct lightning stroke is a short-tail lightning impulse. Experimental study and analysis of insulator breakdown characteristics under short-tail lightning impulses are given in this paper with the contrast of standard lightning impulse. The impacts of crossarm and transmission lines are also considered. The experimental results indicate that the insulator 50% breakdown voltage under short-tail impulse is significantly higher than that of standard wave with both positive and negative polarities, about 25~30%. The 50% breakdown voltage of insulator with negative-polarity short-tail impulse is about 5% higher than with positive-polarity one. The insulator voltage-time characteristics are also given with short-tail and standard lightning impulses.

Keywords: Short-tail lightning impulse, Standard lightning impulse, Insulator, 50% breakdown voltage, Voltage-time characteristics

1. Introduction

Lightning surge analysis is very important in insulation design of transmission lines and substations. Therefore, it is necessary to study the insulator flashover phenomena and breakdown characteristics. Previous studies on insulator breakdown characteristics are usually on the condition of standard lighting impulse. However, recently more researches show that the measured voltage waveform on the insulator by a direct lightning stroke is a short-tail lightning impulse, and more researches on insulator breakdown characteristics and models are based on short-tail lightning impulse [1-4]. Therefore, these results on standard lighting impulse are not applicable to describe an accurate lightning overvoltage process, and the differences may cause errors in the evaluation of lightning performance. The breakdown characteristics of insulator under short-tail impulse, which is similar to actual waveform, are needed to investigate too.

In this paper, voltage waveform on the insulator in lightning surge is analyzed at first. Then the accurate measurements of 50% flashover voltage and voltage-time characteristics of insulators in different voltage levels with short-tail and standard lightning impulse are carried out. Finally, the experimental results are compared and analyzed.

† Corresponding Author: Electric Power Research Institute, Jibei Electric Power Company Limited, Beijing 100045, China (zhao.yuan@ncepri.com.cn)
* Electric Power Research Institute, Jibei Electric Power Company Limited, Beijing 100045, China
** Department of Electrical Engineering, Tsinghua University, Beijing 10084, China

2. Overvoltage waveform on insulator in lightning surge

The insulation design under lightning overvoltages is measured by the international standard 1.2~50 μs voltage wave, which has a virtual front time of 1.2 μs ± 30%, a virtual time to half value of 50 μs ± 20% and a tolerance on the crest voltage of ± 3%. Many previous researches are based on the experimental results with standard lightning impulse. However, it is not clear that the flashover phenomena with the short-tail lightning impulse, which appears by an actual lightning stroke, can be accurately modeled by these approaches. The overvoltage waveform on insulator in lightning surge is analyzed based on measurements simulation in this chapter.

2.1 Measurement results

The measured results of the voltage waveform on insulators are few. But from these results, still can see that it's short-tail voltage waveform on insulator in lightning surge.

A rectangular pulse with a rise time of 20 μs and duration of 4 μs, and a slow-front wave with a rise time of 3 μs are injected into the top of the tower [1]. In Fig. 1 are shown the measured voltage waveforms on the insulators. The insulator voltage waveforms fall in a very short period of time. The simulations in paper [5] agree well with the experimental results.

(a) Upper phase insulator voltage

(b) Middle phase insulator voltage

(c) Lower phase insulator voltage

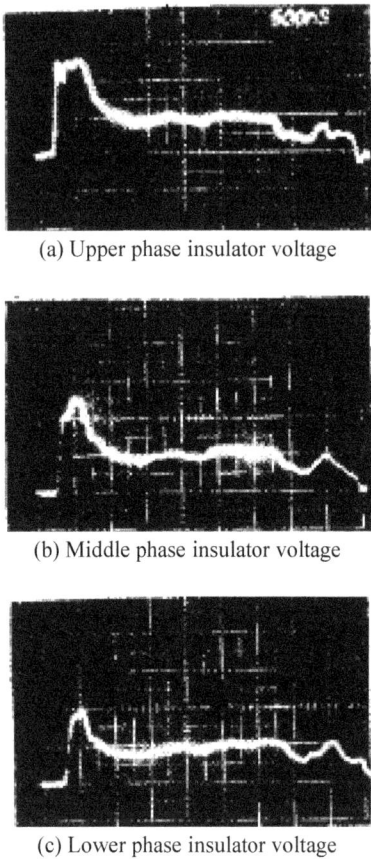

Fig. 1. Measured insulator voltage in paper [5] (Sweep: 0.5 μs/div).

Insulator voltages are recorded when lightning strikes the tower in paper [6], and one is shown in Fig. 2. Although the insulator flashover at about 20 μs, still can see the waveform tail is very short, only about 6 μs.

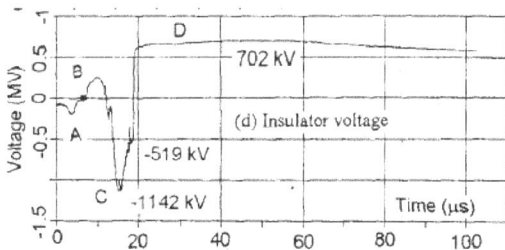

Fig. 2. Measured waveform in paper [6] (Record 89121413).

The paper [7] provides an insulator voltage waveform when a tower with arrester in a lightning surge, and compared with the calculated results, as shown in Fig. 3. Although the installation of an arrester can impact on the waveform, it's very clear that the insulator voltage waveform has a short tail.

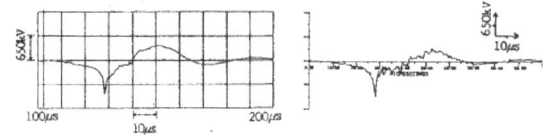

Fig. 3. Measured waveforms and calculated waveforms by EMTP in paper [7].

2.2 Simulation results

To investigate the overvoltage during direct lightning strike, models on PSCAD/EMTDC platform are designed to simulate the transients during lightning direct strike on transmission tower and impulse voltage waves imposed on insulators in paper [4]. The calculated voltage waveforms drop quickly after reaching the peak, which is influenced by reflected waves of grounding body of the tower and adjacent tower. According to the simulation results, the actual impulse voltage is not same as standard one. The wave tail is only about 2 μs.

The short-tail waveforms in lightning surge, considering different lightning current waveforms, grounding resistances and span distances between poles, are analyzed in paper [8]. The insulator overvoltages diverge significantly from the 1.2/50 μs double exponential form. With the increasing of grounding resistance and spans, the tail time of lightning overvoltage become longer. But whichever lightning current waveforms, grounding resistances and spans, its tail time is less than 10 μs

2.3. Analysis of lightning overvoltage on insulator

From above results can see that overvoltages on insulators are short-tail waveforms. Fig. 4 shows the lightning overvoltage waveform on insulator. When the lightning current flows from the top of the tower to the bottom, the voltage drops because of the grounding resistance. The decreasing time is related to the structure and height of the tower, and the reduction of the voltage is related to the grounding resistance. After that, the wave reflection of the adjacent towers makes the voltage drop further, and the decreasing time is related to the span of and towers. Because the height of the tower and the distance between two towers is very short compared with the current speed, the lightning overvoltage on the insulator falls soon after the peak, which has the features of short-tail wave.

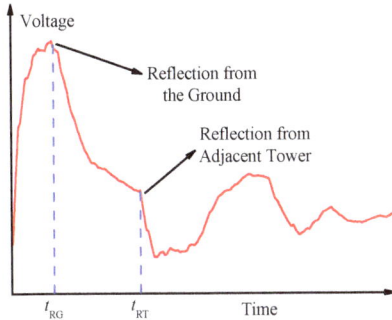

Fig. 4. Analysis of overvoltage waveform on insulator.

3. Experiment results and analysis

Experimental study and analysis of breakdown characteristics of insulators in different voltage level with short-tail and standard lightning impulses are given in this chapter.

3.1 Experimental apparatus

Short-tail impulse wave can be achieved by adjusting wave-tail resistance, and generally require the use of external wave-front resistance [9]. Schematic of experiment for short-tail impulse is shown in Fig. 5, in which,

IG ：Impulse Generator

IUT ：Insulator under Test (with simulated crossarm and transmission lines)

D ：Divider

CP ：Current Probe

DSO ：Digital Storage Oscilloscope

HC ：High-speed Camera

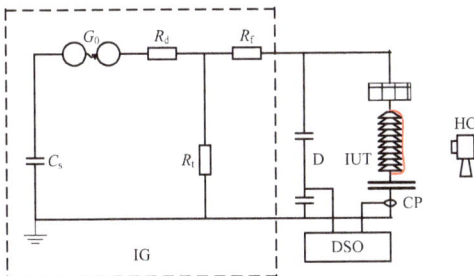

Fig. 5. Schematic of experiment for short-tail impulse.

A short-tail lightning impulse is generated, which the rise time is 1.5 μs and the decay time to the half of peak value is 15 μs. The short tail lightning impulse of the positive or the negative polarity is applied to 110 kV, 220kV and 500 kV voltage levels insulators. The voltage across the insulator and the current flowing the insulator are measured. As a contrast, standard lighting impulse is also applied.

In the experiment, up-down method is used to measure 50% breakdown voltage. All the voltage data has been corrected by atmospheric parameters. Especially, simulated crossarm and line are considered in the experiment.

3.2 Results and analysis

3.2.1 Photos by high-speed camera

processes of insulator flashover are recorded by high-speed camera. The time interval between two photos is 2.31 μs, and the time of exposure is 1.00 μs. Fig. 6 shows a typical insulator flashover under short-tail lightning impulse.

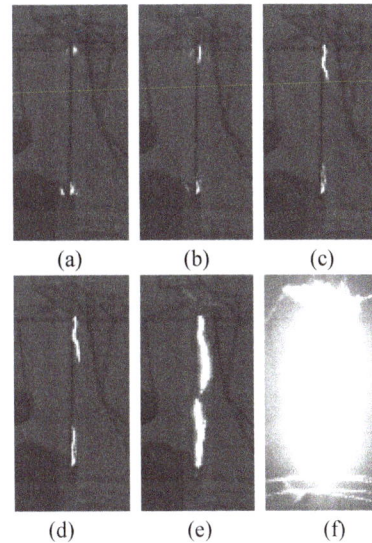

Fig. 6. Insulator flashover under short-tail lightning impulse.

3.2.2 Voltage-time characteristics

The voltage-time characteristics under positive or negative polarity of short-tail and standard lightning impulses are shown in Fig. 7, in which

PSL ： Positive-polarity Standard Lightning impulse waveform

NSL : Negative-polarity Standard Lightning impulse waveform

PST : Positive-polarity Short-Tail lightning impulse waveform

NST : Negative-polarity Short-Tail lightning impulse waveform

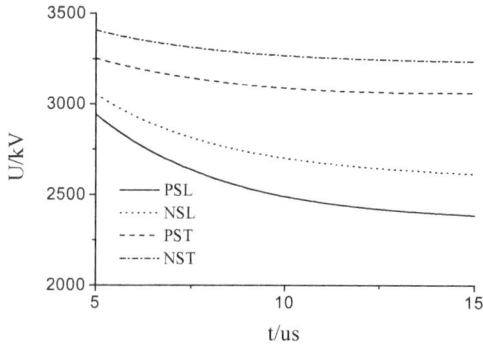

Fig. 7. Voltage-time characteristics of different waveforms in different polarities.

From the experimental data in Fig. 5 can see that in the same breakdown voltage, the breakdown time of insulator with short-tail lightning impulse is longer than with standard impulse. It means that in insulation design with transmission lines in a direct lightning stroke is stringent. Also from Fig. 7 can see that in the same breakdown voltage, the breakdown time of insulator with positive-polarity impulse is shorter than with negative-polarity impulse.

3.2.3 The 50% breakdown voltage

Table 1 shows the 50% breakdown voltages of 500 kV insulators with positive or negative polarity of short-tail lightning impulse on the contrast of results with standard lightning impulse. From the experimental data in Table 1 can see the 50% breakdown voltage with short-tail impulse and is larger compared with the standard wave. The difference of each polarity is about 25~30%.

Table 1. The 50% breakdown voltage of 500 kV insulator under positive or negative polarity of short-tail and standard lightning impulses (kV)

	Positive polarity	Negative polarity
Short-tail impulse waveform	3056	3222
Standard lightning impulse waveform	2331	2542

Fig. 8 shows the 50% breakdown voltage of insulators in different voltage level under short-tail impulse waveform. From the Fig. 8 can see that the 50% breakdown voltage of insulator with negative-polarity impulse is about 5% higher than with positive- polarity impulse.

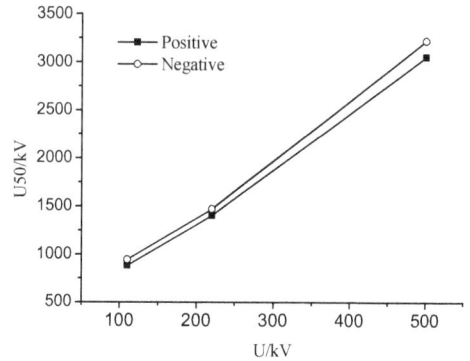

Fig. 8. 50% breakdown voltage of insulators in different voltage level under short-tail impulse waveform.

4. Conclusions

Short-tail lightning impulse appears across the insulators on the transmission line by a direct lightning stroke. The breakdown characteristics of insulators in different voltage level with short-tail lightning impulse, including positive and negative polarity. As a contrast, standard lighting impulse is also applied. The voltage-time characteristics and 50% breakdown voltage are reported in this paper. The main conclusions are summarized as follows;

(1) The breakdown phenomena with short-tail and standard lightning impulse are investigated in detail. The breakdown characteristics of insulators with the short-tail lightning impulse are more accurately than the standard one.

(2) From voltage-time characteristics can see that in the same breakdown voltage, the breakdown time of insulator with short-tail lightning impulse is longer than with standard impulse. It means that in insulation design with transmission lines in a direct lightning stroke is stringent.

(3) The 50% breakdown voltage with short-tail impulse and is 25~30%.larger compared with the standard wave. The 50% breakdown voltage of insulator with negative-polarity short-tail impulse is about 5% higher than with positive-polarity one.

For future study, the experimental data can be used in Leader Propagation Model to calculate the breakdown characteristics of short-tail impulse waveform, and a set of parameter suitable for short-tail impulse can acquired.

References

[1] H. Motoyama, Experimental Study and Analysis of Breakdown Characteristics of Long Air Gaps with Short Tail Lightning Impulse, pp. 972-979, IEEE Transactions on Power Delivery, vol. 11, no. 2, April 1996.

[2] YI Hui, ZHANG Jun-lan, Test and Study of Insulation Characteristic of EHV Transmission Line Insulation under the Short Tail Surge, High Voltage Engineering, pp. 16-17, vol. 28, no. 6, June 2002.

[3] M. Ab Kadir and I. Cotton, Application of the Insulator Coordination Gap Models and Effect of Line Design to Backflashover Studies, pp. 443-449, Electrical Power and Energy Systems, vol. 32, no. 5, June 2010.

[4] WANG Xiao-chuan, ZENG Rong, HE Jin-liang and etc., Experiment and Simulation of Air Gap Breakdown Characteristics Under Short-tail Impulse Waveform, High Voltage Engineering, pp. 925-929, vol. 34, no. 5, May 2008.

[5] Masaru Ishii, Tatsuo Kawamura, Teruya Kouno, Multistory Transmission Tower Model for Lightning Surge Analysis, IEEE Transactions on Power Delivery, pp. 1327-1335, vol. 6, no. 3, July 1991.

[6] Atsuyuki Inoue, Sei-ichi Kanao, Observation and Analysis of Multiple-Phase Grounding Fault Caused by Lightning, IEEE Transactions on Power Delivery, pp. 353-360, vol. 11, no. 1, January 1996.

[7] Y. Matsumoto, O. Sakuma, K. Shinjo, and etc, Measurement of Lightning Surges on Test Transmission Line Equipped with Arresters Struck Natural and Triggered Lightning, IEEE Transactions on Power Delivery, pp. 996-1002, vol. 11, no. 2, April 1996.

[8] William A. Chisholm, New Challenges in Lightning Impulse Flashover Modeling of Air Gaps and Insulators, IEEE Electrical Insulator Magazine, pp. 14-25, vol. 26, no. 2, March/April 2009.

[9] A. Carrus and L. E. Funes, Very Short Tailed Lightning Double Exponential Wave Generation Techniques Based on Marx Circuit Standard Configurations, pp. 782-787, IEEE Transactions on Power Apparatus and Systems, vol. 103, no. 4, April 1984.

8

Design and Analysis of a Magnetless Flux-Switching DC-Excited Machine for Wind Power Generation

Christopher H. T. Lee[†], K. T. Chau*, Chunhua Liu* and Fei Lin*

Abstract – This paper proposes a new outer-rotor magnetless flux-switching DC-excited (FSDC) machine which can offer low-speed operation to directly capture the wind power for power generation. Compared with its permanent-magnet (PM) counterparts, the proposed machine equips no PM such that it enjoys the definite cost benefit. One of the key designs of the proposed machine is the multitoothed per stator pole structure. With the multitoothed structure, the proposed machine not only offers the flux-modulation effect to boost up its torque density, but also is favorable for the low-speed operation. In addition, the external DC-field excitation can be controlled independently to offer the flux-controlled ability to achieve constant voltage output among various wind speeds. By performing the time-stepping finite element method (TS-FEM), the performances of the proposed machine can be analyzed, and hence verifying the design.

Keywords: Magnetless machine, DC-field excitation, Multitoothed, Flux-switching, Wind power generation, Finite Element Method (FEM)

2. Introduction

There is an accelerating pace on the development of the utilization of the renewable energies due to the demand on the environmental protection, and wind power is one of the hottest topics to be well focused [1]-[5]. Generally speaking, wind power generation can be categorized into two major systems, namely the constant-speed constant-frequency (CSCF) and variable-speed constant-frequency (VSCF) systems [6]-[9], respectively. The CSCF system allows the turbine speed to be kept into constant without any connection with the variation of the wind speed. Hence, it ends up with high mechanical stress and low efficiency performance. Meanwhile, with the adoption of power electronics, the VSCF system can capture the maximum wind power by changing the turbine speed along with the wind speed. As expected, it overcomes the problem from the CSCF system and thus resulting with higher efficiency [10]-[11]. However, the natural wind speed (5-10 m/s) confines the turbine speed (around 200 rpm) such that a specific low-speed machine is favorable for the VSCF system [12].

† Corresponding Author: Dept. of Electrical and Electronic Engineering, The University of Hong Kong, Hong Kong SAR, China (htlee@eee.hku.hk)
* Dept. of Electrical and Electronic Engineering, The University of Hong Kong, Hong Kong SAR, China ({ktchau, chualiu, linfei}@eee.hku.hk)

In recent years, the flux-switching permanent magnet (FSPM) machine which offers the bipolar flux-linkage waveforms and higher power density was proposed [13]-[14]. In addition, with the adoption of the multitoothed structure, the FSPM machine is desirable for the low-speed operation [15] and thus it is also highly suitable for the wind power generation. Nevertheless, the PM material cost has been soared drastically while the corresponding supply is limited and fluctuating [16]. Therefore, the focuses of the advanced magnetless machines should be enhanced [17]-[20]. Meanwhile, the concept of the independent DC-field excitation which could offer an external field excitation was proposed [21]-[25] and it could also be implemented into the FSPM machine.

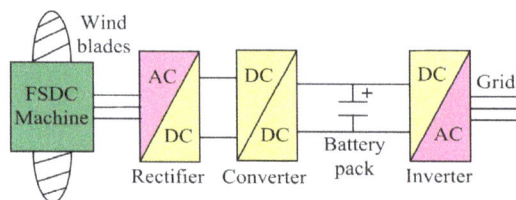

Fig. 1. Wind power generation system.

In this paper, by purposely replacing the independent DC-field winding with the PM materials of the FSPM machine, a FS DC-excited (FSDC) machine is proposed. The proposed FSDC machine adopts the inner-stator outer-rotor 36/32-pole structure. With the controllability of the

DC-field excitation, the proposed machine can offer constant voltage output over wide ranges of wind speeds and loads. By applying the time-stepping finite element method (TS-FEM), the machine designs will be verified.

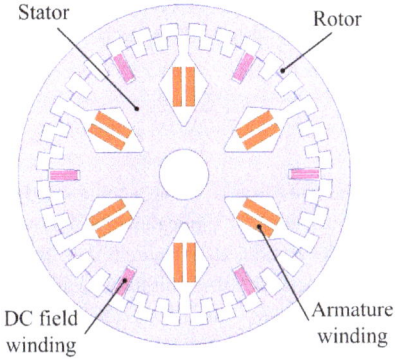

Fig. 2. Proposed 36/32-pole FSDC machine.

2. System and Machine Configurations

Fig. 1 shows the system configuration which the wind blades are directly mounted on the outer-rotor of the proposed FSDC machine. The system holds the similar configuration with the conventional low-speed outer-rotor generator and the generated output voltage is the same as the conventional generator [3]. Meanwhile, the proposed system requires a much larger size and heavier weight with much lower speed design [6], [10].

Fig. 2 shows the topology of the proposed FSDC machine which consists of an inner-stator of 6 salient poles, each fitted with 6 teeth and results with equivalent stator teeth number of 36. This ends with the outer-rotor of 32 salient poles. The proposed machine adopts two types of windings, namely the armature windings and DC-field windings. Both of the two windings adopt the concentrated winding arrangements which are easily constructed and as shown in Fig. 3(a) and Fig. 3(b), respectively.

The design of the pole-pair arrangement of the proposed FSDC machine is based on the following criteria [15]:

$$\begin{cases} N_{sp} = 2mk \\ N_{se} = N_{sp}N_{st} \quad (N_{st} = 2,4...) \\ N_r = N_{se} - N_{sp} \pm 2k \end{cases} \tag{1}$$

where N_{sp} is the number of stator poles, N_{st} the stator teeth,

N_{se} the equivalent stator poles, N_r the rotor poles, m the number of phases and k is any integer. The key design data of the machine is shown in Table 1.

The key features of the proposed FSDC machine are summarized as follows:

1) By adopting the inner-stator outer-rotor structure, the proposed machine enables the rotor to couple with the wind blades directly, leading to eliminate the mechanical transmission losses.

2) Without installation of any PM materials, the proposed machine takes the definite merit of cost benefit.

3) With the structure of mulitoothed per stator pole, the proposed machine can offer the flux-modulation effect to boost up its torque density. This structure allows the FSDC machine behaves similarly with the magnetic-geared machine [26]-[29] such that it is favorable for the high-torque low-speed environment.

4) With the conventional brushless AC (BLAC) conduction scheme as shown in Fig. 4, the proposed machine can utilize all the torque producing zones such that the torque performance can be improved.

The DC-field winding can be controlled independently such that the flux-controlled ability can be naturally offered. Therefore, the proposed machine can maintain the constant output voltage among a wide range of wind speeds.

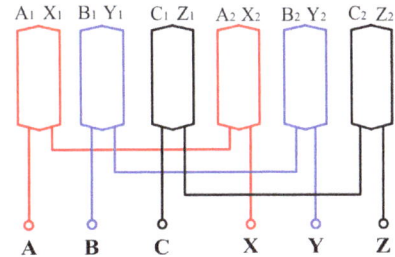

(a)

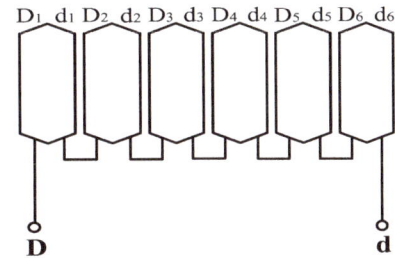

(b)

Fig. 3. Proposed winding arrangements: (a) Armature windings (b) DC-field windings.

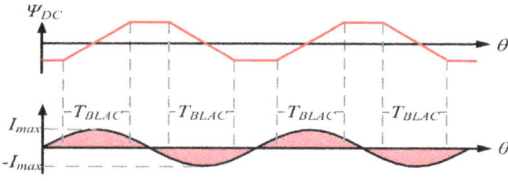

Fig. 4. Operating principle under BLAC operation.

Table 1. Key design data of proposed machine

Item	Value
Rotor outside diameter	270.0 mm
Rotor inside diameter	221.2 mm
Stator outside diameter	220.0 mm
Stator inside diameter	40.0 mm
Air-gap length	0.6 mm
Stack length	80.0 mm
No. of rotor poles	32
No. of stator poles	6
No. of stator teeth	6
No. of equivalent stator poles	36
No. of armature phases	3
No. of turns per armature coil	80

3. Operating Principle

With the DC-field winding, the proposed machine can be operated by the bipolar conduction scheme which is similar as the convention FSPM machine [13]. Based on the same principle, the proposed FSDC machine can also work with the BLAC operation with each phase having 180° conduction angle [25] as shown in Fig. 4. When DC flux-linkage, Ψ_{DC} is increasing, a positive current is applied to the armature winding to produce the positive torque, T_{BLAC}. Meanwhile, a negative current is applied instead when Ψ_{DC} is decreasing and the torque produced will also be positive [8], [25]. Under the normal BLAC operation, the phase currents are three-phase sinusoidal form as given by:

$$\begin{cases} i_a = I_{\max} \sin\theta \\ i_b = I_{\max} \sin(\theta + (2\pi/3)) \\ i_c = I_{\max} \sin(\theta - (2\pi/3)) \end{cases} \tag{2}$$

where I_{max} is the maximum value of the phase current. The electromagnetic torque T_{BLAC} can be expressed as [8]:

$$\begin{aligned} T_{BLAC} &= \frac{P_{BLAC}}{\omega} = 3x \frac{1}{2\pi\omega} \int_0^{2\pi} \left((E_{BLAC} \sin\theta)(I_{\max} \sin\theta) \right) d\theta \\ &= \frac{3E_{BLAC} I_{\max}}{2\omega} \end{aligned} \tag{3}$$

where P_{BLAC} is the electromagnetic power during BLAC scheme, ω the electrical angular velocity, E_{BLAC} is the amplitude of the fundamental component of the back EMF.

Similar as the conventional FSPM machine, the proposed FSDC machine adopts the same speed control algorithm. The operating speed is regulated by the operating frequency, f_{PH} and the value of N_r as shown below [20]:

$$n = \frac{60 f_{PH}}{N_r} \tag{4}$$

where n is the rotor speed which should be designed as around 200 rpm for the application of the wind power generation [12]. It should be noted that the proposed machine is designed with the multitoothed structure, hence resulting with a higher value of N_r. This can achieve the purpose of low-speed operation and thus favoring the application of the wind power generation.

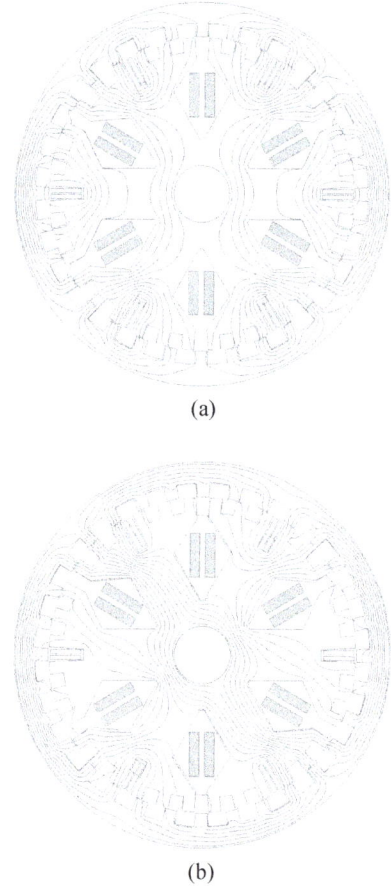

(a)

(b)

Fig. 5. Magnetic field distribution: (a) No-load (b) Full-load.

4. Analysis Approach

In order to describe the machine modeling, TS-SEM is applied and three sets of equations are established. First, the electromagnetic field equation of the machines is governed by [30]-[32]:

$$\begin{cases} \Omega : \dfrac{\partial}{\partial x}\left(v\dfrac{\partial A}{\partial x}\right) + \dfrac{\partial}{\partial y}\left(v\dfrac{\partial A}{\partial y}\right) = -(J_z + J_f) \\ A|_S = 0 \end{cases} \tag{5}$$

where Ω is the field solution region, A and J_z the z-direction components of vector potential and current density, respectively, J_f the equivalent current density of the excitation field, S the Dirichlet boundary, and v the reluctivity. Second, the armature circuit equation of the machine during motoring is given by:

$$u = Ri + L_e\frac{di}{dt} + \frac{l}{s}\iint_{\Omega_e}\frac{\partial A}{\partial t}d\Omega \tag{6}$$

where u is the applied voltage, R the winding resistance, L_e the end winding inductance, l the axial length, s the conductor area of each turn of per phase winding, and Ω_e the total cross-sectional area of conductors of each phase winding. Third, the motion equation of the machine is given by:

$$J_m\frac{\partial \omega}{\partial t} = T_e - T_L - \lambda\omega \tag{7}$$

where J_m is the moment of inertia, ω the mechanical speed, T_L the load torque, and λ the damping coefficient.

By applying the TS-FEM, the no-load and full-load magnetic field distribution of the proposed machine with DC-field current of 500 A-turns is shown in Fig. 5(a) and Fig. 5(b), respectively. Meanwhile, the DC-field current can be independently controllable such that the proposed machine can provide the flux-controlled ability effectively.

5. Performances of the Proposed Machine

By performing the TS-FEM, all important machine performances can be deduced. The machine is designed to avoid the magnetic saturation so that the core loss is minimized. Firstly, the proposed machine is fed purely by the DC-field excitation of 500 A-turns and the flux-linkage of the proposed machine is shown in Fig. 6. It shows that

the flux-linkages of the 3 phases hold the same pattern with no phase shift or severe distortion. This confirms that the arrangement of the DC-field windings match the pole-pair setting of the proposed machine. In addition, the result confirms that the proposed machine obtains the bipolar flux-linkage waveform that is the characteristic of the FSDC machine, hence offering the better torque density than its counterpart does.

Secondly, the steady stages of the no-load EMF waveforms without and with the DC flux control are accessed. The simulated no-load EMF waveforms at 100, 300, and 900 rpm without and with DC flux control are shown in Fig. 7(a) and Fig. 7(b), respectively. Undoubtedly, without any DC flux control, the amplitude of the output voltages vary with the operating speeds as shown in Fig. 7(a); meanwhile, under the flux control, all the output voltages are kept constant over the wide range of speeds as shown in Fig. 7(b). These confirm that the proposed machine can offer flux-controlled ability effectively to achieve efficiency optimization.

Thirdly, the no-load EMF waveform of the proposed machine at the rated speed of 225 rpm is shown in Fig. 8. It can be calculated that the root-mean-square (RMS) value of the no-load EMF is 71.2 V. Meanwhile, it can be observed that the no-load EMF waveform of the proposed machine is not very sinusoidal and this may result with a larger torque ripple when applying the BLAC operation scheme. In order to improve the torque ripple situation, the harmonic-current-injection scheme can be applied [33], but it is out of the scope of this paper.

Fourthly, the air-gap flux density distribution of the proposed machine is shown in Fig. 9. It can be observed that the original flux of each stator pole is modulated into six portions in accordance with the number of teeth per each stator pole. This confirms and verifies the flux-modulation effect of the multitoothed structure. In addition, the peak value of the flux density is 1.12 T, which is also within the acceptable range.

Fifthly, the electromagnetic torque waveform of the proposed machine at the rated condition is as shown in Fig. 10. It can be found that the averaged steady torque of the proposed machine is 30.5 Nm. Meanwhile, the peak magnitude of the cogging torque of the proposed machine is 1.21 Nm. This value is significantly small, which is only about 3.97 % of its rated torque. In addition, to further evaluate the torque performance, the torque ripple factor is defined as following:

$$K_T = \frac{T_{\max} - T_{\min}}{T_{avg}} \times 100\% \tag{8}$$

where T_{max}, T_{min}, and T_{avg} are the maximum, minimum and average torque produced, respectively. It can be found that the torque ripple of the proposed machine is about 36.1 %. This relative large value of torque ripple is due to the mismatch among the BLAC conduction and the asymmetrical no-load EMF patterns.

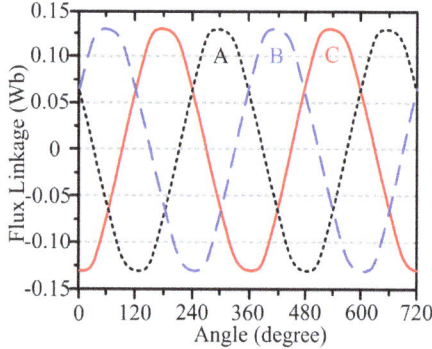

Fig. 6. Flux-linkage waveforms of the proposed machine with DC-field excitation of 500 A-turns.

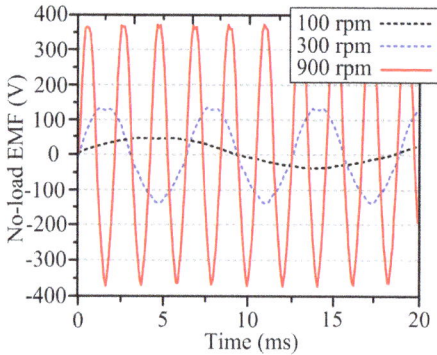

(a)

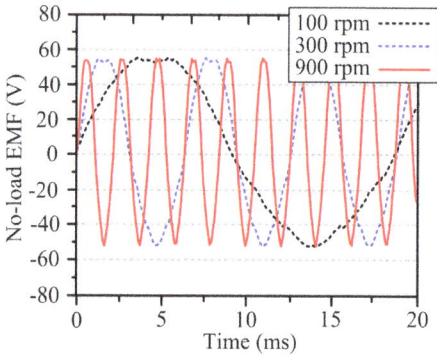

(b)

Fig. 7. No-load EMF waveforms at various operating speeds: (a) Without flux control (b) With flux control.

Finally, the core loss waveform of the proposed machine operating at the rated load is simulated as shown in Fig. 11. It can be observed that the average core loss of the proposed machine is 19.4 W, which is within the acceptable range. In addition, the power level and the core loss of the proposed machine are 700 W and 2.77 %, respectively. Therefore, the core thermal dissipation is not a great problem that can be easily handled by using the air cooling system.

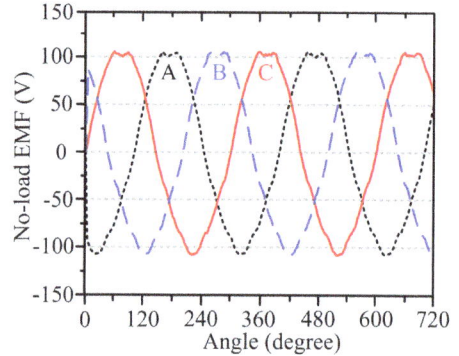

Fig. 8. No-load EMF waveform of the proposed machine at rated speed.

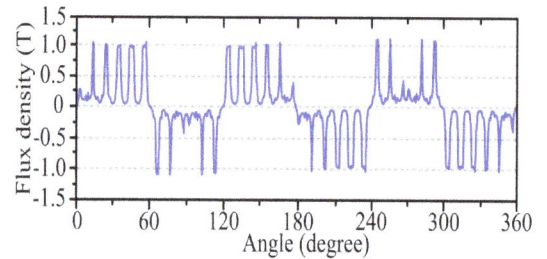

Fig. 9. No-load air-gap flux density distribution.

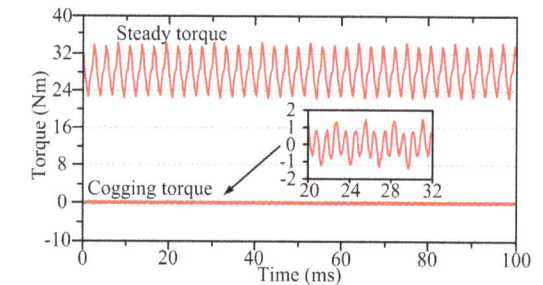

Fig. 10. Output torque waveform at rated speed.

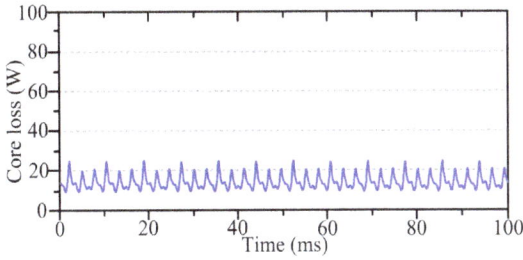

Fig. 11. Core loss waveform at rated load.

Table 2. Machine performances

Item	FSDC
Power	700 W
Rated speed	225 rpm
No-load EMF at rated conditions	71.2 V
Air-gap flux density	1.12 T
Rated torque	30.5 Nm
Cogging torque amplitude	1.21 Nm
Cogging torque / rated torque	3.97 %
Torque ripple at rated load	36.1 %
Core loss	19.4 W

6. Conclusion

In this paper, a new FSDC machine, which can achieve low-speed operation, has been proposed for direct-drive wind power generation. By enjoying the bipolar flux-linkage, the proposed machine can offer higher torque density than its counterparts. Compared with its PM counterparts, the proposed magnetless machine enjoys the merit of better cost-effectiveness. Meanwhile, the independent DC-field excitation offers the flux-controlled ability to the proposed machine, such that it can maintain constant generated voltages among the wide range of wind speeds. In addition, the proposed machine achieves satisfied no-load EMF generation and steady torque-handling capability, hence verifying its suitability for the wind power generation application.

Acknowledgements

This work was supported by a grant (Project No. HKU710612E) from the Hong Kong Research Grants Council, Hong Kong Special Administrative Region, China.

References

[1] K.T. Chau, Y.S. Wong and C.C. Chan, "An overview of energy sources for electric vehicles," *Energy Conversion and Management*, vol. 40, no. 10, pp. 1021-1039, Jul. 1999.

[2] C. C. Chan, and K. T. Chau, Modern Electric Vehicle Technology. *Oxford, U.K.:* Oxford Univ. Press 2001, pp.67-83.

[3] L. Jian, K.T. Chau and J.Z. Jiang, "A magnetic-geared outer-rotor permanent-magnet brushless machine for wind power generation," *IEEE Tran. Ind. Appl.*, vol. 45, no. 3, pp. 954-962, May/Jun. 2009.

[4] K.T. Chau, W. Li, and C.H.T. Lee, "Challenges and opportunities of electric machines for renewable energy," (Invited Paper) *Prog. Electromagn. Res. B*, vol. 42, pp. 45-74, 2012.

[5] Z. Lu, and Y. Qiao, "A consideration of the wind power benefits in day-ahead scheduling of wind-coal Intensive power systems," *IEEE Trans. Power Syst.*, vol. 28. no. 1, pp. 236-245, 2013.

[6] Y. Fan, K.T. Chau and M. Cheng, "A new three-phase doubly salient permanent magnet machine for wind power generation," *IEEE Trans. Ind. Appl.*, vol. 42, no. 1, pp. 53-60, Jan./Feb. 2006.

[7] J. Li, and H. Xu, "Research on control systems, of high power DFIG wind power system," *in Proceedings of IEEE MMIT08 Conference*, Three Gorges, China, Dec. 2008.

[8] C. Yu, K.T. Chau and J.Z. Jiang, "A flux-mnemonic permanent magnet brushless machine for wind power generation," *Journal of Appl. Physics*, vol. 105, no. 7, paper no. 07F114, pp. 1-3, Apr. 2009.

[9] C. Liu, K.T. Chau and X. Zhang, "An efficient wind-photovoltaic hybrid generation system using doubly-excited permanent-magnet brushless machine," *IEEE Trans Ind. Electron.*, vol. 57, no. 3, pp. 831-839, Mar. 2010.

[10] Y. Fan, K.T. Chau and S. Niu, "Development of a new brushless doubly fed doubly salient machine for wind power generation," *IEEE Trans. Magn.*, vol. 42, no. 10, pp. 3455-3457, Oct. 2006.

[11] X. Wang, J.M. Yang, and Y. Hu, "Power electronic technology in wind generation system of variable speed-constant frequency," *in Proceedings of PESA2009 Conference*, Hong Kong, China, May 2009.

[12] J. Li, K.T. Chau, J.Z. Jiang, C. Liu and W. Li, "A new efficient permanent-magnet vernier machine for wind power generation," *IEEE Trans. Magn.*, vol. 45, no. 6, pp. 1475-1478, Jun. 2010.

[13] K.T. Chau., C.C. Chan, and C. Liu, "Overview of permanent-magnet brushless drives for electric and hybrid electric vehicles," *IEEE Trans. Ind. Electron.*, vol. 55, no. 6, pp. 2246-2257, Jun. 2008.

[14] R. Cao, C. Mi, and M. Cheng, "Quantitative comparison of flux-switching permanent-magnet motors with Interior permanent magent motor for EV, HEV, and PHEV applications," *IEEE Trans. Magn.*, vol. 48, no.8, pp. 2374-2384, Aug. 2012.

[15] J.T. Chen, Z.Q. Zhu, and D. Howe, " Stator and rotor pole combination for multi-tooth flux-switching permanent-magnet brushless AC machines," *IEEE Trans. Magn.*, vol. 44, no. 11, pp. 4659-4667, 2008.

[16] J. Li, and K.T. Chau, "Performance and cost comparison of permanent-magnet vernier machines," *IEEE Trans. Appl. Supercond.*, vol. 22, no.3, pp. 5202304, Jun. 2012.

[17] J. Faiz, J. W. Finch, and H.M.B. Metwally, "A novel switched reluctance motor with multiple teeth per stator pole and comparison of such motors," *Electric Power Syst. Res.*, vol. 34, no. 3, pp. 197-203, 1995.

[18] K.T. Chau, M. Cheung and C.C. Chan, "Nonlinear magnetic circuit analysis for a novel stator-doubly-fed doubly-salient machine," *IEEE Trans. Magn.*, vol. 38, no. 5, pp. 2382-2384, Sept. 2002.

[19] C. Shi, J. Qiu, and R. Lin, "A novel self-commutating low-speed reluctance motor for direct-drive applications," *IEEE Trans. Appl. Supercond.*, vol. 43, no. 1, pp. 57-65, Jan. 2007.

[20] C.H.T. Lee, K.T. Chau, C. Liu, D. Wu, and S. Gao, "Quantitative comparison and analysis of magnetless machines with reluctance topologies," *IEEE Trans. Magn.*, vol. 49, no. 7, pp. 3969-3972, Jul. 2013.

[21] K.T. Chau, Y.B. Li, J. Zhang, and C. Liu, "Design and analysis of a stator-doubly-fed doubly-salient permanent-magnet machine for automotive engines," *IEEE Trans. Magn.*, vol. 42, no. 10, pp. 3470-3472, Oct. 2006.

[22] S. Taibi, A. Tounzi, and F. Piriou, "Study of a stator current excited vernier reluctance machine," *IEEE Trans. Energy Convers.*, vol. 21, no. 4, pp. 823-831, Dec. 2006.

[23] C. Yu, K.T. Chau, X. Liu and J.Z. Jiang, "A flux-mnemonic permanent magnet brushless motor for electric vehicles," *Journal of Appl. Physics*, vol. 103, no. 7, paper no. 07F103, pp. 1-3, Apr. 2008.

[24] C. Liu, K.T. Chau, J. Zhong and J. Li, "Design and analysis of HTS brushless doubly-fed doubly-salient machine," *IEEE Trans. Appl. Supercond.*, vol. 21, No. 3, pp. 1119-1122, Jun. 2011.

[25] C.H.T. Lee, K.T. Chau, and C. Liu, "Electromagnetic design and analysis of magnetless double-rotor dual-mode machines," *Prog. Electromagn. Res.*, vol. 142, pp. 333-351, 2013.

[26] L. Jian, K.T. Chau, W. Li and J. Li, "A novel coaxial magnetic gear using bulk HTS for industrial applications," *IEEE Trans. Appl. Supercond.*, vol. 20, no. 3, pp. 981-984, Jun. 2010.

[27] X.Y. Zhu, L. Chen, Q. Li, Y. Sun, W. Hua, and W. Zheng, "A new magnetic-planetary-geared permanent magnet brushless machine for hybrid electric vehicle," *IEEE Trans. Magn.*, vol. 48, no. 11, pp. 4642-4645, 2012.

[28] S. Niu, N. Chen, and S.L. Ho, "Design optimization of magnetic gears using mesh adjustable finite-element algorithm for improved torque," *IEEE Trans. Magn.*, vo. 48, no. 11, pp. 4156-4159, Nov. 2012.

[29] C.H.T. Lee, K.T. Chau, C. Liu, and C. Qiu, "Design and analysis of a new multitoothed magnetless doubly-salient machine," *IEEE Trans. Appl. Supercond.*, vol. 24, no. 3, pp. 5200804, Jun. 2014.

[30] S.J. Salon, Finite Element Analysis of Electrical Machines. *Kluwer Academic Publishers*, Boston USA, 1995.

[31] Y. Wang, K.T. Chau, C.C. Chan, and J.Z. Jiang, "Transient analysis of a new outer-rotor permanent-magnet brushless dc drive using circuit-field-torque time-stepping finite element method," *IEEE Trans. Magn.*, vol. 38, no. 2, pp. 1297-1300, Mar. 2002.

[32] Y. Zhang, K.T. Chau, J.Z. Jiang, D. Zhang and C. Liu, "A finite element – analytical method for electromagnetic field analysis of electric machines with free rotation," *IEEE Trans. Magn.*, vol. 42, no. 10, pp. 3392-3394, Oct. 2006.

[33] W.X. Zhao, M. Cheng, K.T. Chau, and C.C. Chan, "Control and operation of fault-tolerant flux-switching permanent-magnet motor drive with second harmonic current injection," *IET Electric Power Appli.*, vol. 6, no. 9, pp. 707-715, 2012.

9

Influences of Channel Size and Operating Conditions on Fluid Behavior in a MHD Micro Pump for Micro Total Analysis System

Kosuke Ito*, Toru Takahashi**, Takayasu Fujino[†] and Motoo Ishikawa*

Abstract – The present paper discusses experimental and numerical studies on the fluid behavior in a magnetohydrodynamic(MHD) micro pump for micro total analysis system(μ-TAS). In the experiment, the MHD micro pump has a length of 26.0 mm, a width and a height of 0.5 mm. As a working fluid, Phosphate Buffered Saline(PBS) solution is used under the assumption of blood analysis using μ-TAS. A neodymium permanent magnet with the maximum magnetic flux density of 0.32 T is used for applying a magnetic field to a channel in the MHD pump. Experimental and numerical results show that Hartmann flow is not observed in the channel because the MHD interaction is very weak, so that Poiseuille flow is maintained in the channel. The numerical study also examines the influences of the channel height and the strength of externally applied magnetic flux density on the fluid temperature in the channel. The numerical results indicate that an increase in fluid temperature by operating the MHD pump is less than 1 K when the magnetic flux density and the channel height are more than 0.3 T and 0.3 mm, respectively.

Keywords: MHD, μ-TAS, Micro pump, Microfluidics

1. Introduction

The development of a blood analytical technology with micro total analysis system(μ-TAS) is strongly desired to find urgent patients in disaster area and developing country. The μ-TAS is capable of analyzing and processing the blood simultaneously with only about 10 μl blood. A magnetohydrodynamic(MHD) micro pump has been proposed as one of the components of AC current injection [1]-[3]. Fig. 1 shows the principle of MHD micro pump, which is composed of a channel with a pair of electrodes and a externally applied magnetic field. The driving force of MHD micro pump is Lorentz force induced by the interaction between the externally applied magnetic field and the electric current flowing in fluid.

The MHD micro pump has neither moving parts nor valves. Therefore, the geometry design and the fabrication of MHD micro pump are relatively simpler than those of other mechanical micro pumps. A driving voltage of MHD micro pump is lower than one of other non-mechanical micro pumps [4]. The electrolytic bubble generation and the Joule heating are regarded as drawbacks of MHD micro pump. Lemoff and Lee [5] studied an MHD micro pump

driven by AC currents, which used an electromagnet. Their experimental results showed that choosing high operation frequencies such as the order of kHz prevents the electrolytic bubble generation.

So far there have been few studies on the influence of Joule heating on the fluid temperature in a channel. If the operation of MHD micro pumps leads to a considerable increase in fluid temperature, the MHD micro pumps cannot be utilized for μ-TAS for blood analysis.

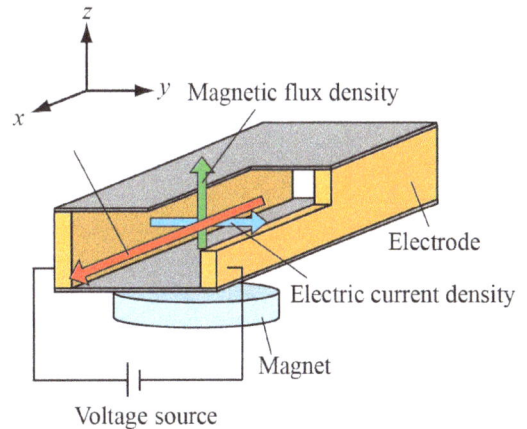

Fig. 1. Principle of MHD micro pump.

This work carries out three-dimensional MHD numerical simulations of MHD micro pumps to clarify the influence

---
†   Corresponding Author: University of Tsukuba, Japan (tfujino@ kz.tsukuba.ac.jp)

*   University of Tsukuba, Japan.

**   National Institute of Advanced Industrial Science and Technology, Japan.

of operating the MHD micro pumps on the fluid temperature for several channel geometries with different channel height and for several magnetic fields with different strength of magnetic flux density. In addition, for the validation of the three-dimensional MHD numerical program developed by us, the distributions of fluid velocity obtained in the numerical simulations are compared with the ones measured in the demonstration experiments of MHD micro pump conducted by us.

2. Experimental and Numerical Procedure

2.1 Experimental Set up

Fig. 2 illustrates a schematic drawing of an MHD micro pump. The MHD micro pump is composed of a base with a micro-channel, a pair of electrodes, a neodymium permanent magnet, capillary tubes and tube connectors. The base is made of acrylic plastic. A copper tape with a length of 26 mm and a thickness of 50μm is used for each of the electrodes. The channel has a length of 50 mm, a width of 0.5 mm, and a height of 0.5 mm.

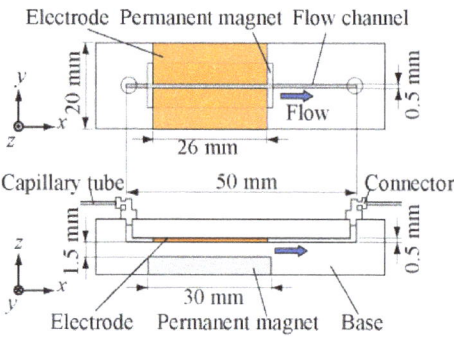

Fig. 2. Schematic drawing of MHD micro pump.

A dual-tracking multi-output DC power supply (PMM18-2.5SDU, KIKUSHI) is used for the driving power of MHD micro pump. The power supply, the MHD micro pump, and a shunt resistor (10 mΩ) are connected in series. A digital multi-meter and the shunt resistor are connected in parallel for measuring voltages between the electrodes. Currents between the electrodes are calculating from the measured voltages and the value of shunt resister.

A phosphate buffered saline(PBS) solution is employed as an electrically conducting fluid. Table 1 summarizes the property of PBS.

The neodymium permanent magnet has a dimension of

30 mm x 10 mm x 10 mm. Fig. 3 shows the distribution of magnetic flux density measured by a Gauss meter (Model421, Lake Shore). A distance between the bottom surface of micro-channel and the top surface of neodymium permanent magnet is 1.5 mm. The maximum magnetic flux density in the micro-channel is 0.32 T.

Table 1. Physical property of PBS

Electrical conductivity σ_0 [S/m]	1.59
Viscosity μ [Pa-s]	0.7×10^{-3}
Mass density ρ [kg/m³]	1020
Specific heat C [J/kg-K]	2.92
Thermal conductivity k [W/m-K]	0.6

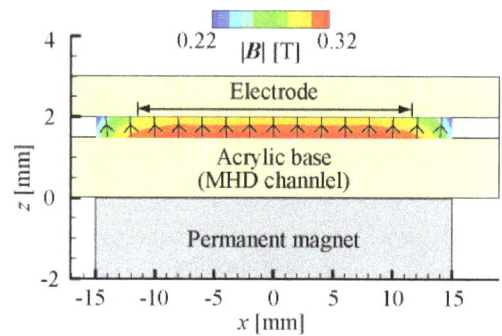

Fig. 3. Distributions of magnetic flux density in the experiment.

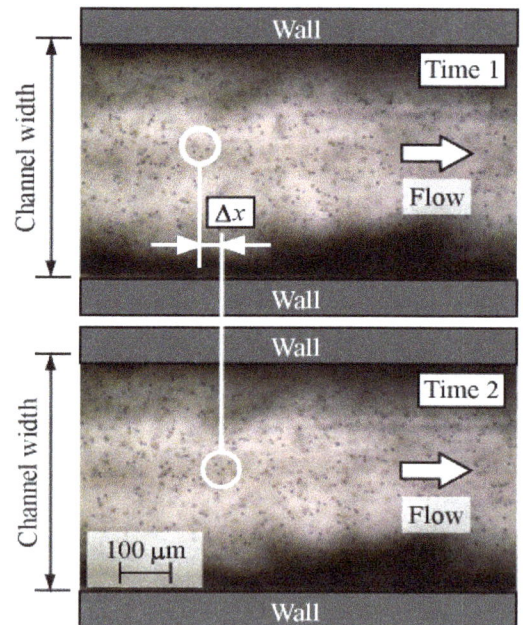

Fig. 4. Captured images of tracer particles in PBS solution flow.

A digital optical microscope (EV5680, EXEMODE) and a high speed camera (VW-9000, KEYENCE) are used to observe and record moving tracer particles in the micro-channel. The microscope and the camera are positioned above the micro-channel. A region visualized is 0.5 mm × 1.0 mm just behind the exit of MHD micro pump. Fig. 4 shows captured images of tracer particles in the PBS solution flow at two different times. The tracer particles have a diameter of about 5 μm. The material of tracer particles is polystyrene. In this work, the flow velocity of PBS solution is assumed to be the same as that of tracer particles.

2.2 Numerical Procedure

The basic equations for fluid dynamics are the continuity equation, the momentum conservation equations, and the energy conservation equation as follows:

Continuity equation:

$$\nabla \cdot \boldsymbol{u} = 0 \tag{1}$$

Momentum conservation equations:

$$\frac{\partial \boldsymbol{u}}{\partial t} + (\boldsymbol{u} \cdot \nabla)\boldsymbol{u} = -\frac{1}{\rho}\nabla p + \nu\nabla^2 \boldsymbol{u} + \frac{1}{\rho}(\boldsymbol{j} \times \boldsymbol{B}) \tag{2}$$

Energy conservation equation:

$$\rho C\left(\frac{\partial T}{\partial t} + \boldsymbol{u} \cdot \nabla T\right) = k\nabla^2 T + \frac{|\boldsymbol{j}|^2}{\sigma} \tag{3}$$

where $\boldsymbol{u}$ is the flow velocity, t the time, ρ the mass density, p the static pressure, ν the kinetic viscosity, $\boldsymbol{j}$ the electric current density, $\boldsymbol{B}$ the magnetic flux density, C the specific heat, T the static temperature, k the thermal conductivity, and σ is the electrical conductivity.

The dependency of electrical conductivity on fluid temperature is determined as follows [7]:

$$\sigma = \sigma_0(1 + 0.02T) \tag{4}$$

The flow is assumed to be laminar since the Reynolds number is 10^0 to 10^1. The continuity equation and the momentum conservation equations are solved by the SMAC method [8]. The energy conservation equation is solved by the CIP method [9].

The basic equations for electrodynamics are the steady Maxwell equations and the Ohm's law. The induced magnetic field is neglected because the magnetic Reynolds number is much smaller than unity.

Steady Maxwell equations:

$$\nabla \times \boldsymbol{E} = \boldsymbol{0} \tag{5}$$

$$\nabla \cdot \boldsymbol{j} = 0 \tag{6}$$

Ohm's law:

$$\boldsymbol{j} = \sigma(\boldsymbol{E} + \boldsymbol{u} \times \boldsymbol{B}) = \sigma(-\nabla\phi + \boldsymbol{u} \times \boldsymbol{B}) \tag{7}$$

where $\boldsymbol{E}$ and ϕ is the electric field and the electric potential, respectively.

The partial differential equation on the electric potential ϕ is derived from Eqs. (5)-(7). It is solved by the MAC method [10].

The boundary conditions are given in Table 2.

Table 2. Boundary conditions

Inlet boundary conditions:	
$\dfrac{\partial u_{x,in}}{\partial x} = u_{y,in} = u_{z,in} = 0$	(8)
$P_{in} = 0$	(9)
$T_{in} = T_0$	(10)
$\left(\dfrac{\partial \phi}{\partial n}\right)_{in} = 0$	(11)
Wall boundary conditions:	
$\boldsymbol{u_{wall}} = \boldsymbol{0}$	(12)
$\dfrac{\partial P_{wall}}{\partial n} = 0$	(13)
$T_{wall} = T_0$	(14)
$\left(\dfrac{\partial \phi}{\partial n}\right)_{wall} = 0$	(15)
Outlet boundary conditions:	
$\dfrac{\partial \boldsymbol{u_{out}}}{\partial t} + u_{x,out}\dfrac{\partial \boldsymbol{u_{out}}}{\partial x} = \boldsymbol{0}$	(16)
$P_{out} = 0$	(17)
$\dfrac{\partial T_{out}}{\partial x} = 0$	(18)
$\left(\dfrac{\partial \phi}{\partial n}\right)_{out} = 0$	(19)
Electrode surface boundary conditions:	
$\phi_{anode} = V_{input}$	(20)
$\phi_{cathode} = 0$	(21)

Table 3 lists the numerical conditions. PBS is employed as the working fluid. The distribution of magnetic field illustrated in Fig. 3, which was used in the experiment, is adopted as the normal condition of magnetic field in the numerical simulations. In this study, we carry out numerical simulations for several magnetic field distributions: those distributions are obtained by multiplying the strength of magnetic field (normal case) shown in Fig. 3 by an arbitrary value.

Table 3. Numerical conditions

Working fluid	PBS
Magnetic flux density B_z [T]	0.32-1.0
Applied DC current I [mA]	0.8-210
Channel height h [mm]	0.1-0.5
Electrode length l [mm]	26.0
Time step Δt [s]	5.0×10^{-6}
Initial temperature T_0 [K]	300

3. Result and Discussion

3.1 Fluid Behavior in MHD micro pump

In this section, experimental and numerical results for the maximum magnetic flux density of 0.32 T(normal case) and the channel height of 0.5 mm will be shown.

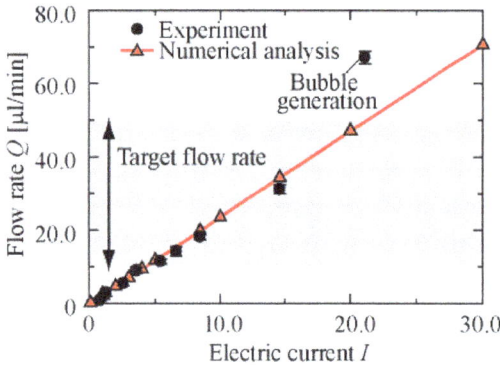

Fig. 5. Relation between flow rate and electric current.

Fig. 5 shows the relation between flow rate Q and electric current I. The experimental and numerical results are in good agreement for the electric current of 20 mA or less. On the other hands, there is a discrepancy between the experimental and numerical results above the electric current of about 20 mA. This is probably because the numerical simulations neglect electrolytic bubble generation observed in the experiment. The flow rate in a range of 10 µl/min to 50 µl/min is required for blood analysis by µ-TAS. Fig. 5, therefore, indicates that the numerical program developed by us is usable for understanding the fluid behavior in MHD micro pumps for blood analysis using µ-TAS.

Fig. 6 shows velocity profiles in the y-direction at x=+15 mm behind the exit of MHD micro pump for the following two different conditions of electric current: 14.6 mA and 8.4 mA. The experimental and numerical results are in good agreement. Hartmann flow is not observed due to weak MHD interaction, and therefore Poiseuille flow is maintained in the MHD micro pump. This velocity profile is obtained for all the conditions of magnetic field and electric current listed in Table. 3.

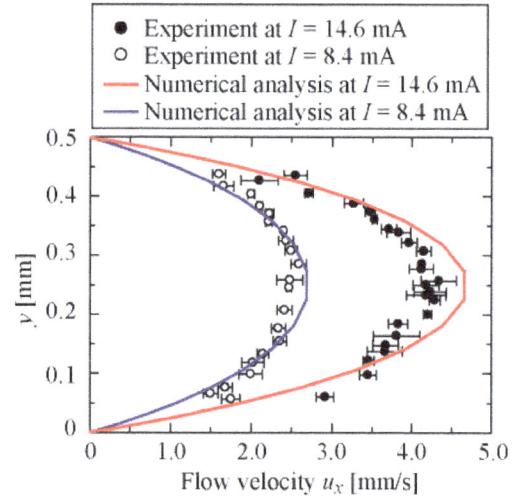

Fig. 6. Velocity profiles in the y-direction at x=+15 mm behind the exit of MHD micro pump.

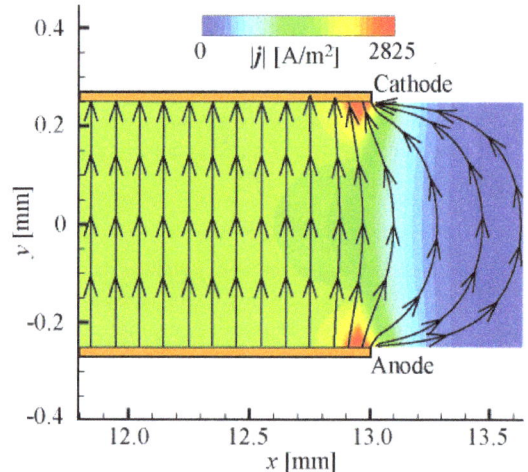

Fig. 7. Distribution of electric current density and streamlines of electric current (electric current: 20 mA, flow rate: 47.1 µl/min).

Fig. 7 shows the distribution of electric current density and the streamlines of electric current at I=20.0 mA and Q=47.1 μl/min. Fig. 8 depicts the temperature distribution under the above-mentioned conditions. The electric current concentrates at the edge of anode and cathode electrodes, where the electric current density is about twice as much as at the central point of the MHD micro pump (x=+12 mm, y=0 mm). Because the Joule heating increases in proportion to the square of electric current density, the Joule heating at the edge of electrodes is about four times more than at the central point. There is, however, no excessive increase of fluid temperature at the edge of electrodes, as can be seen from Fig. 8. The fluid temperature at the edge of electrodes is much lower than that at the central point, and also it is almost equal to the wall temperature. This is because a micro channel generally has an intense heat exchange between fluid and wall.

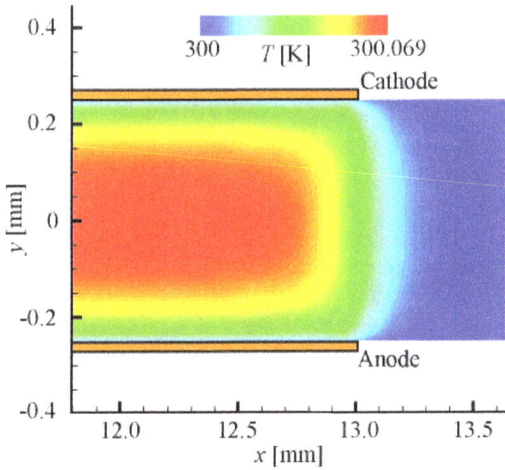

Fig. 8. Temperature distribution (electric current: 20 mA, flow rate: 47.1 μl/min).

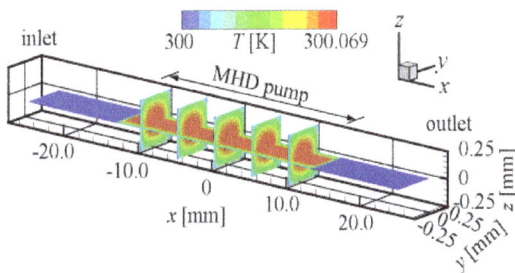

Fig. 9. Three dimensional distribution of temperature (electric current: 20 mA, flow rate: 47.1 μl/min).

Fig. 9 illustrated the three-dimensional distribution of fluid temperature at I=20.0 mA and Q=47.1 μl/min. The increase of fluid temperature is observed only in MHD micro pump. The fluid is rapidly cooled just behind MHD micro pump.

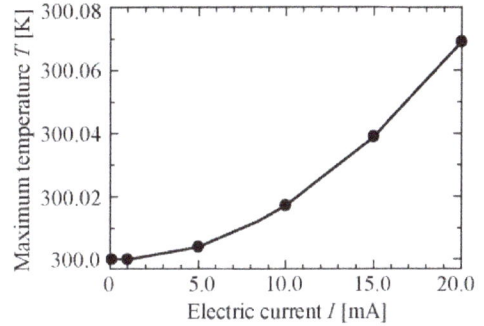

Fig. 10. Relation between electric current and maximum fluid temperature.

Fig. 10 shows the relation between electric current and maximum fluid temperature. The fluid temperature rises with increasing electric current. At I=20 mA, the increase amount of fluid temperature is only 0.069 K. In this case, the flow rate is 47.1 μl/min and its value meets the requirement of flow rate for the blood analysis using μ-TAS.

3.2 Relation between Channel Size and Fluid Temperature

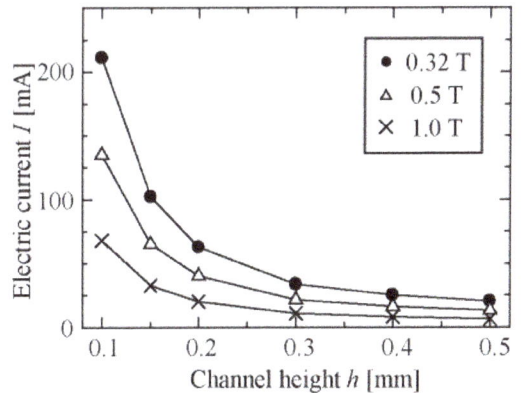

Fig. 11. Relation between channel height and electric current (flow rate: 47.1 μl/min).

In this section, we discuss the relation between channel size and fluid temperature in MHD micro pump at the constant flow rate of Q =47.1 μl/min. Fig. 11 illustrates the relations between channel height and electric current for several magnetic field distributions. Under any magnetic field condition, the electric current increases exponentially with decreasing channel height when the flow rate is kept constant. Fig. 12 shows the relation between channel height and maximum fluid temperature for several magnetic field distributions. An amount of the increase in fluid temperature by operating the MHD pump can be kept less than 1 K for the channel height more than 0.3 mm for all the magnetic flied conditions assigned in this work. An excessive increase of temperature is observed at the channel height of 0.1 mm. In this case, the increase amount of fluid temperature is much larger than 1 K for all the magnetic field conditions.

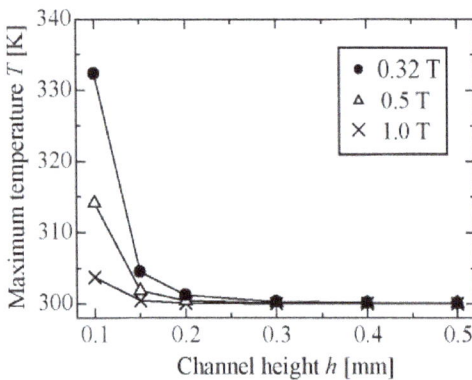

Fig. 12. Relation between channel height and maximum fluid temperature (flow rate: 47.1 μl/min).

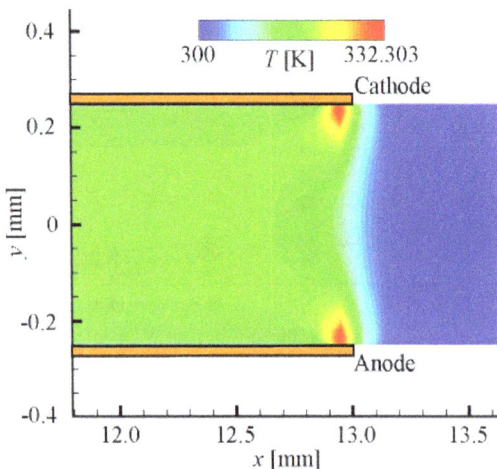

Fig. 13. Temperature distribution (channel height: 0.1 mm, magnetic flux density: 0.32 T, flow rate: 47.1 μl/min).

Fig. 13 depicts the temperature distribution at the channel height of 0.1 mm, the maximum magnetic field of 0.32 T, and the flow rate of 47.1 μl/min. An excessive increase of temperature at the edge of electrodes is observed in Fig. 13 unlike the temperature distribution shown in Fig. 8 which is obtained for the channel height of 0.5 mm. The is because the electric current for the channel height of 0.2 mm becomes ten times as large as that for the channel height of 0.5 mm so as to keep the flow rate constant.

4. Conclusions

The authors carried out the experimental and numerical studies on the fluid behavior in an MHD micro pump using a permanent magnet. As the working fluid, Phosphate Buffered Saline (PBS) solution was utilized under the assumption of blood analysis using μ-TAS. The main results are summarized as follows:

(1) The comparison of flow velocity between the experiments and the numerical simulations indicates that our numerical program is very reliable for simulating the fluid behavior in the MHD micro pump.
(2) In a range of the flow rate of 10-50 required for blood analysis using μ-TAS, the increase amount of fluid temperature by operating the MHD micro pump can be suppressed below 1 K in at the magnetic flux density of more than about 0.3 T and the channel height of more than 0.3 mm.
(3) Poiseuille flow profile is maintained in the MHD micro pump because of weak MHD interaction.

Acknowledgements

This work was supported by Hatakeyama Culture Foundation.

References

[1] J. Jang and S. S. Lee, Theoretical and experimental study of MHD (magnetohydrodynamic) micropump, *Sensors and Actuators A*, Physical 80, pp. 84-89 (2000)
[2] A. Homsy. V. Linder, F. Lucklum, and N.F. de Rooij, Magnetohydrodynamic pumping in nuclear magnetic resonance environments, *Sensors and Actuators B*, Chemical 123, issue 1, pp. 636-646 (2007)
[3] D. Sen, K. M. Isaac, and N. Leventis, Simulation of Electrochemical MHD induced Flow in a Microfluidic

Cell without Channels, *AIAA Theoretical Fluid Mechanics Conference*, (2011)

[4] A. Nisar et al., MEMS-based micropumps in drug delivery and biomedical applications, *Sensors and Actuators B*, Chemical 130, pp. 917-942 (2008)

[5] A. V. Lemoff and A. P. Lee, An AC magnetohydrodynamic micropump, *Sensors and Actuators B*, Chemical 63, pp. 178-185 (2000)

[6] K. P. Travis, B. D. Todd, and D. J. Evans, Departure from Navier-Stokes hydrodynamics in confined liquids, *Phys. Rev. E*, Vol. 55, pp. 4288-4295 (1997)

[7] V. Patel and S. K. Kassegne, Electroosmosis and thermal effects in magnetohydrodynamic (MHD) micropumps using 3D MHD equations, *Sensors and Actuators B*, Chemical 122, pp. 42-52 (2007)

[8] A. A. Amsden and F. H. Harlow, A Simplified MAC Technique for Incompressible Fluid Flow Calculations, *Journal of Computational Physics*, Vol. 6, pp. 322-325 (1970)

[9] H. Takewaki, A. Nishiguchi, T. Yabe, Cubic interpolated pseudo-particle method (CIP) for solving hyperbolic-type equations, *J. Comput. Phys.*, Vol. 61, pp. 261-268 (1985)

[10] A. Bottaro, Note on Open Boundary Conditions for Elliptic Flows, *Numerical Heat Transfer*, Part B, Vol. 18, pp. 243-256 (1990)

10

An Optimal Combined SVM Model for Short-term Wind Speed Forecasting

Bai Dan-dan[†], He Jing-han*, Tian Wen-qi*, Wang Xiaojun* and Tony Yip*

Abstract – A high precise wind speed forecasting method is one of current wind power research hotspots. This paper presented a combined wind speed forecasting model based on support vector machine (SVM) optimized by particle swarm optimization (PSO) using historical data of wind speed at the site. The model took the results of back propagation neural network (BPNN), radial basis function neural network (RBFNN), genetic neural network (GNN) and wavelet neural network (WNN) as the inputs, and adopted the actual wind speed as the output. Meanwhile, particle swarm optimization was used to optimize model parameters. Apply this model in hourly prediction of wind speed using historical data from a wind farm in Shanxi Province. It is observed that its prediction accuracy was not only higher than that of any of its single network but higher than traditional linear combined forecasting model and neural network combined forecasting model.

Keywords: Wind speed forecasting, Combined forecasting model, Neural network, Support vector machine, Particle swarm optimization

1. Introduction

Wind power has attracted a great deal of people's attention because of the advantages such as the small investment, clean pollution-free and resources widely distributed. However, the uncertainty of wind resources results in randomness and volatility of wind farm power output. So an accurate prediction of wind speed and wind power is urgent need to ensure system safety, reliability and controllability after large-scale wind parks integration [1].

To deal with wind speed forecasting, many methods have been developed such as physical method, which uses lots of physical considerations like temperature, wind direction, topography, roughness to reach the best forecasting precision and other is statistical method, which specializes in finding the relationship of the measured data [2]. Wind speed can be predicted by using time series analysis, Kalman Filter method, gray theory, artificial neural network, also by support vector machine [3-6]. Every forecasting model is the abstraction and simplification of the physical things containing variables and parameters requiring decision-making [7]. In order to improve prediction precision, researchers have done a lot of exploration mainly from three aspects: The first is parameter optimization. Literature [8] used particle swarm optimization algorithm

for least squares support vector machine model parameters. Secondly, optimize input sample. Literature [9] used time series model to select the input variables before neural network iteration, [10] turned wind speed into a series of original sequences with different frequency and waveform characteristics based on multi-scale morphology, and then imported them into support vector regression machine. The last is to establish a combined forecasting model using different single forecasting models which are built for the same prediction objects. With the purpose of optimal weighting coefficient of each single method, [11] made use of solving programming problem established by minimum variance.

Combined forecasting model is a hot issue for inclusiveness and multidimensionality. But current wind speed combined forecasting model is easy to fall into local optimum by compliance with traditional empirical risk minimization principle, While support vector machine, based on the structural risk minimization principle [12], guarantees sample empirical risk minimization and reduces the confidence range at the same time. Therefore, this paper puts forward a kind of support vector machine combined forecasting model based on particle swarm optimization, which takes prediction results of BP neural network, RBF neural network, wavelet neural network and genetic neural network as the inputs of SVM, and the actual wind speed values as the output. Particle swarm optimization, moreover, is used to optimize SVM parameters, thus fully dig the characteristics of original wind speed data. Apply wind

† Corresponding Author: Dept. of Electrical Engineering, Beijing Jiaotong University, China (11121572@bjtu.edu.cn)
* Dept. of Electrical Engineering, Beijing Jiaotong University, China

speed data acquired from an actual wind farm in Shanxi Province to verify above proposed method, and the result shows that the forecasting accuracy is not only higher than any used single model, but also higher than traditional linear combined forecasting model and neural network combined forecasting model.

2. Wind Speed Combined Forecasting Model Based on SVM Optimized by PSO

2.1 Wind Speed Variable Weight Combined Forecasting Model

As a result of the different learning rules of all sorts of single wind speed forecasting model, prediction effect may change over time. If the weight coefficient has been the same all the time, the best prediction effect wouldn't be got. Therefore research on nonlinear variable weight combined forecasting model is significant and necessary.

If there are m numbers of single forecasting models, named $f_1, f_2, \ldots, f_m$, then the combined model will be:

$$f(t) = \omega_1(t)f_1(t) + \omega_2(t)f_2(t) + \ldots + \omega_m(t)f_m(t) \qquad (1)$$
$$t = 1, 2, \ldots, N$$

Where, $f(t)$ is the wind speed value at time t, $\omega_i(t)$ is the weight coefficient of model i at time t.

$f_i(t)$ ($i = 1, \ldots, m$) is a constant value at time t ($t = 1, 2, \ldots, N$), so above question turns to an estimation problem of optimal weight coefficient $\omega_i(t)$ ($i = 1, \ldots, m$, $t = 1, 2, \ldots, N$), which turns the wind speed variable weight combined forecasting model to variable coefficient regression model.

Support vector machine has unique advantages on the search for global optimal solution for being rooted in statistics of VC dimension theory and structural risk minimization theory [12], and it can improve convergence speed and precision of nonlinear problems by transforming them into linear problems in a high dimensional space through kernel function. Hence, this article will use support vector machine combined forecasting model optimized by particle swarm optimization to approach the above variable coefficient regression model. The model of the regression equation is expressed as follows:

$$f(x) = \sum_{i=1}^{l}(\alpha_i - \alpha_i^*)K(x_i, x) + b \qquad (2)$$

That is to solve the following programming problem:

$$\max W = -\frac{1}{2}\sum_{i=1}^{l}\sum_{j=1}^{l}(\alpha_i - \alpha_i^*)(\alpha_j - \alpha_j^*)K(x_i, x)$$
$$+ \sum_{i=1}^{l}[\alpha_i(y_i - \varepsilon) - \alpha_i^*(y_i + \varepsilon)] \qquad (3)$$
$$s.t. \sum_{i=1}^{l}(\alpha_i - \alpha_i^*) = 0 \qquad 0 \le \alpha_i, \alpha_i^* \le C \qquad i = 1, 2, \ldots, l$$

Where, α_i and α_i^* are Lagrange multipliers, C is the punish coefficient, ε is the loss function, $K(x_i, x)$ is the kernel function, here adopts Gauss kernel function:

$$K(x_i, x) = \exp(\frac{\|x_i - x\|}{2\sigma^2}) \qquad (4)$$

2.2 Combined Forecasting Model Based on SVM and Optimized by PSO

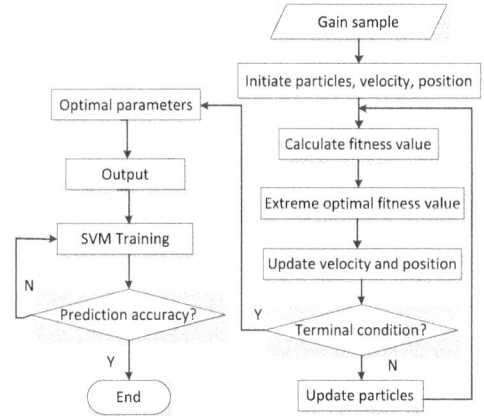

Fig. 1. Flow chart of SVM model Optimized by PSO.

The parameter selection of kernel function and SVM model is the crux of the algorithm, so parameter optimization technique can effectively improve learning ability and prediction accuracy of the forecasting model. This paper uses practical swarm optimization algorithm to search for optimal parameters: Practical swarm optimization algorithm firstly initializes a group of particles, which are indicated by position, velocity, and fitness value, in the solution space, and each particle represents a potential optimal solution for optimal parameters of SVM model. Then update individual positions after tracking by individual extreme value and group extreme value from comparison of fitness values. Several iterations later, the optimal position of fitness value can be searched, and that is the optimal parameters of the forecasting model. Afterwards, train SVM network to get the prediction model. Fig. 1 describes the algorithm implementation process.

2.3 Wind Speed Combined Forecasting Model Based on SVM Optimized by PSO

The wind speed forecasting model takes BP neural network, RBF neural network and wavelet neural network and genetic neural network as single forecasting model, that is taking their outputs as the inputs of SVM model and the actual wind speed gathered from wind farms as the output. Particle swarm algorithm is used to optimize model parameters, so as to improve prediction accuracy. The process is realized as shown in Fig. 2.

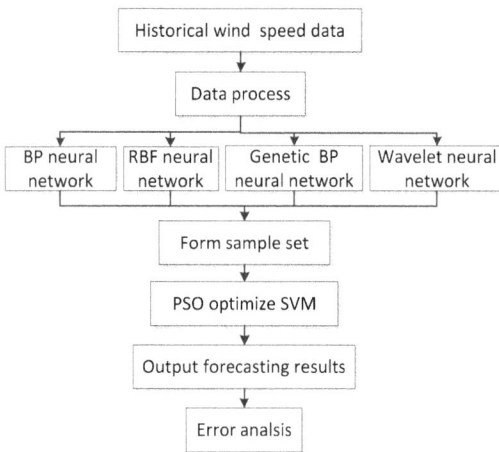

Fig. 2. Flow chart of optimal combined forecasting.

That is:

Step1: Take results of four single methods as the input $X_i(i = 1, 2, 3, 4)$, actual wind speed as the output $Y = (y_1, y_2, ..., y_l)$;

Step2: Solve for optimal coefficient $\alpha_i, \alpha_i^* (i = 1, 2, ..., l)$ according to equation (3);

Step 3: Take $\alpha_i, \alpha_i^* (i = 1, 2, ..., l)$ into regression equation (2) to calculate prediction values.

3. Results and Analysis

3.1 PSO Data Description and Model Parameters

The wind speed data studied in this paper is taken from a wind farm in Shanxi Province containing the whole 2011 year wind speed datas with an hour interval measured in fan hub height. Choose wind speed sequence from 8th January to 10th February as research sample data. Take 24days data from 1th August to 31thJanuary of them as the training and testing samples, and 10 days data from 1th to 10th Feburary as the evaluating samples.

3.2 Evaluation of performance of proposed model

Apply BP neural network model, RBF neural network model, wavelet neural network model and genetic neural network model, SVM combined model (SCM), PSO optimized SVM combined model (PSOSCM), traditional linear combined model (LCM) and neural network combined model (NNCM) for wind speed forecasting based on above wind speed data. Results are shown as Fig. 3 and Fig. 4.

Evaluate models separately in terms of Mean Absolute Deviation (MAD), Mean Forecast Error (MFE), Mean Square Error (MSE) and Mean Absolute Percentage Error (MAPE) described in Table 1 and Table 2. In Table 1 SVM combined model has a smaller MAD, MSE, MAPE compared with other four kinds of single forecasting model results, and the MFE is close to zero, which verifies that SVM combined model can not only optimize prediction accuracy of traditional single forecasting model but also meet unbiasedness requirement. Fig. 5 shows the results of traditional linear combined model, neural network combined model and PSO optimized SVM combined model. It shows that the optimal model has improved prediction accuracy compared with other combined models and the MFE is the most close to zero, which amply demonstrate PSO algorithm has searched the optimal parameters for SVM combined model and has promoted nonlinear learning ability for wind speed change.

Combined forecasting model gathers advantages of all kinds of single forecasting models, but restricted to the prediction accuracy of a single forecasting model at the same time. So improvement of the prediction accuracy of a single model will directly benefit to the optimization of combination forecasting model. If add external environment such as temperature, wind direction, and pressure conditions in the proposed prediction model, the change rule will be more clear and forecasting precision will get

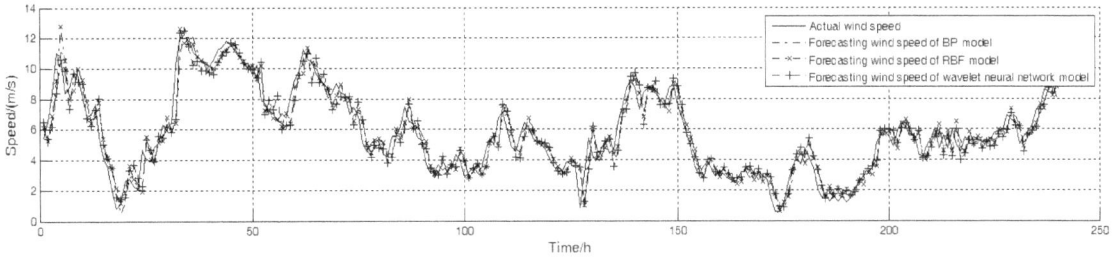

Fig. 3. Prediction of BP and RBF and Wavelet neutral network.

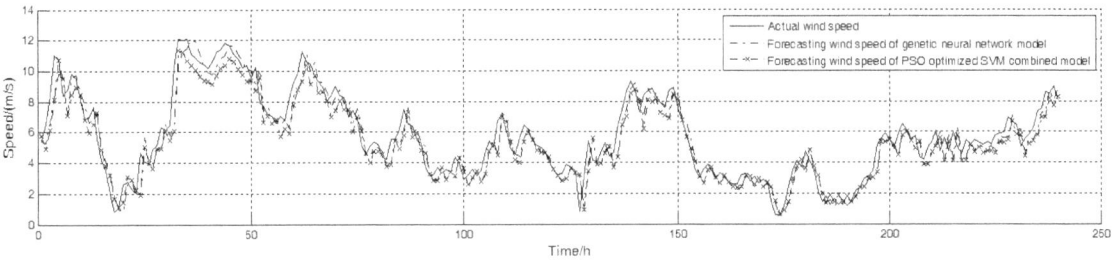

Fig. 4. Prediction of GNN and PSOSVM combined forecasting.

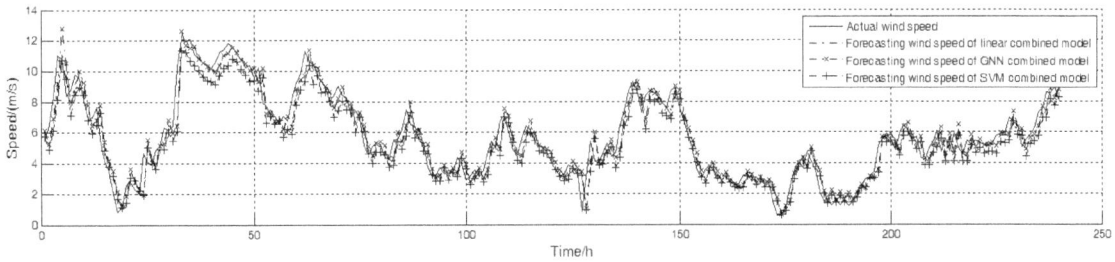

Fig. 5. Prediction of linear and GNN and PSOSVM combined forecasting.

higher. These will be one of key problems for our further study on wind speed forecasting.

Table 1. Forecasting results of different model

Model	MAD	MSE	MFE	MAPE
BP	0.7137	0.8385	-0.0311	17.2727
RBF	0.6566	0.7535	0.0171	16.2297
WNN	0.6690	0.7752	0.0314	16.3517
NNCM	0.7720	0.9965	-0.4727	16.5011
SCM	0.6026	0.7225	-0.0254	15.5847

Table 2. Forecasting results of different combined model

Model	MAD	MSE	MFE	MAPE
LCM	0.7507	0.9514	-0.4423	16.2205
NNCM	0.6469	0.7283	-0.0520	15.6306
SCM	0.6026	0.7225	-0.0254	15.5847
PSOSCM	0.6013	0.7184	0.0012	15.0017

4. Conclusions

Particle swarm optimized combined support vector machine model for wind speed forecasting is proposed on the basis of four kinds of single neural network forecasting model. The presented model is applied to a wind farm in Shanxi Province. This paper uses genetic algorithm to search the best model parameters of BP neural network, which proves that the combination of different algorithms can maximize their advantages. At the same time, practical swarm optimization algorithm has tracked the optimal parameters of combined model based on SVM, which shows that combined forecasting model can also reduce the prediction error by parameters optimization. The comparison results with traditional linear combined model and neural network combined model identifies the highlight prediction precision of the wind speed combined forecasting model based on SVM optimized by PSO as a nonlinear variable weight combined forecasting model for effectively integrating the learning advantages of four neutral networks.

References

[1] GU Xing-kai, FAN Gao-feng. Summarization of wind power prediction technology[J]. Power System Techno-logy, 2007, 31(2) : 335-338.

[2] Foley A M, Leahy P G, McKeogh E J. Wind Power Forecasting & Prediction Methods[J]. IEEE International Conference on Environment and Electrical Engineering, 2010, 61-64.

[3] PAN Di-fu, LIU Hui, LI Yan-fei, et al. A wind speed forecasting optimization model for wind farms based on time series analysis and Kalman filter algorithm[J]. Power System Technology, 2008, 32(7) : 82-86.

[4] LI Jun-fang, ZHANG Bu-han, et al. Grey predictor models for wind speed-wind power prediction[J]. Power System Protection and Control, 2010, 38(19) : 151-159.

[5] FAN Gao-feng, WANG Wei-sheng, LIU Chun, et al. Wind power prediction based on artificial neural network[J]. Proceedings of the CSEE, 2008, 28(34) : 118-123.

[6] ZHANG Hua, Zeng Jie. Wind speed forecasting model study based on support vector machine[J]. Acta Energiae Solaris, 2010, 31(7) : 928-931.

[7] HAN Shuang, YANG Yong-ping, et al. Application study of three methods in wind speed prediction[J]. Journal of North China Electric Power University, 2008, 35(3) : 57-61.

[8] YANG Hong, GU Shi-fu, et al. The short-term wind speed forecast analysis based on the PSO-LSSVM predict model[J]. Power System Protection and Control, 2011, 39(11) : 44-48, 61.

[9] CAI Kai, TAN Lun-nong, LI Chun-lin , et al. Short-Term Wind Speed Forecasting Combing Time Series and Neural Network Method[J]. Power System Technology, 2008, 32(8) : 82-85.

[10] CHEN Pan, CHEN Hao-yong, YE Rong. Wind speed forecasting based on multi-scale morphological analysis[J]. Power System Protection and Control, 2010, 38(21) : 12-18.

[11] ZHANG Guo-qiang, ZHANG Bo-ming, et al. Wind speed and wind turbine output forecast based on combination method[J]. Automation of Electric Power Systems, 2009, 33(18) : 92-96.

[12] Deng Nai-yang, Tian Ying-jie. Support vector machines[M]. Beijing : Science Publishing Company, 2009.

11

A new Calibration Device of Electronic Transformer based on IEEE1588 Time Synchronization Mode

Zhang Ji[†], Shao Hanqlao*, Peng Changyong* and Du Zhi*

Abstract – Time synchronization mode is very important for error calibrating of electronic transformer in the digital substation. This paper firstly introduces the technical principles of IEEE1588 time synchronization protocol, then it introduces the design scheme and technical features of a new calibration device of electronic transformer based on IEEE1588 time synchronization mode. Then, it compare and analysis the results of error calibration of electronic transformer using the calibration device，and compare the differences among the three time synchronization modes, including IEEE1588,B code and PPS. The test results verifies the IEEE5588 time synchronization mode has more advantages in the error calibrating of electronic transformer, and provides the references of IEEE1588 time synchronization protocol using in the smart substation.

Keywords: IEEE1588, Time synchronization, Electronic transformer, Error calibration

1. Introduction

With the development of smart grid by the State Grid Corporation of China (SGCC), represented by electronic transformers, a great quantity of intelligent electronic devices (IED), have been applied in digital substations.

After sampling process, electronic transformers send digital signals including both output signals of current transformer and voltage transformer to IED in isolation layer by merging unit. If and only if digital signal acquisition and transmission base on a uniform timing sequence and clock standard, could the precision, reliability and validity of information be guaranteed, which ensures the normal operation of digital substations. Merging unit time synchroni-zation technology plays an important role in digital substations, which generally use GPS time signal as external time reference signal and PPS, IRIG-B code or SNTP, etc. as clock synchronization modes. Compared with IEEE1588, these clock synchroni-zation modes have several disadvantages, such as less reliability, lower time hack accuracy, lack of compatibility with clock interface, etc. As a network time protocol, IEEE 1588 could realize sub-microsecond-level synchronization accuracy, which makes the digital substation network time synchronized. This paper will introduce the technical features and applications of this precision time protocol.

†   Corresponding Author: Technology&Economic Research Institute of Hubei Electric Power Company, China (ceeezj@qq.com)

*   Technology&Economic Research Institute of Hubei Electric Power Company, China

2. Technology Principle

2.1 Brief discriptions

IEEE 1588, namely precision time protocol (PTP), which could reach sub-microsecond-level synchroni-zation accuracy, was a high precision time synchronization protocol released in 2002 to be used in the areas of industrial control and measurements. All types of synchronous messages that defined in the protocol are all sent based on UDP/IP protocol. Hence, it's very suitable for LAN distributed applications since existing Ethernet could be sued as timing channels. Three main technologies, i.e. Network communication, local computation and distributed objects technologies are adopted to realize precision timing synchronization in measure and control systems. A simple time synchronization system with IEEE 1588 protocol includes at least one master clock and multiple slave clocks[3]. If more than one potential master clocks exist, the active master clock is decided by optimized master clock algorithm[1].

2.2 Time synchronization principles

Time synchronization of IEEE 1588 is carried out in two steps, firstly clock offset measurement and then network delay measurement.

First, the master clock sends synchronous messages in priority periodically. The master clock records the time mark T_1 when synchronous messages are sent, as shown in Fig. 1. And slave clocks record the time mark T_2 when synchronous messages are received. Then the master clock sends following messages, if the network time delay is T_{delay}, then the offset time lag T_{offset} [2] could be obtained in Equal 1.

$$T_{offset} = T_2 - T_1 - T_{delay} \qquad (1)$$

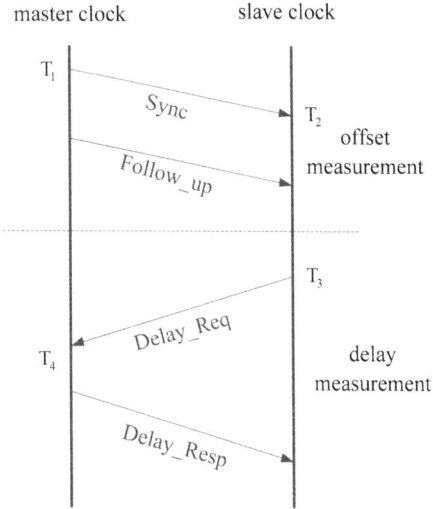

Fig. 1. Time synchronous principles of IEEE1588 Protocol.

In the systems with fixed network structure and network which load does not vary much, the delay measurement is normally sent randomly, as shown in Fig. 1. The slave clocks send delay_req message, and time mark T_3 is recorded. When the master clock receive delay_req message, time mark T_4 is recorded, and delay-resp message is sent to slave clocks in turn. Then the network time delay is given by Equal 2.

$$T_{delay} = T_4 - T_3 + T_{offset} \qquad (2)$$

Combining Equal 1 and Equal 2,

$$T_{offset} = \frac{(T_2 - T_1) - (T_4 - T_3)}{2} \qquad (3)$$

$$T_{delay} = \frac{(T_2 - T_1) + (T_4 - T_3)}{2} \qquad (4)$$

With T_{delay}, T_{offset}, slave clock can be revised in a uniform time standard with the master clock [3][4].

3. Design and Technology Features

3.1 Hardware Design

Solar The calibration device of electronic voltage/ current transformer(ECT/ EVT) based on IEEE 1588 protocol is composed of IEEE1588 ethernet switch, IEEE1588 clock, standard voltage/current converter, standard channel data acquisition unit, synchronous pulse transmission module, measured channel date acquisition unit, data processing unit and GPS, as shown in Fig. 2.

The procedure of the calibration device is listed in following
steps:

- IEEE1588 ethernet switch receive the IEC1850 and IEEE1588 messages sent by merging unit of measured electronic current transformer with digital output, then the messages are transmitted to IEEE1588 clock and digital input date acquisition of measured channel unit by multicast .
- IEC61580 message is sent to data processing error computation and display unit after being resolved in digital input date acquisition unit of measured channel.
- Output PPS signal is sent to synchronous pulse transmission module when IEEE1588 clock time hack is completed.
- When PPS signal received, synchronous pulse transmission module generate sampling pulse to data acquisition unit of standard channel
- Data acquisition unit of standard channel proceeds data sampling with signal generated by standard voltage/current converter, meanwhile the sampling results are transmitted to data processing error computation and display unit.

3.2 Software design of time synchronization

The system software flow chart is shown in Fig. 3. First, working mode of the master clock is set; second, time synchronization model is set to ETE or PTP; then, IEEE1588 parameters are set to ensure success of time hack; and finally, start error tests and get test results.

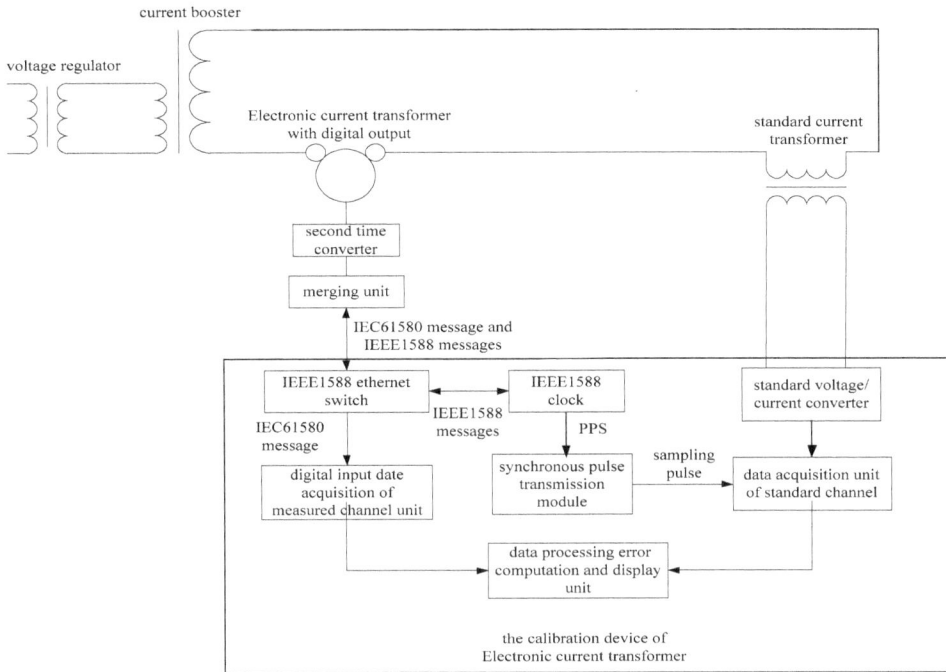

Fig. 2. Block diagram of the Calibration Device.

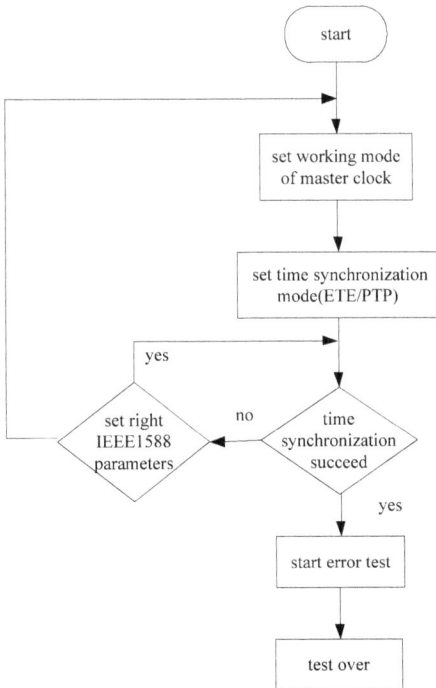

Fig. 3. Flowchart of system software.

3.3 Technology features

Technology features of this calibration device of electronic transformer based on IEEE1588 protocol are summarized as follows:

- The high precision DAQ card PCI5922 from NI contains voltage reference, and has self-calibration ability, which could eliminate the error caused by temperature variations, and very suitable for field test thanks to its compactness [5].

- Provided with IEEE1588 clock, which could be served as master clock as well as slave clock; and it has the ability to output PPS only when the time synchronization successes.

- Rack-mounted web-managed industrial gigabit Ethernet switch SICOM3024PT, which could supports both time synchronization modes of ETE and PTP, is adopted, giving the ability to consider the influence of time synchronization mode on the error of voltage/current electronic transformer . If data frames of IEEE1588 and IEC61580 in measured merging unit share the same EMAC, the IEEE1588 ethernet switch could be expand to two channels, one for time synchronization and another for IEC61580 data sampling[6].

4. Tests

4.1 Test based on IEEE1588 mode

This calibration device of electronic transformer based on IEEE1588 time synchronization mode is used to test the errors of electronic current transformer. The error of electronic current transformer in the calibration device are shown in Table 1.

Table 1. Error of electronic current transformer based on IEEE1588 mode

	Ascend		Descend	
	Ratio Error(%)	Angular Error(')	Ratio Error(%)	Angular Error (')
1%	0.10	-11.7	0.08	-15.5
5%	0.10	-5.0	0.11	-6.3
20%	-0.01	-10.6	-0.01	-10.3
100%	0.02	-14.0	0.02	-14.1
120%	0.03	-14.2		

4.2 Test based on PPS mode

The IEEE1588 clock and IEEE1588 ethernet switch of this calibration device are replaced by the PPS time synchronization module.Then the new calibration device is used to test the error of electronic current transformer based on PPS mode. The error of electronic current transformer in the calibration device are shown in Table 2.

Table 2. Error of electronic current transformer based on PPS mode

	Ascend		Descend	
	Ratio Error(%)	Angular Error(')	Ratio Error(%)	Angular Error (')
1%	0.10	-15.2	0.16	-17.0
5%	0.13	-6.1	0.12	-5.4
20%	-0.02	-10.3	0.01	-9.0
100%	0.02	-14.1	0.03	-13.9
120%	0.03	-14.3		

4.3 Test based on IRIG-B code mode

The IEEE1588 clock and IEEE1588 ethernet switch of this calibration device are replaced by the time synchronization module of IRIG-B code.Then the new calibration device is used to test the error of electronic current transformer based on IRIG-B code mode.The error of electronic current transformer in the calibration device are shown in Table 3.

Table 3. Error of electronic current transformer based on PPS mode

	Ascend		Descend	
	Ratio Error(%)	Angular Error(')	Ratio Error(%)	Angular Error (')
1%	0.08	-14.7	0.15	-16.4
5%	0.14	-4.3	0.10	-6.3
20%	-0.04	-10.0	0.01	-10.6
100%	0.01	-13.9	0.01	-14.2
120%	0.02	-14.1		

4.4 Analysis of three time synchronization modes

The error curves of three different time synchronization modes, including IEEE1588, PPS and IRIG-B code,are shown in Fig. 4 and Fig. 5.

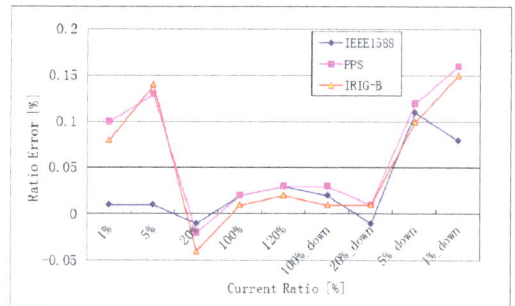

Fig. 4. Ratio difference curves of three time synchronization modes.

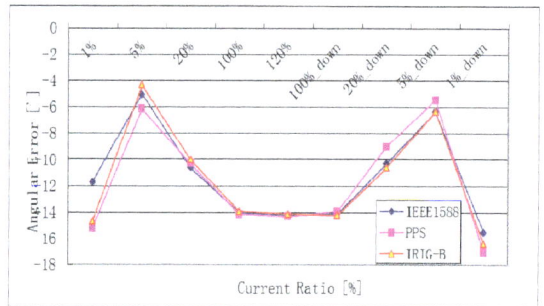

Fig. 5. Angular difference curves of three time synchronization modes.

The error curves above show good error consistency with ratio error and angular error of electronic current and voltage transformer.The error curves of PPS mode and IRIG-B code mode are almost consistant,but the error curve of IEEE1588 mode have little differences with the other two due to the different time synchronization principle.As the

PPS mode and IRIG-B code must raly on the master colck outside,the error test will be influened by the factors outside.However, IEEE1588 mode can set the IEEE1588 clock as master clock,which can guarantee excellent reliability and flexibility in the test [7],and meet the requirements for calibration of electronic transformer.

5. Conclusions

Authors It is sure that IEEE1588 time synchronization mode will be the development orientation of time synchronization technology in the digital substation, because it has so many characteristics such as high synchronization precision and network structures. This paper introduces the design principles and technical features of a calibration device of electronic transformer based on IEEE1588 time synchronization mode, and get test result of error calibration of electronic transformer by it. It verifies this device can meet the field error test of electronic voltage/current transformer (ECT/ EVT) as a basis for further study.

References

[1] ZHAO Shanglin, HU Minqiang, DOU Xiaobo, et al. Research of time synchronization in digital substation based on IEEE 1588[J]. Power System Technology, 2008, 32(21) : 97-102

[2] YU Pengfei, YU Qiang, DENG Hui, et al. The research of precision time protocol IEEE 1588I-J]. Automation of Electric Power Systems, 2009, 33(13) : 99-103.

[3] YIN Zhiliang, LIU Wanshun, YANG Qixun, et al. A new IEEE 1588 based technology for realizing the sampled values synchronization on the substation process bus[J]. Automation of Electric Power Systems, 2005, 29(13) : 60-63.

[4] R. Holler, T. Santer, N. Kero. Embedded Syn UTC and IEEE 1588 clock synchronization forIndustrial Ethernet. Proceedings of Emerging Technologies and Factory Automation, 2003 : 422~426.

[5] HUANG Guofang, XU Shimingt ZHOU Bin, et al. Optimization design of a new synchronized measuring and control device for substations[J]. Automation of Electric Power Systems, 2009. 33(19) : 77-79.

[6] LIU Huiyuan, HAO Houtang, LI Yanxin, et al. Research on a synchronism scheme for digital substations[J]. Automation of Electric Power Systems, 2009, 33(3) : 55-58.

[7] Li Hongbin, liu Yanbing, Zhang Mingming. Key Technology of Electronic Current Transformer[J]. High Voltage Enginnering, 2004, 30(10) : 4-6.

12

Integrated Modeling between Smart Substation and Dispatch Automation System

Deng Yong[†], Mi Weimin, Ren Xiaohui*, Xu Dandan** and Chen Zhengping***

Abstract – Based on technical supporting systems for smart grid dispatch and smart substation system and according to the functional demand of dispatch master station to the substation, an integrated modeling scheme for coordinated sharing of models/images and seamless communication between dispatch master station and smart substation is proposed, by which practicable converting rules between IEC 61850-based model and IEC 61970-based model are drafted; a unified data model-based mapping approach for IEC 61850-based model and IEC 61970-based model is put forward; by means of splitting/merging technique of models and self-converting technique of Ids of graphs and so on, the integrated modeling between dispatch master station and smart substation is implemented; utilizing message bus service of dispatch master station and manufacturing message specification (MMS) protocol stack, the IEC 61850 protocol-based communication at dispatch master station side is attained, thus the coordinated sharing of models/images and seamless communication between smart substation and dispatch master station is implemented, and the capacity of information interaction between smart substation and dispatch master station is increased and the workload of model coordination in heterogeneous systems is reduced.

Keywords: IEC 61970, IEC 61850, Model coordination, Distributed and Integrated modeling, Model splitting/merging

1. Introduction

At present, IEC61850 standard is applied in smart substation while IEC61970 standard is used in smart grid dispatch support system. Dispatch center and substation automatic systems follow different standard hierarchies, in which lead to difficulties to share information between dispatch substations and master stations systems, to satisfy the demands of grid dispatch integrated maintenance and longitudinal through, as well as the demands of safe and stable operation of interconnected large power system. Therefore, it is a resolving issue to achieve coordinating share of model/graph and seamless communication between smart substation and dispatch master station systems.

References [1-3] made conclusions and outlooks of the status of smart grid technology; reference [4-8] studied smart dispatch modeling system and its key technology; reference [9] presents system hierarchy of the second edition of IEC61850 and the contents added or modified from the first edition; reference [10] studied relevant

technologies to achieve substation communication gateway configuration by applying proxy service mode; reference [11] discussed relevant technologies of substation state monitoring system and the coordination of substation model and dispatch master station model; reference [12-15] described IEC61970 and IEC61850 standard.

This paper presents an integrated modeling scheme of smart substation and dispatch master station; the modeling range contains coordination and share of grid model, integrated maintenance of grid graph and seamless communication between smart substation and dispatch master station. This scheme follows IEC61970 and IEC61850 standards, combined with CIM/E (common information model based efficient model exchange format), CIM/G（common information model/graph）and power grid equipment common model naming specification to formulate model/graph sharing standard of smart substation and dispatch master station.

2. Overview

Integrated modeling of smart substation and dispatch master station including the coordination and sharing of grid model, integrated maintenance of grid graph and

† Corresponding Author: Fujian Electric Power Dispatching and Controling Center, China (huang_jiashu@foxmail.com)
* Fujian Electric Power Dispatching and Controling Center, China
** State Grid Electric Power Research Institute, China

seamless communication between smart grid and dispatch master station. Main functions include formulating transform rules from IEC 61850 model to IEC 61970 model; applying mapping from IEC 61850 model to IEC 61970 model based on unified data model; integrated modeling; seamless communication mechanisms between substation and dispatch master station based on IEC 61850 protocol. General structure is as shown in Fig. 1. System configuration module on smart grid side is responsible for building public information models to achieve the transformation from IEC 61850 to IEC 61970 model. Offline model management module in dispatch center is built to complete the integration of dispatch master station model and single substation model. IEC 61850 online client of dispatch master station implements seamless communication between substation and dispatch master station based on IEC 61850 protocol.

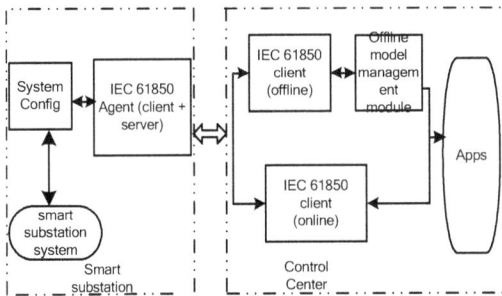

Fig. 1. General structure.

SCD (substation configuration description) file is able to completely describe primary equipment, secondary equipment and their relationships in detail. In the past, dispatch master stations were in charge of building and maintaining primary equipment model while substations only concerned about secondary equipment. In that case, the descriptions of primary equipment in SCD files are quite inadequate. In this scheme, as a substation standard, it requires a model includes complete primary equipment and its topology relationships to build a new smart substation system; it also requires visualization tools being provided by modified substation system to complete primary equipment information from the original model.

The process of building smart substation and dispatch master station model is shown in Fig. 2. Smart substation system creates SCD model according to information provided by system configuration module, and then generates CIM/E and CIM/G files based on SCD file to communicate with dispatch master station client by

substation proxy server. Dispatch master station client includes online client and offline client. Online client achieves the communication between substation and dispatch master station based on IEC61850 protocol. Offline client uses file service to receive CIM/E and CIM/G files and new information model will by created by triggering model merging process after the files have been verified. When model merging is complete, key information of CIM/G could be converted automatically according to key information mapping table generated during merging process. At last, in order to provide model/graph to smart grid dispatch support system, edition manage module must be triggered to present new information model to subscribers.

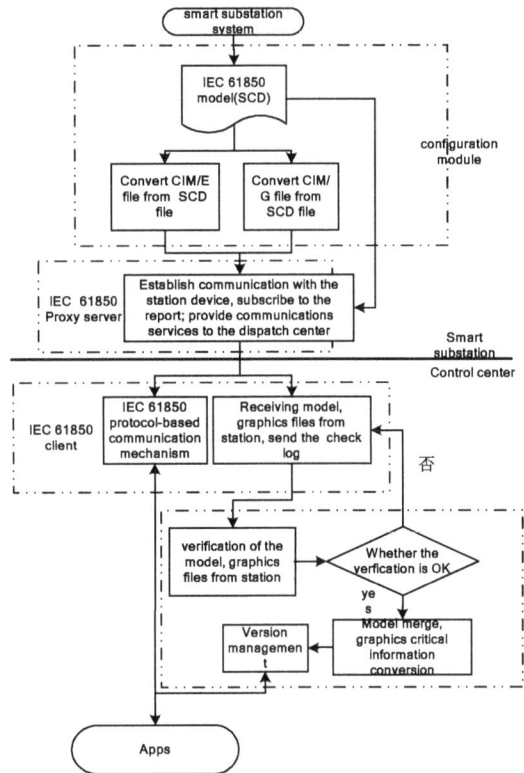

Fig. 2. Modeling flow.

3. Main Functions

3.1 Model transfer rules

Based on IEC61850 and IEC61970 standards and combined with CIM/E, CIM/G and power grid equipment common model naming specification to practice similarity and difference analysis for these 2 models as well as bonding the realistic business requirements of dispatch master station towards substation in order to make model transfer rules. It can be specified into general rules and special rules.

General rules. For types and attributes have been defined in IEC61850 and IEC61970 standards, it is easy to apply mapping and the mapping relationships are model transfer general rules.

Special rules. For special demands of dispatch master station model, it is possible that there is only 1 standard definition between 2 standards or both standards have not been defined. The model should be expanded. Expanded types and attributes must be stored into data model dictionary by maintenance tool to make it is convenience to build expanded model while exporting unified data model.

3.2 Model mapping measures basing on unified data model

IEC61850 and IEC61970 standard has different ways to build model, but they are quite in common by using object oriented measures to build power system models.

Due to analysis for IEC 61850 and IEC 61970 model, presenting a method which is based on unified data model to make IEC 61850 model convert to IEC 61970 model. Analysis content is shown below：

IEC 61850 model analysis. IEC 61850 is international standard for substation communication network and system, which is used to build unified communication service and information transmission for automation equipment and related system of substation, which it is an abstract description base on communication expression for real function and equipment, its model belong to transmitting information model. It is use unified communication as condition among equipment for substation automation. There is not the relation with application function as long as the progress of message transmission has identical meaning. The progress of modeling is using communication service as precondition for information sealing progress，it does not have complicate model analysis and structure，which is based on XML to use SCL language to describe.

IEC 61970 model analysis. IEC 61970 is EMS application progress interface[12,14], which is used for unified application progress interface and information application to promote different manufacturers develop independently to integrate every application progress, which model belong to application information model. The modeling process to be able to reflect the relationship as a precondition of the information model, object-oriented

analysis and construction techniques, with emphasis on the relationship between information objects using UML (Unified Modeling language) language to describe.

In summary, the IEC 61850 model and IEC 61970 model belongs to the different levels of model, the coordination between the models is necessary. The coordination between these two models is not only the coordination of the model structure, the coordination of the semantic space is inevitable.

Unified data model uses model transformation rules based on XML language modeling. The model is based on the IEC 61970 model, described in the IEC 61850 model and IEC 61970 model class and parameter mapping between the mapping information of the measurement model, device naming rules and data type and other information. Compatible with IEC 61850 and IEC 61970 standard data models, you can easily set up the mapping between them; For a standard data model defined which should be based on the IEC 61970 standard for substation model expansion. The unified data model fragments are shown in Fig. 3. The system provides visualization tools to establish a unified data model. Dispatch center maintenance staff uses a visual tool, select the classes and attributes of the IEC 61970 model, modeling tools automatically matched the SCD file corresponding to establish a unified data model according to the needs of the master model.

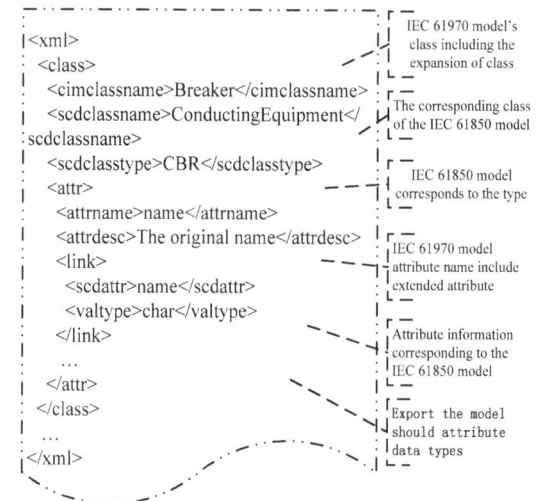

Fig. 3. A part of data model.

The method takes into account the two models of semantic space and the coordination of the model structure, device naming convention as one of the conversion rules to solve unified named for station terminal and master station, which is realize IEC 61850 model to IEC 61970 model mapping.

3.3 Integrative Modeling

The integrated modeling completed by smart grid dispatch technology of master station support system off-line model management center. Mainly through the completion of the model split / merge, version management module. Integrated modeling this scenario is the master station model with a single plant station model stitching, just consider the dispatch center and a station in the boundary of the station model, compared to the traditional model of combined boundary, easy to maintain, but there may be boundary information does not overlap. To solve this problem, the Station-side model for boundary devices modeled in detail, smart border maintenance coerced approach to border station end model equipment, the master station model boundaries equipment supplemented, access master station   model. The result of the merger into offline database incrementally, by Unicom check and without disturbance to invest online operating system.

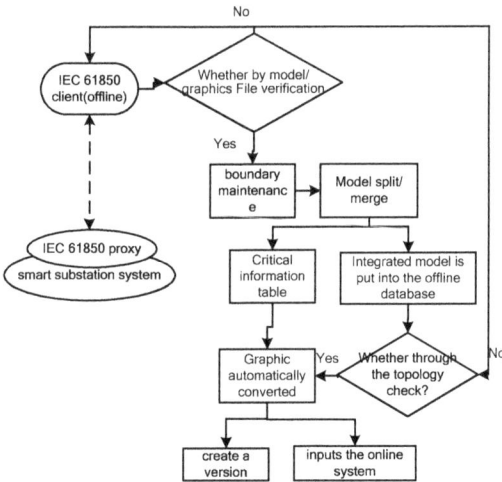

Fig. 4. Flow of integrated modeling.

The modeling process is shown in Fig. 4. The master station IEC 61850 Offline client transmission substation graphics and model file to the offline model management module. After received new file, the offline model management center startup file check process, if the file does not pass the check, they will check file error message return IEC 61850 Offline client; otherwise enter the boundary configuration process. The station boundary of

the model is generated together with the model file into the model split/merge process. The integration of the model into the offline database, start Unicom check process. If it is not by Unicom check the error log return to the offline client; by Unicom verification the combined model merging process generates key information table conversion substation graphics. Finally, put into the online system through the calibration model and graphical version.

3.4 Sealed communication dispatch center based on the IEC 61850 protocol Mechanism

At present, most of the communication between the substation and dispatch center with the traditional IEC 60870-5-101 or IEC 60870-5-104 rules, this requires RTU protocol conversion, for information unable to protocol conversion has to be discarded.

The program is based on the existing functionality of the smart grid dispatch of master station technology support systems and intelligent substation system, starting from the dispatch center on the functional requirements of the substation, give full consideration to the system scalability, reliability, standardization and maintainability IEC 61850 client is responsible for scheduling the master station and substation system established in the dispatch center docking.

IEC 61850 client package for dispatching master station seals substation communication with the master station. Service bus the dispatching master customer end and substation server-side manufacturing quoted text specification (manufacturing message specification, MMS) protocol stack, a network packet, the scheduling master side and substation end of the information exchange; take advantage of smart grid scheduling technology support system , and scheduling master information exchange.

Based on the IEC 61850 protocol transport mechanism for sealed communication of intelligent substation and dispatch center, to replace the existing variety of line and network layer protocols, scheduling master data and intelligent substation direct mining, a communication protocol of the target.

4. Real Projects

The intelligent substation and dispatch center integrated modeling technique has been successfully used in dispatch center of some province electric power company and some 500 kV Substation.

Master station IEC 61850 client is registered as a process of smart grid dispatching technology support system, its operation and exit by the system management. The system configuration is shown in Fig. 5. Master station IEC 61850 Offline client is responsible for model and graphic file transfer. The master station IEC 61850 online client the CIM / E provided by the station side, the SCD file and the corresponding relationship to establish a data mapping table, the IEC 61850 data model and master station data model associated initialization. IEC 61850 server communication master station online client using the IEC 61850 protocol substation side, completed the telemetry data updated in real time remote signaling and remote control commands.

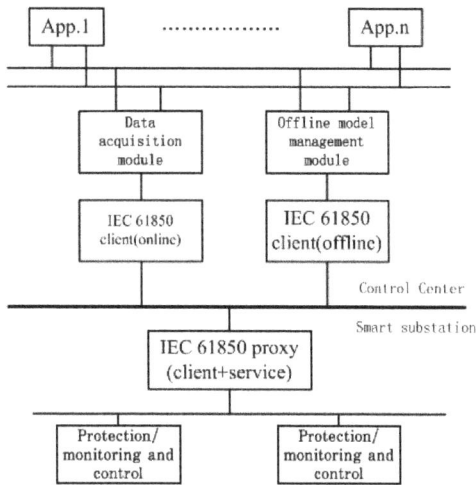

Fig. 5. System configuration.

The use of intelligent substation and dispatch center integrated modeling techniques, the province achieved the integrated modeling of the dispatch center and a 500 kV Substation. The model of integration to put into the online system with 240 s, graphical key conversion 3 s and telemetry data refresh every 3 s, remote signaling data refresh every 1 s.

Through the application of this technology, the province's power dispatch center eliminates the need for dispatching the master station side of the 500 kV substation model artificial modeling and graphics rendering, the end of the new processing station master maintenance-free, greatly reducing the time and effort of the new processing station into master station system; dispatching master station and substations by the IEC 61850 protocol to communicate, eliminating the complexity of the Statute of the conversion work, the dispatch center and substation interoperability and information exchange.

5. Conclusions

The program is proved that they have the following characteristics through the practical application:

Based on the unified data model of the intelligent substation model (IEC 61850) to the mapping of the dispatching master station model (IEC 61970), coordinated sharing smart substations model and dispatching master station model. Through the application of the technology in practice, further validate the feasibility and practicality of the conversion method, substation and dispatch center model conversion and mapping technology from theoretical research towards practical application.

Dispatch master station implements with IEC61850 client to achieve the exchange of system model, graph files between dispatch master station and smart substation as well as real-time information exchange based on IEC61850 protocol. Smart substation communicates with dispatch master station through unified protocol. IEC61850 standard is applying as the only communication protocol from process level to interval level in smart substation, from interval level to station level, from substation to dispatch master station. Complicated protocol transfer could be saved while unchanged information lost caused by different communication protocols could be avoid.

It is can be proved that the model transfer rules in this scheme is applicable and the demands of exchanging information between substation (IEC61850) and dispatch master station (IEC61970) can be satisfied by using model mapping measures from IEC61850 to IEC61970 based on unified data model. The application of this scheme make it is direct to transfer IEC61850 model of substation into IEC61970 model of dispatch center which increased the abilities of information exchange between substation and dispatch center. It achieves 'zero maintenance' of system graph gallery in dispatch master station and telesignalisation, telemetering, telecommand and teleadjusting are free from debugging.

References

[1] Chen Shuyong, Song Shufang, Li Lanxin, et al. Survey on smart grid technology[J]. Power System Technology, 2009, 33(8) : 1-7(in Chinese).

[2] Yang Dechang, Li Yong, Rehtanz C, et al. Study on the structure and the development planning of smart grid in China[J]. Power System Technology, 2009, 33(1720) : 13-20(in Chinese).

[3] Zhang Boming, Sun Hongbin, Wun Wenchuan, et al.

Future development of control center technologies for smart grid[J]. Automation of Electric Power Systems, 2009, 33(17) : 21-28(in Chinese).

[4] Lin Jinghuai, Mi Weimin, Ye Fei, et al. Several Issues in Modeling Technique for Intelligent Dispatch[J]. Power System Technology, 2011, 35(6) : 1-4(in Chinese).

[5] Huang, Haifeng, Cao Yang, Song xin, et al. Power network model management system fitted for smart dispatch[J]. Preceedings of the CSEE, 2009, s1 : 7-10 (in Chinese).

[6] Qian Jing, Fan Guangmin, He Lei, et al. Implementation of distributed coordination modeling in power dispatch data platform[J]. Power System Technology, 2009, 32(10) : 136-141(in Chinese).

[7] Liu Tao, Zhu Cuixia, Tang Weidong, et al. Distributed power network modeling and real DATA sharing scheme for central china power grid[J]. Automation of Electric Power Systems, 2011, 35(2) : 48-88(in Chinese).

[8] Mi Weimin, Wei Lingxiao, Qian Jing, et al. Application of CIM & XML base combination method of power network models in dispatch system of Beijing electric power corporation[J]. Power System Technology, 2008, 32(10) : 33-37(in Chinese).

[9] Li Yongliang, Li Gang. An introduction to 2nd edition of IEC 61850 and prospects of its application in smart grid[J]. Power System Technology, 2010, 34(4) : 11-16(in Chinese).

[10] Chen Ailin, Yue Quanming, Feng Jun, et al. Application of IEC 61850 proxy server in seamless communication between smart substation and control center[J]. Automation of Electric Power System, 2012, 34(20) : 99-102(in Chinese).

[11] Zhang Jinjiang, Guo Chuangxin, Cao Yijia, et al. Substation equipment condition monitoring system and IEC model coordination[J]. Automation of Electric Power System, 2009, 33(20) : 67-72(in Chinese).

[12] IEC. IEC 61970 energy management system application program interface (EMS-API) part 1 : guidelines and general requirements[S]. IEC, 2003.

[13] IEC. CPSM minimum data requirements in terms of the EPRI CIM[S]. IEC, 2004.

[14] IEC. IEC 61850-6 Ed.2 : 2004(E) Conmaunication networks and systems for power utility automation-part6 : configuration description language for communication in electrical substations related to IEDs[S]. IEC, 2004.

Analysis and Corresponding Measures on Subsynchronous Resonance of Series Compensated Transmission for Large-scale Coal-fired Power Base in China

Du Ning[†], Song Rui-Hua*, Liu Chuan-Wen*, Chen Zhen-Zhen*, Xiang Zu-Tao*
and Ban Lian-Geng*

Abstract – This paper investigates the risk assessment of subsynchronous resonance for series compensated UHV transmission of Ximeng coal-fired power base, which exemplifies a common problem in the development of the power grid in China. To avoid the problem of huge computational amount of offline analysis on subsynchronous resonance, a new method of online analysis and forewarning based on time-domain simulation is proposed. In addition, various suppression measures are summarized and discussed, as well as combination of multiple measures.

Keywords: Large-scale coal-fired power base, Subsynchronous resonance (SSR), Online analysis and forewarning, Suppression, Combination

1. Introduction

As the energy resources and loads are quite uneven distributed in China, nowadays, State Grid Corporation of China has been developing UHV transmission technology in order to meet the rapidly increasing requirements for electric power of the society. Generally speaking, series compensation is a common and effective measure to make full use and exploit economic benefits of long transmission lines. Therefore, series compensation technology is extensively applied in 1000 kV UHV AC transmission pilot project, which has been put into operation since 2009, and other UHV transmission projects which are still in plan and under construction in China.

Nevertheless, it is certainly known that application of series compensation technology in power systems may cause problem of subsynchronous resonance (SSR) between coal-fired power plants and series compensated transmission systems [1]-[2]. However, there have been many research results on SSR problem so far, which are mainly for cases of a single power plant with a grid-to-grid network. Different from that, due to the special distribution of energy resources and loads as previously mentioned, there will be many large-scale coal-fired power bases in

China, and the electric power generated by all plants in a base might be intensively delivered by UHV transmission lines with a point-to-grid structure of network. Due to the scale of energy base which consists multiple generators as well as multiple oscillation modes, and characteristics of the power network structure, it may lead to more serious and more complicated SSR problem.

This paper focuses on the SSR problem of of series compensated transmission for large-scale coal-fired power base, and takes the planning UHV transmission project for Ximeng energy base of China as the studying object. Then, an new method of online analysis and forewarning will be proposed, and corresponding measures will be discussed.

2. SSR Analysis of Ximeng Energy Base

According to the electric power system planning of China, a large-scale coal-fired power base will be built in Ximeng area, Inner Mongolia, and a UHV AC transmission system will be constructed from Inner Mongolia to Northern and Eastern China. As illustrated in Fig. 1, the transmission lines from XM to BJ, and from BJ to JN, are both about 350 km. In order to increase the capability of transmission lines, it is planned to apply 40% series compensation, which may lead to SSR problem.

Frequency scan is a common method to estimate SSR risk. By frequency scan analysis, the impedance-frequency

---

[†] Corresponding Author: China Electric Power Research Institute, China (duning@epri.sgcc.com.cn)

* China Electric Power Research Institute, China

Table 1. Typical frequency scan results for transmission system of Ximeng energy base (Plant 1)

No.	Number of elements in operation							Series resonance frequency (Hz)	Reactance dip	
	Plant 1	Plant 2	Plant 3	Plant 4	XM-BJ line	SC at XM	BJ-JN line		Frequency (Hz)	Dip
1	2	2	3	3	2	2	2	--	16.8-17.8	28%
2	2	2	3	3	2	1	2	--	24.9-25.8	8%
3	2	2	3	3	1	1	2	--	20.6-21.7	29%
4	1	2	3	3	2	2	2	--	16.4-17.5	40%
5	1	1	1	1	2	2	2	--	14.5-15.4	77%
6	1	0	0	0	2	2	2	12.5	--	--
7	1	0	0	0	2	1	2	--	22.9-24.1	43%
8	1	0	0	0	1	1	2	15.5	--	--

characteristics of transmission system of a plant with frequency will be obtained, which is irrelevant to the characteristics of generators. Part of frequency scan results for Ximeng base is illustrated in Fig. 2 and Table 1. It is observed that there are generally two types of conditions.

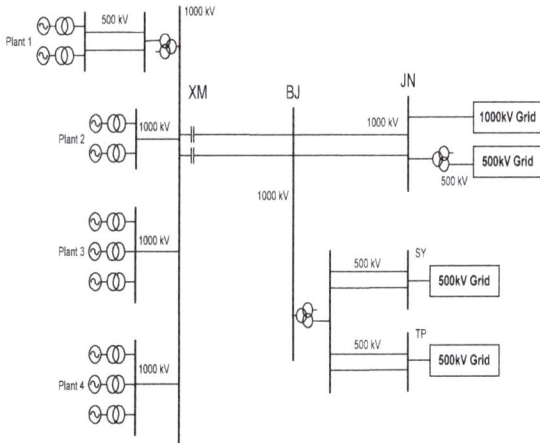

Fig. 1. Diagram of transmission system of Ximeng energy base.

2.1 Reactance dip

As shown in Fig. 2(a), the curve of reactance-frequency drops and then rises when the frequency is around 17 Hz, however the reactance keeps positive within the entire subsynchronous frequency range. Commonly, this phenolmena is quantified by reactance dip [3], which is defined as:

$$\textbf{reactance dip} = \frac{Y - X}{Y}$$

There is no precise threshold value to classify SSR risk nowadays. However, experience shows that it requires attention when reactance dip is larger than 20%.

2.2 Series resonance

As shown in Fig. 2(b), the reactance falls below zero and then rises above zero when the frequency is around 11 Hz, which indicates series resonance. In this case, there might be high SSR risk.

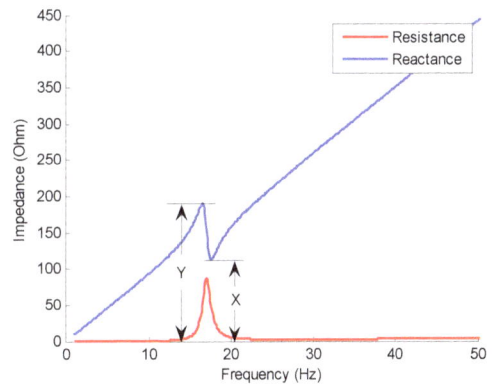

(a) Reactance dip

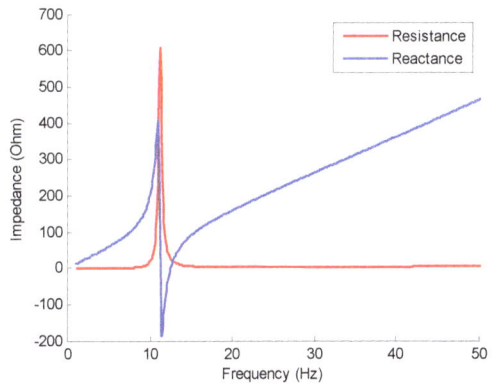

(b) Series resonance

Fig. 2. Typical impedance-frequency characteristics of transmission of Ximeng energy base (Plant 1).

It is observed from Table 1 that, if all elements of the system, including lines, series capacitors and generators of all plants, are in operation, the reactance dip is 28%. When one of series capacitors or one of XM-BJ lines quits, the value will decrease to 8% and keep almost the same, respectively. If one or more of generators quit, the reactance dip will rise up, it may even turn into series resonance for islanding conditions.

In summary, comparing with normal series compensated transmission, there might be worse SSR problem for large-scale power base, and corresponding studies on both SSR risk estimation and suppression should be carried out.

3. SSR Online Analysis and Forewarning

As for offline analysis of subsynchronous resonance, theoretically, all possible conditions should be considered, which includes the status of generators, transmission lines, series capacitors, transformers, as well as output of generators and distribution of power flow all over the power grid. Simply, suppose that there are N elements separate from each other, considering each one being on or off, the total amount of conditions could be 2^N. Thus, the calculation amount will be 1,024 when N is equal to 10, and the number could be as large as 32,768 when N is equal to 15. It is obviously difficult to take all conditions into consideration in offline analysis, instead, only several typical conditions are considered in practice,

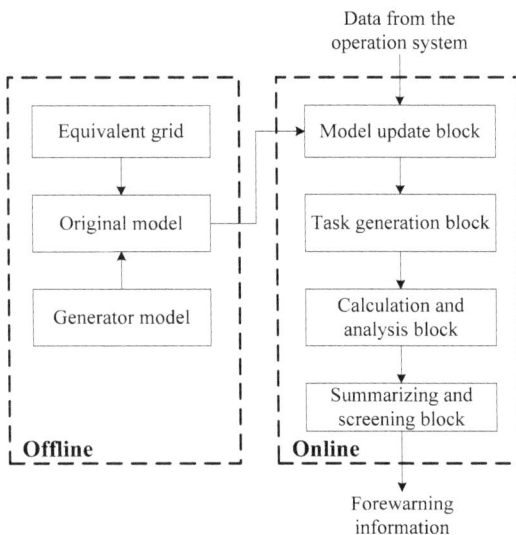

Fig. 3. Scheme of SSR online analysis and forewarning.

To avoid large amount of calculation in SSR risk assessment, we have tried another analysis method that using online analysis and forewarning. The main idea of this method is to analyze contiguous conditions regularly, and reveal conditions under which subsynchronous resonance may take place.

The scheme of this method is illustrated in Fig. 3. This method contains two main parts: offline part and online part.

In offline part, the original model for SSR analysis is stablished, which includes the model of generators and equivalent model of the grid.

There are four blocks in online part.

3.1 Model update block

From the operation system, the current grid topology and power flow distribution could be obtained through necessary data format conversion and calculation. And then, the original analysis model stablished in offline part will be updated according to the newest data, and the model of current condition will be built.

3.2 Task generation block

In this block, the status of every element is examined first, so we can learn how could its status change from current condition. Through permutation and combination of status of each elements, computational tasks with various conditions will be generated.

It is important to note that, as the online analysis repeats regularly, within such as several minutes, it is not necessary to consider all conditions like offline analysis. Generally, a group of $N\pm1$ and $N\pm2$ conditions should suffice for online analysis, and the amount for a grid with N separate elements is

$$\mathbf{C}_N^1 + \mathbf{C}_N^2 = N + \frac{N\cdot(N-1)}{2} = \frac{1}{2}\left(N^2 + N\right)$$

A comparison of computational amount between offline and online analysis is showed in Table 2. It is obvious that the amount of online tasks increases much slowly than offline analysis when the number of elements rises.

Table 2. Comparison of computational amount between offline and online analysis of SSR

Number of elements (N)	Offline analysis	Online analysis
10	1,024	55
15	32,768	120

3.3 Calculation and analysis block

In this block, all these computational tasks will be executed. Each task is written into an input data file for time-domain electromagnetic analysis, and submitted to the computing platform like a cluster compute system. According to Table I, the amount of tasks will be hundreds, which is not a great number for a cluster compute system of this time.

After numerical calculation of each task, the information of subsynchronous signals should be extracted from the results, including the frequency and corresponding magnitude and damping factor. FFT and ESPRIT algorism [4] can be used during frequency analysis.

3.4 Summarizing and screening block

In this block, all time-domain calculation and frequency analysis results should be summarized, and the conditions with large magnitude and negative damping factor could be recognized, thus the forewarning information will be produced.

4. SSR Suppression for Large-scale Coal-fired Power Base

Table 3. Comparison of main SSR suppression measures

Measures	Advantages	Disadvantages
Pole-face amortisseur windings	Good effect, low cost	Reduce efficiency of generator
Block filter	Simple structure	High cost, large area, high loss, the effect changes with environment temperature
SEDC	Low cost, easy to install	Limited by capacity of generator, incompetent at large turbulence
SVC	Good effect, low cost	High steady loss
TCSC	Increase stability of system and power transfer limit	High cost, high requirements on reliability
STATCOM	Good effect, low steady loss	High cost

To solve the SSR problem caused by application of series compensation technology in AC transmission systems, researchers out of the world have been carrying out many studies on suppression measures with different mechanism [5]-[8], for example, using supplementary excitation damping controller (SEDC), blocking filter, thyristor controlled series compensation (TCSC), static var compensator (SVC), static synchronous compensator (STATCOM), and so on,

among which the former four measures have been practically applied in China. According to the actual operation situation, these measures can take some effects on SSR suppression, they may also have disadvantages on suppression capacity, controllability, flexibility, etc. Table 3 gives a comparison of main SSR suppression measures.

For series compensated transmission of large-scale coal-fired power base, as there might be high SSR risk, further studies should be carried out on suppression measures.

Recently, researches have studied on some new mitigating methods such as using induction machine [9] or double fed machine [10], by which a co-axial machine will be fixed at the shaft terminal of the generator. However, it may take years for practical application.

On the other hand, researches have tried to employ two or more measures together for better suppression effect, such as a combination of SEDC and TCSC [11], and one of SEDC and STATCOM [12]. In this way, we could benefit from advantages of both measures. However, it needs further studies on the coordination of all measures involved.

5. Conclusion

This paper investigates the SSR problem of series compensated transmission for large-scale coal-fired power base in China. It indicates that there might be worse SSR problem for large-scale power base, thus corresponding studies on both SSR risk estimation and suppression should be carried out. As the large amount calculation for offline analysis of SSR, an online analysis and forewarning method is proposed. In addition, suppression measures for the SSR problem in series compensated transmission for large-scale coal-fired power base are discussed, including both new methods and combination of multiple methods.

Acknowledgements

This work was supported by the National High Technology Research and Development Program of China (863 Program) (No. 2011AA05A119) and the Major Projects on Planning and Operation Control of Large Scale Grid of the State Grid Corporation of China (No. SGCC-MPLG001-2012).

References

[1]   Cheng Shi-jie, Cao Yi-jia, Jiang Quan-yuan. Theory and

method of subsynchronous oscillation in power system (in Chinese). Beijing: Science Press, 2009.

[2] Anderson P M, Agrawal B L, Van Ness J E. Subsynchronous resonance in power systems. New York: IEEE Press, 1990.

[3] Agrawal B L, Farmer R G. Use of frequency scanning techniques for subsynchronous resonance analysis. IEEE Transactions on Power Apparatus and Systems, 1979, PAS-98(2): 341-349.

[4] Roy R, Kailath T. ESPRIT - Estimation of signal parameters via rotational invariance techniques. IEEE Transactions on Acoustics, 1989, 37(7): 984-995.

[5] Khaparde S A, Krishna V. Simulation of unified static var compensator and power system stabilizer for arresting subsynchronous resonance. IEEE Transactions on Power Systems, 1999, 14(3): 1055-1062.

[6] Song Rui-hua, Xiang Zu-tao, Ban Lian-geng. Theory & project application of mitigating SSR with SVC. IEEE PES General Meeting 2010, Minneapolis, USA, 25-29 July 2010.

[7] Padiyar K R, Prabhu N. Design and performance evaluation of sub-synchronous damping controller with STATCOM. IEEE Transactions on Power Delivery, 2006, 21(3): 1398-1405.

[8] Bongiorno M, Angquist L, Svensson J. A novel control strategy for subsynchronous resonance mitigation using SSSC. IEEE Transactions on Power Delivery, 2008, 23(2): 1033-1041.

[9] Purushothaman S, De León F. Eliminating subsyn-chronous oscillations with an induction machine damping unit (IMDU). IEEE Transactions on Power Systems, 2011, 26(1): 225-232.

[10] Xu Jin-feng, Song Rui-hua, Xiang Zu-tao, etc. Simulation studies on mitigating subsynchronous resonance with coaxial double fed motor. 2012 International Conference on Power Systems Technology, POWERCON 2012, Auckland, New Zealand, 30 Oct - 2 Nov 2012.

[11] Wu Xi, Jiang Ping. Research on sub-synchronous oscillation mitigation using supplementary excitation damping controller and thyristor controlled series capacitor (in Chinese). Transactions of China Electrotechnical Society, 2012, 27(4): 179-184, 239.

[12] Sun Yan-long, et al. Study on Damping scheme of subsynchronous resonance in a multi-machine system by combination of SEDC with STATCOM (in Chinese). Power System Technology, 2013 (in press).

14

A Study on Cooperative Control Method to Secure the Reactive Power in the Power System

Hyun-Chul Lee*, Ki-Seok Jeong*, Ji-Ho Park** and Young-Sik Baek[†]

Abstract – This paper proposes the cooperative control method between reactive power compensation devices for securing the reactive reserve power of the FACTS in the power system. The Cooperative control consists of the two different methods, such as concurrent and sequence control. This proposed method is used different method according to the normal and abnormal state. The concurrent control is applied for maintaining the bus voltage in desired level when the load of power system has been remarkably increased or decreased. This control method is operated at once by calculating necessary active power capacity based on the voltage difference to each substation voltage and reference voltage in the power system. The sequence control has been used to secure momentary reactive power. The momentary reactive power is processed with the fast control characteristics. Therefore, this control has been managed for improving the voltage stability in the power system. Also, it has been demonstrated by the PSS/E and the Python programs.

Keywords: Cooperative control, Reactive power compensation devices, Voltage stability

1. Introduction

The power system structure was changed by the various loads and the increased power demand. The domestic power system has been appeared as more large and complicated power system. Therefore, the power system has been operated the various reactive power compensation devices to supply with necessary power demand [1-3].

So, the power systems has been operated various reactive power compensation devices to supply with necessary power demand. As one of the devices, the FACTS (Flexible AC Transmission System) have been applied to the power system. This device has got to the fast voltage recovery ability. In the power system, the voltage compensation devices needed to be cooperative control for effective operation [3-5].

This paper was studied regarding cooperative control method in the voltage compensation devices between substation and near substation. The control was proposed concurrent control and sequence control method. First, the concurrent control was controlled the reactive power compensation devices by calculating the required reactive power. Second, the sequence control was controlled the

reactive power compensation devices according to situation of the power system. The compensation devices was used the shunt element (Shunt Capacity and Shunt Reactor) for the voltage and reactive power, in addition to use the FACTS (SVC, STATCOM), that is power semiconductor devices by developing the technology. The simulation was performed to use PSS/E (analysis program of the real power system), and the cooperative control method was implemented to used Python (compatible external macro program). A case study was applied to the domestic power system through the proposed operating algorithm by using the shunt elements and the FACTS devices. The simulation result was satisfied range of the maintained voltage depending on the operating algorithm.

2. Reactive Power compensation devices

2.1 Characteristic of the compensation devices

It is a well-known fact that shunt compensation can be used to provide reactive power compensation. In following formula, shunt elements were expressed in supplied current and reactive power to the power system [5-6].

$$I_L = -B_L V_i, \qquad Q_L = B_L V_i^2 \qquad (1)$$
$$I_L = B_C V_i, \qquad Q_C = -B_C V_i^2 \qquad (2)$$

Where, B_L and B_C is inductive and capacitive

† Corresponding Author: Dept. of Electrical Engineering, Kyungpook National University, Korea (ysbaek@knu.ac.kr)
* Dept. of Electrical Engineering, Kyungpook National University, Korea (oneye@knu.ac.kr, biotronic@naver.com)
** Dept. of Electrical Engineering, Koje Colleage, Korea (pjh@ee.knu.ac.kr)

susceptance respectively. V_i is the i-th bus voltage.

Traditional shunt capacitors or newly introduced FACTS controllers can be used for this purpose. Shunt capacitors are relatively inexpensive to install and maintain. Installing shunt capacitors in the load area or at the point that they are needed will increase the voltage stability. However, shunt capacitors have the problem of poor voltage regulation and beyond a certain level of compensation; a stable operating point is unattainable. Furthermore, the reactive power delivered by the shunt capacitor is proportional to the squre of the terminal voltage, during low voltage conditions Var support drop, thus compounding the problem. Fig. 1 is shown the characteristic of the shunt capacitor.

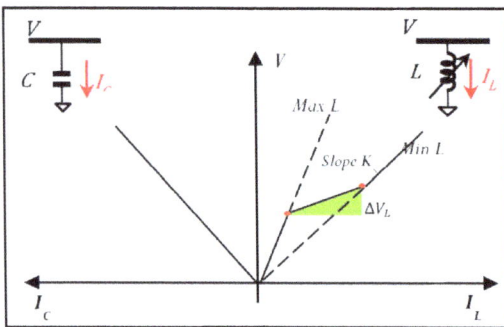

(a) shunt elements

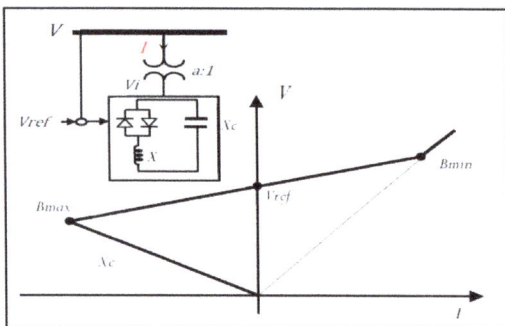

(b) SVC (Synchronous Var Compensator)

Fig. 1. Characteristics of reactive power compensation devices.

2.2 Control of the reactive power

The proposed method was controlled reactive power compensation devices in the power system. This method is used profile to voltage sensitivity of each bus. The reactive power had been supplied calculated capacitance. It was voltage magnitude control of the connected reactive power devices in the bus. However, compensation devices types

have been differ from controlled response speed for reactive power. It is necessary for cooperative control to get different response speed because of reactive power compensation devices characteristics. SVC and STATCOM should be secured momentary reactive power in steady state. The bus voltage has been recovered to fast response speed by the FACTS devices in transient state. Other devices were consisted in mechanical reactive power equipment (shunt reactor and capacitor). It has been operated slow response speed but it has got the large reactive power capacity. The bus voltage has been maintained by operating the MSR/MSC in transient state. It might be controlled power system by analyzing central and local database.

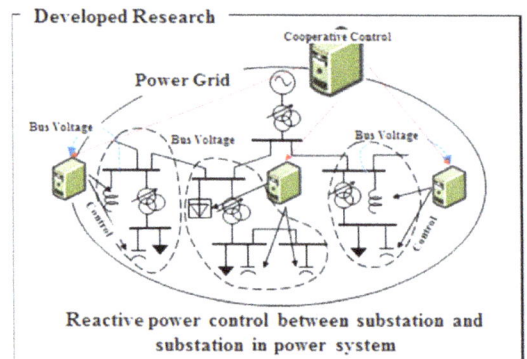

Fig. 2. Concept of reactive power control in the power system.

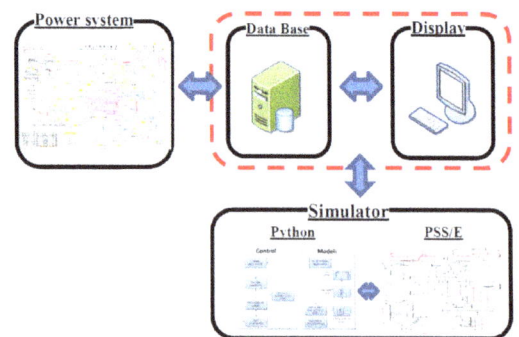

Fig. 3. Data flow for the simulator development in the power system.

The database could be stored the bus voltage, active/reactive power, and reactive power devices state. The simulator has been monitored by using the PSS/E and Python about the voltage sensitivity and reactive power capacity. Fig. 3 is presented the data flow about the simulator.

2.3 Algorithm of cooperative control

Fig. 4 was shown the cooperative control algorithm. This black diagram has shown the securing momentary reactive reserve to maintain the bus voltage in the power system.

First, the power system had been set the control area and calculated the voltage sensitivity due to reactive power capacity in each bus. And then, checking the bus voltage in exceeds reference voltage or not. In case of exceed reference voltage, FACTS device should be operated the fastest response. It was controlled the necessary reactive power by calculating voltage sensitivity.

Case within range reference voltage, FACTS device should be operated for obtaining momentary reactive power reserve. It shouldn't be ended the control until FACTS devices has been securing the reactive power reserve.

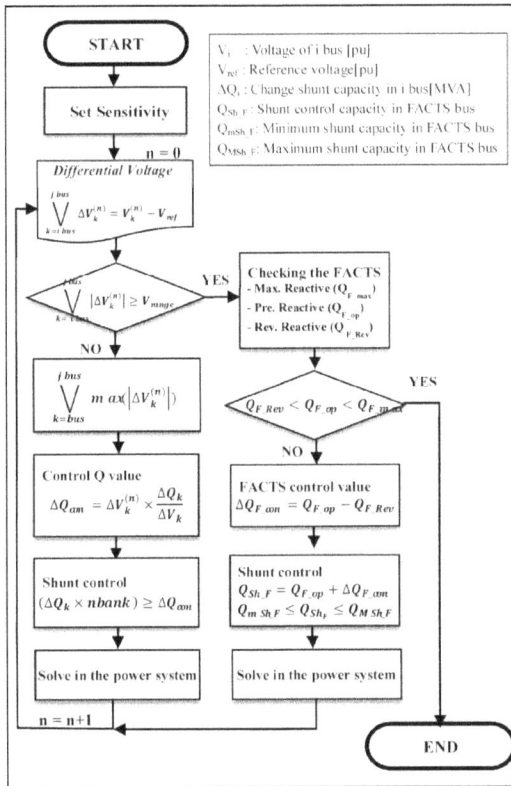

Fig. 4. Flowchart of reactive power control for voltage stability in the power system.

The reactive power is controlled by using compensation devices in the substation inside unit. However, if the margin value couldn't be found the reactive power, reference voltage should be difficult to maintain. This method is used to maintain the bus voltage by controlling reactive power in near substation. At this time, compensation devices were controlled concurrent at once by calculating all bus voltage sensitivity. The FACTS device was operating to the fastest response when bus voltage was exceeding the reference voltage in power system. Therefore, the device got the reactive power in order to fast voltage control from transient state in power system. This method was called sequentially control. It has been securing the reactive power by operating the MSR/MSC (relatively slow response than the FACTS device).

The concurrent control method was operated stable bus voltage from the unstable the power system by changing the transmission line and load capacity. Formula 3 was shown the capacity to be demanded for reactive power.

$$\Delta Q_{req} = \frac{V_{ref} - V_i}{\Delta V_i / \Delta Q} \tag{3}$$

Where, the V_{ref} is reference voltage, V_i is i-th bus voltage, and $\Delta V_i / \Delta Q$ is i-th bus voltage sensitivity by ratio of reactive power change.

Reactive power compensation derives has been operated bus voltage sensitivity according to change reactive power compensation device. If violated bus voltage was lacked of reactive power, it was used reactive power compensation device in near substation. Formula 4 was shown capacity to be demanded by using concurrent control method for reactive power.

$$\Delta Q_{req} = \frac{1}{\Delta V_i / \Delta \Delta_i} \left[max \left\{ \forall_{bus} \left(V_{ref} - V_k \right) \right\} \right] \tag{4}$$

To manage reactive power, the bus voltage should be steady sate in power system. The momentary reactive was set the reserve power according to devices capacity. Formula 5 and Formula 6 were capacity to be demanded by using sequence control method for reactive power.

$$\Delta Q_{i\_Freq} = Q_{i\_Fpre} - Q_{i\_Fres} \tag{5}$$

Where, Q_{i_Fbus} is reactive power control capacity in the i-th bus installed FACTS devices, Q_{i_Fpre} is present reactive capacity, and Q_{i_Fres} is reactive reserve setting capacity in FACTS devise. Q_{i_Fbus} was changed reactive capacity by controlling shunt devices. ΔQ_{i_Freq} was demanded reactive power on the FACTS devices.

$$Q_{i\_FBus} = \begin{cases} Q_{i\_Fpre} + \Delta Q_{i\_FBus} & if\left(\Delta Q_{i\_Freq} > 0\right) \\ Q_{i\_Fpre} & if\left(\Delta Q_{i\_Freq} \leq 0\right) \end{cases} \quad (6)$$

3. Simulation Result

The case study is confirmed low bus voltage as bus range condition 1.02 ± 0.02 [p.u]. In simulation results, the concurrent control method has been cooperative control to operate FACTS devices by maximum capacity mode and shunt elements by calculating demanded reactive power.

3.1 simulation condition

The test power system has been targeted at the domestic metropolitan area. This area has been load concentration problem. This test was simulation on two cases. First case is simulation about the transmission line trip in power system. Second case is simulation about the increased load capacity. Table 1 was shown the simulation result by using control method.

Table 1. Simulation control result of compensation devices

MVAR	Set	1st control Concurrent	2nd control Sequence
CASE 1	FACTS	+101.6	-45.3
	Shunt	2set	4set
		+200	+270
CASE 2	FACTS	+98.4	-74.1
	Shunt	3set	4set
		+300	+350

3.2 Case Study

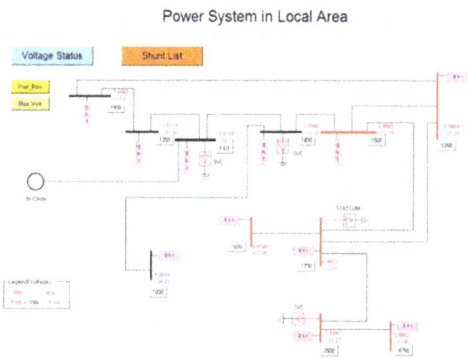

Fig. 5. Power system for simulation (Case I).

The violated bus was founded difference largest bus

voltage (1600) with the reference voltage. The violated voltage bus was controlling reactive power derives. However, calculated reactive power has been lacked for recover the bus voltage. The near bus (1700) of the violated bus was operated both the STATCOM (+101[MVA]) and shunt elements (+200[MVA]). And then, the sequence control had been operated to secure reactive power of the FACTS device by controlling shunt elements (+270[MVA]). Fig. 5 is shown power system diagram of the developed simulator.

Fig. 6 and Fig. 7 is shown the controlled bus to before and after control of reactive power compensation devices.

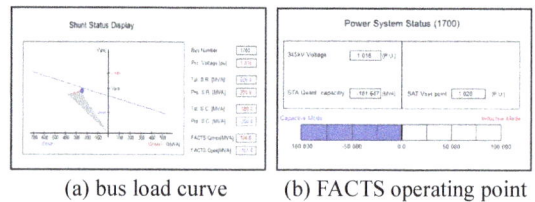

(a) bus load curve   (b) FACTS operating point

Fig. 6. Initial power system state (load curve and FACTS capacity).

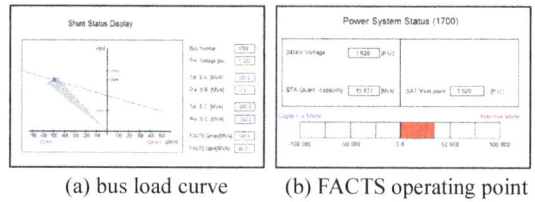

(a) bus load curve   (b) FACTS operating point

Fig. 7. After simulated power system (load curve and FACTS capacity).

The case 2 has been simulated the violated voltage bus (1800). This bus was controlled reactive power devices. The bus voltage had been recovered by calculating required reactive power. Reactive power devices of near bus were controlled due to insufficient capacity of reactive power compensation devices in the violated bus. So, near bus (1400) of the violated bus was operated both the STATCOM (+98.4[MVA]) and shunt elements (+300[MVA]). Sequence control was operated to secure reactive power of the FACTS by controlling shunt element (+350[MVA]).

4. Conclusion

This paper is algorithm development. It had been unstable the power system condition by transmission line fault or load capacity change. So, compensation devices of the bus used to make stable power system.

The amount of reactive power at the maximum loading point from the synchronous compensator was found to 100~200[MVAR]. This has been good starting point for different controller. This method of determining the compensation devices has been find the relationship between the clanging load capacity that the devices can diver without casing voltage collapse. Other compensation devices of reactive power was found to be 50~700 [MVAR]. But, this compensation device couldn't be continuous control. Therefore, it could be improved voltage stability through the cooperative control of FACTS devices and other compensation devices.

The control method has been considered effective operated power system by securing momentary reactive reserve and maintaining the reference voltage.

Acknowledgements

This study was supported MEST(Ministry of Education, Science and Technology) in 2013, Thank to the relevant authorities. (grand code: 2013006460)

References

[1] Geun-Joon Lee, Jong-Soo Yoon, Byung-Hoon Jang, Sung-Won Jung, Sun-Ho Yoon, Hyun-Chul Lee, "Local Coordinative Power control Method to Manage Voltage/Momentary Reactive Reserve", Conference KIEE, pp.115-116, 2008.

[2] Ki-Seok Jeong, Byun-Hoon Chang, Hyun-Chul Lee, Geun-Joon Lee, Yong-Sik Baek, "A Study on Local coordinative Reactive Power Control between STATCOM and Other Reactive Power Controllers for Voltage Stability Improvement at Substation", Trans. KIEE, Vol.59, No.3, pp.523-530, MAR. 2010.

[3] Ferry A. Viawan, Deniel Karlsson, "Combined Local and Remote Voltage and Reactive Power Control in the Presence of Induction Machine Distributed Generation", IEEE Trans. on PWRS, Vol.22, No.4. pp.2003-2012, Nov. 2007.

[4] Tariq Masood, R.K. Aggarwal, S.A. Qureshi, R.A.J Khan,"STATCOM Model against SVC Control Model Performance Analyses Technique by Matlab",ICREPQ 2010, MAR, 2010.

[5] Ji-Ho Park, Sang-Duk Lee, Tae-Yong Jyung, Ki-Seok Jeong, Yong-Sik Baek, Gyu-Seok Seo, " Coordination of UPFC and Reactive Power Source for Staedy-state Voltage Contorl, Trnas. KIEE, Vol.60, No.5, pp. 921-928, MAY,2011.

[6] Arthit Sode-Yome, N. Mithulananthan, "Comparison of shunt capacitor, SVC and STATCOM in static voltage stability margin enhancement", IJEEE, pp.158-171. APR. 2004.

Development of Optimization Parameter System for AVC in Hebei Low Voltage Grid

Yang Xiao†, Gao Zhiqiang* and Fan Hui*

Abstract – Proposing the optimal AVC coordinating parameters and comparing effects of different grid planning for reactive balance in an scientific manner is one of the functional characteristics of smart AVC. In this paper Optimization Parameter (OP) system is developed in Hebei low voltage grid for this purpose. The data are taken from EMS. The state estimation is carried out periodically after obtaining real-time data. Taking state estimation results as the basic power flow and simulating AVC operations, the voltage variations and line loss analysis can be calculated for comparison. The system can compare optimization parameters for the real grid or the virtual grid. It's depending on whether considering the actual capacity of reactive power compensation equipments. The structure, hardware configuration and data exchange of the system are illustrated. Finally, several analysis tools of the OP system are introduced. With the system, it can be evaluated that effects of reducing line loss benefited from the AVC operations. It can propose coordinating priority of capacitors or on-load regulating transformers. Also it can provide the optimal AVC coordinating parameters and the optimal grid plans, which lead to better voltage control and better layout of reactive power compensation equipments.

Keywords: Automatic voltage control, Low voltage grid, Optimization parameter, Line loss, AVC coordinating parameter, Planning grid

1. Introduction

The problems of power system became important with the increase of grid capacity and voltage level upgrades, such as the reactive power balance and voltage control. The voltage of the key point is much high in valley load period in recent years. Therefore, the insulation of transformations, transmissions and consumer equipments must bear over-voltage, greatly shortening the life of these equipments and probably affecting the safe operation of them [1-3]. Automatic Voltage Control (AVC) system has been applied as a means for controlling voltage and adjusting reactive power because it has many special excellences [4-8]. However, in low voltage grid or rural grid, AVC system only have reactive compensate effect, for not enough reactive capacity. It has been a problem to get better plan for the layout of reactive power compensation equipments. Meanwhile, there are few ways to compare effects of different AVC coordinating parameters on the voltage regulation. It is an urgent need to develop Optimization

Parameter (OP) system.

The main function of the OP system can be described as follow: (a) It can check whether the available AVC coordinating parameters are optimization. (b) It can evaluate reducing line loss benefited from the operations. (c) The OP system can provide the decision-supporting function for planning new power capacitors or on-load regulating transformer of substation, which can reduce the line loss of low voltage grid or rural grid.

The main motive of this paper is to provide the purpose, system structure, hardware configuration and data exchange of the OP system. Blueprint of developing OP system to smart AVC system is shown, though still in the research. Furthermore, a number of important concepts about the coordinating priority, such as the voltage-capacitor sensitivity and the voltage-tap sensitivity, are also given. Finally, several analysis tools of the OP system are introduced.

2. The structure of the OP system

With the real-time data taken from EMS, the OP system is the on-line monitoring and analyzing system for Hebei low voltage grid. The state estimation is carried out periodically after obtaining real-time data. Taking state

† Corresponding Author: Hebei Electric Power Research Institute, Shijiazhuang 050021, Hebei Province, China (yangholmes. student@sina.com)

* Hebei Electric Power Research Institute, Shijiazhuang 050021, Hebei Province, China {(hepri-gaozq, hepri-fh)@he.sgcc.com.cn}

estimation results as the basic power flow, the theoretical voltage variations and line loss analysis can be carried out within the virtual equipments such as capacitors and transformers. It can calculate the voltage-capacitor sensitivity, which is voltage variation of one key bus dividing the reactive variation when regulating a unit of capacitor. Also the OP system can calculate the voltage variation of one key bus while regulating the transformer tap, which is voltage-tap sensitivity. These sensitivities decide the coordinating priority of capacitors or transformers. All the voltage variations can be viewed with Hebei grid geographical diagram.

The OP system in Hebei grid contains five subsystems. The flowchart of it is illustrated in Fig. 1.

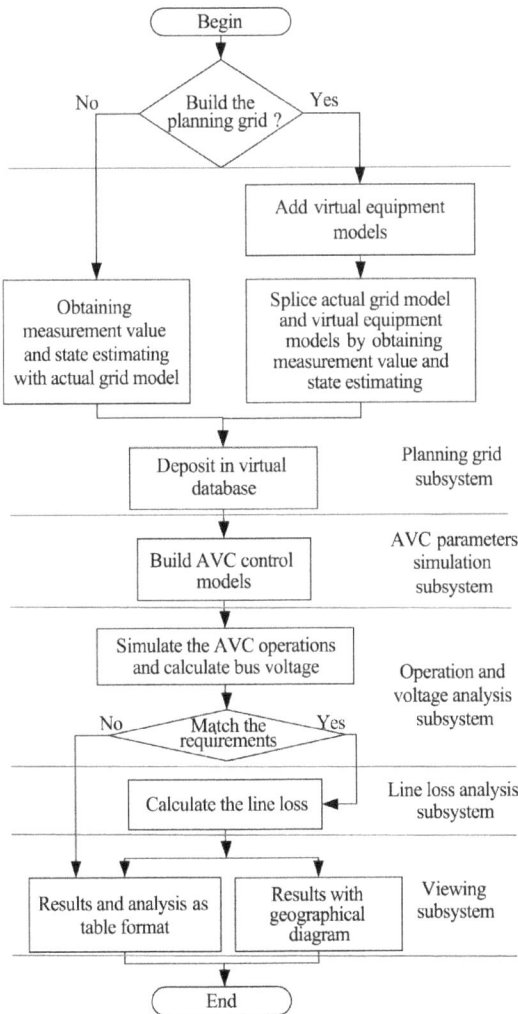

Fig. 1. Flowchart of OP system.

2.1 Planning grid subsystem

The planning grid subsystem can establish and save different planning grid models. The planning grid models include actual grid model and virtual equipment models (transformer, line, bus, generator, capacitor, etc). Based on two data sources (real-time data and historical data), the virtual equipment models can be added via user interface. The subsystem can splice actual grid model and virtual equipment models by obtaining measurement value of actual grid model and carrying state estimation with the planning grid model. If establishing the virtual equipment models is not needed, the subsystem only obtains measurement value and carries state estimating with actual grid model. The models deposit in virtual database.

2.2 AVC parameters simulation subsystem

It is used to build AVC control models and save different AVC coordinating parameters such as capacitors coordinating velocity and min capacitors switching amount by one operation. They can affect the reactive power balancing and voltage coordinating. Every AVC control model can deposit a number of parameter schemes.

2.3 Operation and voltage analysis subsystem

It is used to simulate the AVC operations. Firstly users should define when the operations taking place in one day or in any time segment. And via the subsystem users can use real-time EMS data or historical EMS data for simulation. The operations include switching of capacitors or adjusting on-load regulating transformer taps. Then the accurate results of power flow and voltage variations are calculated. These data are stored in virtual database for the line loss analysis subsystem. If the AVC coordinating parameters can't match the coordinating requirements, the next step will go to viewing subsystem and propose the incompetent parameters in the form of table. Otherwise, it will enter into line loss analysis subsystem.

2.4 Line loss analysis subsystem

It can calculate line loss of the actual grid or the planning grid. Firstly users should establish line loss models and parameters. It can define or modify range of line loss statistics areas via user interface. The line loss models and parameters of added virtual equipments are stored in database and initialized automatically if updated. According to the aforesaid power flow stored in virtual database, the integral calculation for line loss can be carried out. The results can be shown with Hebei grid geographical diagram.

2.5 Viewing subsystem

It is used to provide analysis tables and comparison curves for users. It also can show bus voltage or line loss with the geographical diagram of low voltage grid.

3. Hardware configuration and data exchange

The hardware configuration of OP system is shown as Fig. 2. The EMS server, EMS workstation and EMS database are built in power dispatch or control center of county in Hebei province. A data receiving server in Hebei Electric Power Research Institute is configured to receive models, graphics and data of Hebei low voltage grid. Two network cards are used in data receiving server. One network card connects with the EMS server to get power system data. The other one connects with OP system.

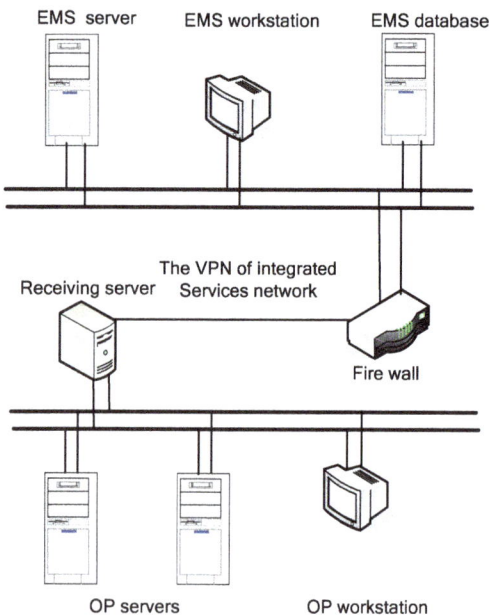

Fig. 2. hardware configuration of OP system.

Two OP servers are configured for storing five subsystems and the virtual database, they are used for calculation and analysis functions. And the system has data backup and recovery function. The files of it can be automatically forwarded to other storage facilities for data

mirroring, disaster recovery or backup applications. A workstation is used for monitoring and operating. With the workstation, users can login OP servers, modify parameters and view the results.

Fig. 3 shows data exchange between five subsystems and EMS server. After state estimation, the EMS server sends correction value to OP system. As it is shown, the virtual database is very important in the system. It not only receives EMS real-time data or historical data, and stores voltage analysis and line loss results. Broken line means the date sending from OP system to AVC system. The purpose is sending the optimization coordinating parameters to AVC substations with control orders via AVC system. Moreover, it could modify AVC coordinating parameters on-line when large load variation happened or structure of low voltage grid changed. This is one of the functional characteristics of smart AVC [9–13].

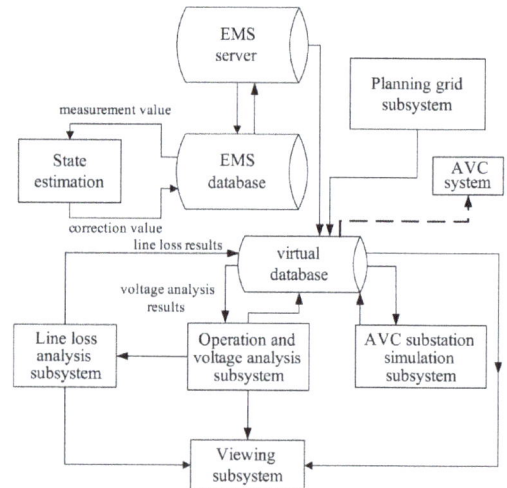

Fig. 3. data exchange between subsystems and EMS server.

4. The introduction of analysis tools and application

4.1 Sensitivity and priority calculator

It can calculate the voltage-capacitor sensitivity. It should be firstly defined the reactive variation when switching one unit of capacitors in one substation. Then the system calculates reactive output and voltage variation of different key points. Using the same reactive output variation, the tool simulates every substation with different load situations. The load situations include spring load, summer load and winter load, for their distinct load characteristics in Hebei low voltage grid. When all voltage-capacitor sensitivities are given, user can define capacitor coordinating priority for some key point in different load situations. Capacitors with bigger voltage-capacitor sensitivity have higher coordinating priority. The process of providing voltage-tap sensitivity and on-load regulating transformer coordinating priority is similar.

4.2 Comparison curve

The main function of the tool is to show comparison curves for different types of data. A series of comparison curves have built in system such as active power, reactive power, voltage, line loss and so on. Users can display these comparison curves easily. Through the comparison, it is illustrated which planning new capacitors of substation or new on-load regulating transformers have better effects on controlling voltage or reducing line loss. The system also provides a user-define interface for comparison curve. Users can define various comparison curve scheme according to their requirements.

4.3 Table report

The main function of table report is to provide customization reports for users. The stored data in system can be proposed in the tables such as sensitivities data, priorities data, capacitors data, on-load regulating transformers data, integral time section of line loss, line loss and so on. Via the report, better layout planning for reactive equipments is revealed. The arithmetic of data in tables can be investigated by users. The format of tables can be defined flexibly. The customization report can be generated and published automatically.

5. Conclusion

In this paper Optimization Parameter system is developed. The structure, hardware configuration and data exchange of it are introduced. With the OP system, the traditional reactive compensate equipments can be united and developed to smart AVC system for rural grid. It is an important component of smart grid. It can propose coordinating priority of capacitors or on-load regulating transformers. It can evaluate reducing line loss benefited from the AVC coordinating parameters or the AVC operations. Also it can provide the optimal AVC coordinating parameters and the optimal grid plans, which

lead to better voltage controlling and better layout of reactive power compensation equipments.

Acknowledgements

This work was supported by the Science and Technology Research Foundation of Hebei Power Grid Corporation (Grant No.kj2009-114 and NO.kj2010-105).

References

[1] Dmitriev M V, Evdokunin G A, Gamilko A. EMTP simulation of the secondary arc extinction at overhead transmission lines under single phase automatic reclosing. IEEE Conference on Power Technology. Russia: IEEE, 2005: 1-6

[2] Malewski R,Douville J,Lavallee L. Measurement of switching transients in 735kV substations and assessment of their severity for transformer insulating. IEEE Transactions on Power Delivery, 1988, 3(4): 1380-1386

[3] Qiu J Y, Ren J J, Hu D C.Lightning overvoltage analysis of 110kV transformer neutral point (in Chinese). High Voltage Engineering, 2007, 33(1): 99-101

[4] Guo Q L, Wang B, Ning W Y, et a1.Applications of automatic voltage control system in North China power grid (in Chinese).Automation of Electric Power Systems, 2008, 32(5): 95-98

[5] Guo Q L, Sun H B, Zhang B M, et a1.Study on coordinated secondary voltage contro1 (in Chinese). Automation of Electric Power Systems, 2005, 29(23): 19-24

[6] Ruiz P A, Sauer P W. Voltage and reactive power estimate for contingency analysis using sensitivities. IEEE Transactions on Power Systems, 2007, 22(2): 639–647

[7] Chowdhury F D, Crow B H, Acar M L,et al. Improving voltage stability by reactive power reserve management. IEEE Transactions on Power Systems, 2005, 20(1): 338–345

[8] Wu D, Li D C, Dong R, et al.Actualization and effect of An-hui automatic voltage control system (in Chinese).Power System Equipment, 2005, 6(5): 63-67

[9] Yu Y X. Technical composition of smart grid and its implementation sequence (in Chinese). Southern Power System Technology, 2009, 3(2): 1-5

[10] Zhang Z Z, Li X Y, Cheng S J. Structures, functions and implementation of united information system for smart grid (in Chinese). Proceedings ofthe CSEE, 2010, 30(34): 1-7

[11] Hu X H. Smart Grid- A Development Trend of Future Power Grid (in Chinese). Power System Technology, 2009, 33(14): 1-5

[12] Cui S Y, Sun S F, Li L X, et al. Survey on Smart Grid Technology (in Chinese). Power System Technology, 2009, 33(8): 1-8

[13] Yu Y X, Liang W P .Smart grid (in Chinese). Power System and Clean Energy, 2009, 25(1): 7-11

16

Advanced Technique for Moisture Condition Assessment in Power Transformers

Jialu Cheng[†], Diego robalino*, Peter werelius** and Matz ohlen**

Abstract – The presence of moisture in a transformer needs to be monitored throughout its service life. Moisture deteriorates transformer insulation by decreasing both electrical and mechanical strength. High moisture content accelerates solid insulation aging, reduces the breakdown strength and makes the transformer vulnerable to the overload conditions due to high temperature spot bubbling. In addition to that, partial discharge can occur in a high voltage region because of the moisture disturbance. Traditional indirect estimation of the moisture concentration in the solid insulation of power transformers includes testing the oil samples as well as measurement of the insulation resistance and loss tangent (50/60 Hz) of the transformer. However, these methods usually give limited information and may lead to wrong conclusions. Direct measurement is not viable and may not be representative to take a paper sample from the surface insulation because moisture distribution is not homogeneous along the insulation geometry. Dielectric Frequency Response, DFR was introduced more than 20 years ago and has been thoroughly evaluated and proven. Several documents have been published summarizing the research work and field tests all over the world. DFR is a practical non-intrusive and non-destructive technique for moisture condition assessment in power transformers, a breakthrough compared with traditional methods. Scientists, researchers and utility operators have shown great interest in the development and application of DFR technique. In this paper, the limitation of traditional methods is presented at first. Later, a comprehensive review of DFR method demonstrates its advantages over traditional methods. Finally, latest research about the mathematical model, temperature correction and test voltage is included to answer common questions regarding the application of DFR method.

Keywords: Moisture, Power transformer, Insulation resistance, Dissipation factor, Dielectric frequency response, Recovery voltage, Temperature correction

1. Introduction

Transformers are by far, one of the most mission critical components in the electrical grid. With the increasing amount of power transformers installed each year, the need for reliable diagnostic methods drives the world's leading experts to evaluate new technologies that improve reliability and optimize the use of the power network.

Moisture is one of the factors deteriorating the insulation of the power components. The presence of moisture in power transformers decreases both electrical and mechanical strength. Heat and moisture together put the insulation under accelerated aging. The aging of cellulose is greatly increased with the existence of moisture. It also reduces the breakdown strength of the power transformers

and makes the transformers vulnerable to the overload conditions due to high temperature spot bubbling. The inception voltage of partial discharge also becomes lower with the presence of water [1]. Therefore, moisture inside power transformers must be periodically monitored throughout their service life in order to minimize the risk of unexpected failure.

A variety of methods are available for indirect estimation of moisture in the solid insulation, including: dielectric response in time domain and frequency domain as well as water content test of oil. These methods are widely used and correct understanding of limitations is required.

From all above listed techniques, dielectric response in the frequency domain has been proven to be an effective solution for power operators. DFR accurately determines the bulk moisture concentration in the solid insulation allowing manufacturers and operators to take assertive corrective or preventive actions that permit extension of the service life of power transformers.

---

†   Corresponding Author: received his BS degree in electrical engineering in 2009 from the Southeast University, Nanjing, China (Jialu.cheng@megger.com)

*   Megger USA

**   Megger Sweden

2. Review of the prevailing methods

2.1 Water content test in oil

The insulation system of power transformers consists of oil and paper/cellulose. The moisture contents in paper/cellulose and in oil are related with each other. Researchers have established a correlation between moisture in oil vs. moisture in paper under the assumption of equilibrium between oil and paper at different temperatures named moisture equilibrium curves as shown in Fig. 1 [1]. The figure indicates that if the water contained in oil is known, the moisture content of paper could be estimated based on the moisture equilibrium curves. Mineral oil from transformers is frequently sampled in field and afterwards the water content is measured by a reliable state of the art method – Karl Fischer Titration [2]. Then the moisture content of the solid insulation could be estimated.

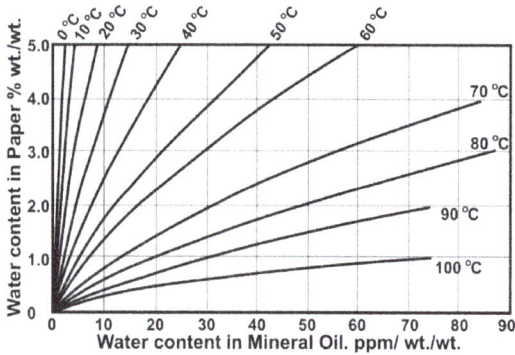

Fig. 1. Water equilibrium curves in oil/paper system [1].

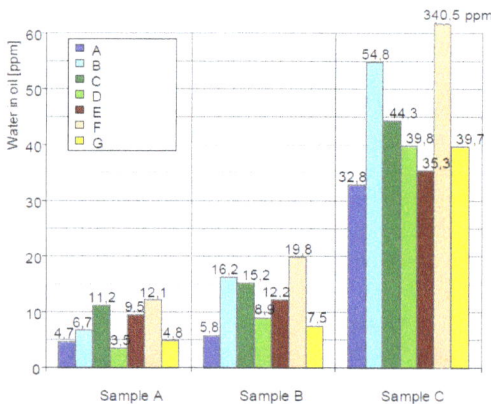

Fig. 2. Moisture in oil in ppm measured by seven laboratories [3].

ASTM D923-07 provides guidelines for sampling electrical insulating liquids. However, due to the low solubility of water in oil at room temperature (20 □C), factors like moisture ingress during oil sampling and transportation, different handling methods in the lab significantly reduce the reliability of the result. As for an example, tests on the same oil sample by seven laboratories across Europe revealed surprising differences as shown in Figure 2 [3]. None of them reasonably agrees with each other. To make things even worse, free water appears when the moisture is saturated in oil and deposits to the bottom which is not included in the oil samples.

The solubility of water in oil is dependent on temperature. The moisture distribution in oil-paper system is usually not homogeneous along the insulation geometry of power transformers. All of these uncertainties impede this method to be a reliable estimation of moisture condition in transformers.

2.2 Insulation resistance measurement

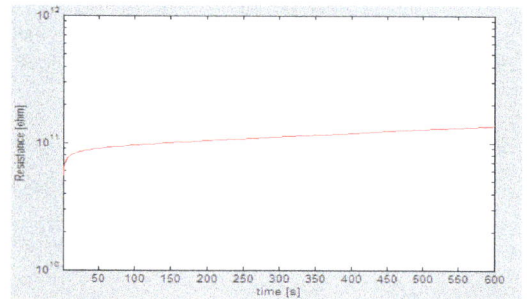

(a)

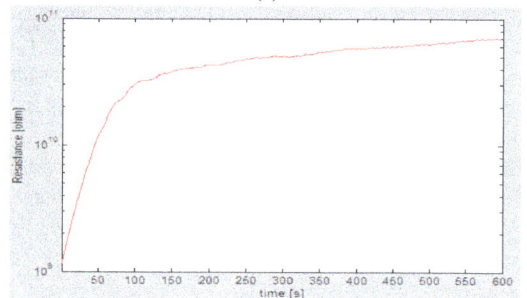

(b)

Fig. 3. The measured resistance curve of the power transformer with (a) new oil (b) aged oil.

The typical insulation resistance, Polarization Index or Dielectric Absorption Ratio tests have been regarded as useful tools in evaluating the condition of an insulation material. It is believed that if the insulation is contaminated with moisture or impurities, the polarization index will decrease. However, the successful application of the test method is limited to solid insulations such as motor insulation. Experiments have been carried out to prove the influence of mineral oil on the measured polarization index. The insulation resistance test on a transformer is made as usual. Then the aged mineral oil inside the transformer is drained out and replaced with new oil. The resistance curves are shown in Fig. 3. The polarization index of the transformer with new oil is 1.49 while the one with aged oil is 4.55. It indicates that the polarization index increases with the aging of transformers which is totally misleading!

2.3. Dissipation factor (tan δ) test

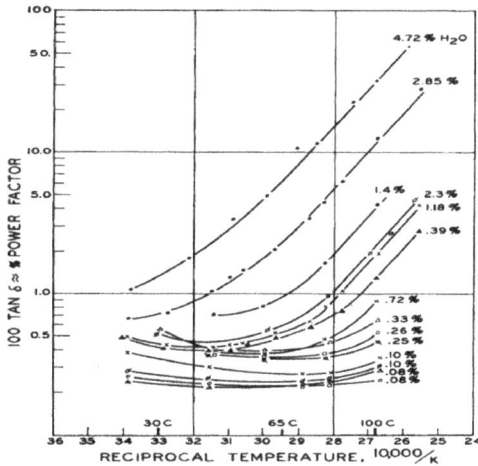

Fig. 4. Power factor at 60 Hz for oil-paper insulation with various moisture contents as a function of temperature (°C) [7].

The widely accepted and most common insulation diagnostic test is carried out measuring the capacitance and dissipation factor of the insulation at line frequency (50/60 Hz). This test is performed whenever there is a need for insulation property investigation. Analysis is based on the comparison historical values against factory values. Since insulation properties are temperature dependent, temperature compensation has to be used for measurements not performed at 20 □C which is normally achieved using temperature correction table factors for certain classes of devices [4] recommended by the international standards.

However, the dissipation factor of insulation materials in good condition is usually quite small and could be vulnerable to the unexpected surface current leakage path

[5] and contaminations [6]. In some extreme cases the dissipation factor shows a negative value. In addition to that, the recommended factors for temperature correction are only average values and therefore subject to error. The classic results presented by Blodget [7] in Figure 4 show the temperature dependent power factor of oil-paper insulation with different moisture contents. The power factor of wet insulation is more temperature dependent than that of the dry insulation. Transformers with different geometry (volume ratio of oil and paper) also have different correction factors.

The 50/60 Hz dissipation factors obtained at different temperatures is not a perfect trending method because of the temperature correction limitation.

2.4. Recovery voltage method

The recovery voltage method, RVM technique has been, by far, a very popular method for insulation condition assessment. It is an advanced diagnostic method used to estimate the moisture content of the insulation from the voltage relaxation measurement.

However, there has been much controversy surrounding the technique [8]. Moisture determinations are often much higher than obtained by other methods. The reason is that it does not take the influence of oil into account. In other words, it lacks the modeling of the insulation system and its interpretation scheme is simplistic as described in [9].

3. Review of the Dielectric Frequency Response method

The Dielectric Frequency Response, DFR method is the extension of tanδ measurement at power frequency. The measurement principle and setup is very similar to traditional 50/60 Hz power factor testing with the difference that a lower measurement voltage is used (200 Vp), insulation properties are measured over a frequency range, typically from 1 kHz to 1 mHz . In frequency domain each material is characterized by a complex permittivity ε_r^* :

$$\varepsilon_r^* = \frac{C_x}{C_0} = \varepsilon_r^{'} - j\varepsilon_r^{''} \qquad (1)$$

where C_x is the measured capacitance and C_0 is the geometrical capacitance. The loss tangent can be expressed by:

$$\tan \delta = \frac{\varepsilon_r^{"}}{\varepsilon_r^{'}} \qquad (2)$$

The results are normally presented as capacitance and loss/tan delta/power factor versus frequency such as shown in Fig. 5. The curve represents the unique characteristics of the insulation.

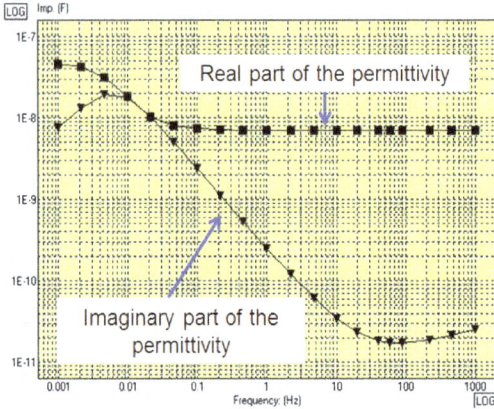

Fig. 5. Typical DFR measurement results.

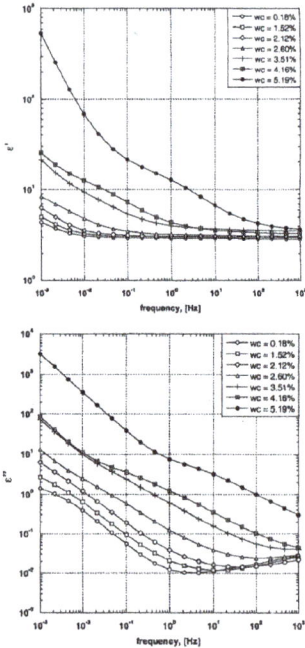

Fig. 6. Real and imaginary part of the permittivity of the Kraft paper at different moisture contents [10].

The characteristic curve of power transformers at various frequencies $\varepsilon_{trans}^{*}(\omega)$ is a function of cellulose property $\varepsilon_{ce}^{*}(\omega)$, oil property $\varepsilon_{oil}^{*}(\omega)$, geometry G (the way cellulose and oil are combined) and temperature T, which could be written as:

$$\varepsilon_{trans}^{*}(\omega) = \mathbf{F}\left(\varepsilon_{ce}^{*}(\omega),\ \varepsilon_{oil}^{*}(\omega),\ G,\ T\right) \qquad (3)$$

ε_{ce}^{*} is the complex permittivity of cellulose which is mainly affected by moisture. Typical moisture dependent characteristic curves of paper/cellulose are shown in Fig. 6 [10].

For transformer oil, the real part permittivity is constant ($\varepsilon_r = 2.2$). The imaginary part is dominated by the DC conductivity (σ) which could be represented by a straight line in frequency domain as shown in Fig. 7 [11].

$$\varepsilon_{oil}^{*}(\omega) = 2.2 - j\frac{\sigma}{\omega\varepsilon_0} \qquad (4)$$

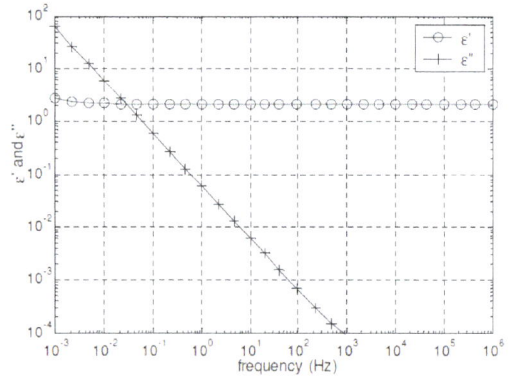

Fig. 7. Real and imaginary part of the permittivity of the mineral oil [11].

The geometry of the transformer insulation is used to describe the relative amount of oil and cellulose and how they are combined. The most widely used model is called XY model [12] which is constructed based on the structure of the cooling duct. Parameter X is defined as the ratio of the sum of all barriers in the duct, lumped together, and divided by the duct width. The spacer Y is defined as the total width of all the spacers divided by the total length of the periphery of the duct as shown in Fig. 8.

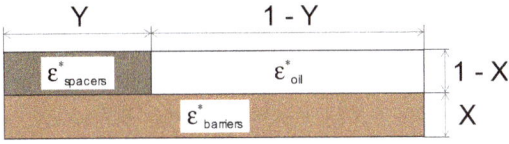

Fig. 8. XY model as described in [12].

Knowing the permittivity of cellulose and oil, as well as their amount and the way they are combined, the permittivity of a power transformer is given by:

$$\varepsilon_{trans}^{*}(\omega) = \frac{Y}{\dfrac{1-X}{\varepsilon_{spacer}^{*}}+\dfrac{X}{\varepsilon_{barrier}^{*}}} + \frac{1-Y}{\dfrac{1-X}{\varepsilon_{oil}^{*}}+\dfrac{X}{\varepsilon_{barrier}^{*}}} \qquad (5)$$

Where ε_{spacer}^{*} and $\varepsilon_{barrier}^{*}$ are the permittivity of cellulose which could be expressed by ε_{ce}^{*}.

The temperature dependence of cellulose and oil are different. DFR method gives a reasonable and logical mathematical approach regarding the thermal behavior of the system. In general, the increase/decrease of temperature makes the dielectric spectrum shift towards higher/lower frequencies at the logrithmic scale while the shape remains unchanged. Fig. 9 is an example of the dielectric frequency response of cellulose at different temperatures. The shifted distance is determined by the activation energy of the material. The activation energy of the Kraft paper is about 1-1.05 eV while mineral oil has activation energy of 0.4-0.5 eV [13]. The higher the activation energy, the further DFR curves shift at a given temperature increase.

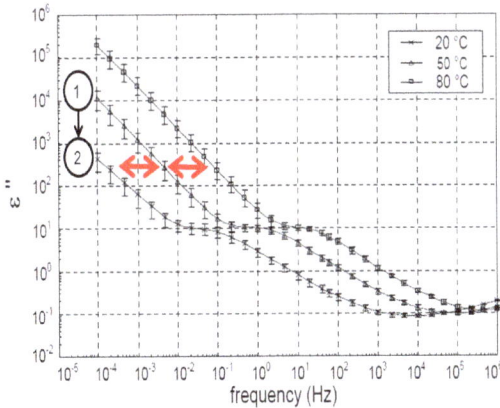

Fig. 9. DFR curves of cellulose at different temperatures [11].

After solving the function containing these parameters, the DFR method gives by far the most accurate results compared with other methods. Figure 10 shows the moisture determination of a oil-paper system using the DFR method (MODS software) at different temperatures compared with the Karl Fisher titration method [14]. It supports the argument that the DFR method could give the reliable moisture determination of insulation systems at different temperatures.

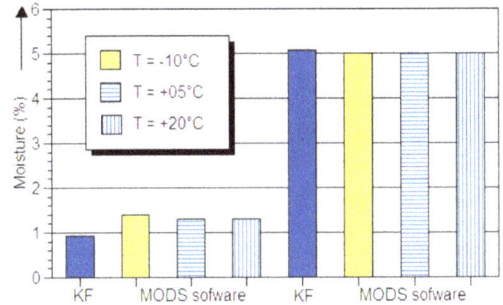

Fig. 10. Comparison of DFR analysis with Karl Fisher method for moisture determination in the oil-paper insulation [14].

In practice, the insulation temperature needs to be carefully measured and put into the post-analysis software By varying the remaining three variables in the function F, a best fit to the measured characteristc curve could be obtained. Thus the moisture content of cellulose, oil conductivity/loss tangent, relative amount of cellulose and oil from the database represent the measured transformer as shown in Fig. 11.

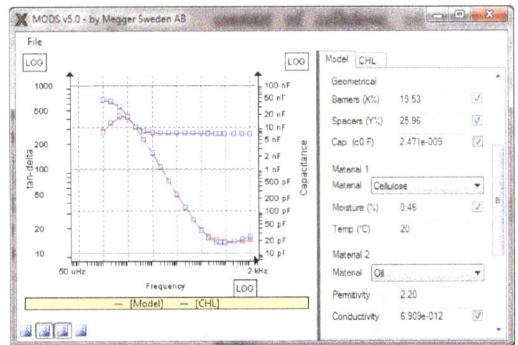

Fig. 11. MODS® moisture and oil analysis for a 20 MVA transformer.

4. Latest discoveries

The most controversial topic is whether X and Y should be fixed according to the designed parameters or let vary to get the best fit during post analysis. More detailed transformer model is analyzed by the Finite Element method [15] so that the differences of two methods are known. Fortunately, the matching reuslts are not very sensitive to the insulation geometry. Results obtained by two methods are listed in [15] and it shows that the moisture content of cellulose, oil conductivity and tan delta of the transformer given by different methods just have slight differences with each other.

Power factor/dissipation factor testing is usually performed at 10 kV or the readings are converted to 10 kV equivalent. One reason that industries standardized on a 10 kV test voltage is for immunity against electrostatic interference. Power transformers with oil-paper type insulating systems in good condition exhibit a flat response when power factor/dissipation factor (50/60Hz) is measured as a function of the test voltage [16].

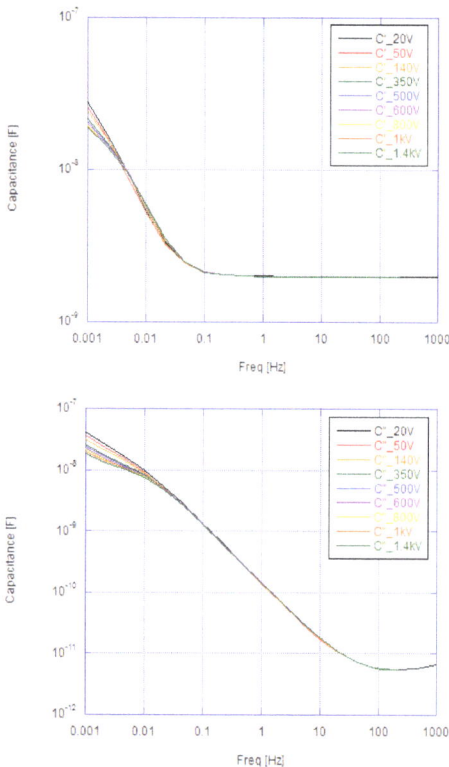

Fig. 12. DFR curves of a 20MVA power transformer at different voltages.

The DFR method uses a 140V RMS test voltage signal that could be amplified up to 30 kV with an external amplifier. However, the response at lower frequencies shows certain voltage dependence . The DFR curves of a 20 MVA power transformer at various test voltages ranging from 20V to 1.4kV are shown in Fig. 12. The distortion at lower frequencies is because of the non-linear characteristics of the oil-paper system. When these two media are put into contact (forming interfaces) under test voltage, charge accumulation occurs at the interfaces due to the differences between their electrical properties. This kind of polarization is the Maxwell-Wagner [9] effect or interfacial polarization. The effect of the polarization is determined by the field strength formed by the test voltage.

5. Conclusions

This article reviews the widely used technologies for transformer moisture estimation. Limitations, merits and guidelines for better interpretation are provided throughout the contet of the document..

The DFR method is demonstrated to be an excellent tool for the off-line assessment of power transformer insulation. Moisture content of cellulose, oil conductivity of power transformers could be reliably estimated from the measurement at all temperatures. DFR performance in the field using high voltages shows a deviation at the very low frequencies, this is explained under the Maxwell-Wagner effect.

References

[1] Yanqing Du, et al., Moisture equilibrium in transformer paper-oil systems, IEEE Electrical Insulation Magazine, vol. 15, no. 1, pp. 11–20, 1999

[2] Peter A. Bruttel, et al., Water determination by Karl Fischer Titration, Metrohm AG

[3] Maik Koch, et al., Reliability and Improvements of Water Titration by the Karl Fischer Technique, 15th ISH 27-31 August 2007 Ljubljana, Slovenia

[4] IEEE Guide for Diagnostic Field Testing of Electric Power Apparatus; Part 1: Oil Filled Power Transformers, Regulators, and Reactors, IEEE 62-1995

[5] Huang Minru, et al., Analysis and Treatment on tanδ Measuring Error of Oil Paper Capacitive Bushing, High Voltage Apparatus, vol. 44, no. 5, pp. 483 – 485, 2008

[6] Jialu Cheng, et al., Improvements of the Transformer

Insulation XY Model Including Effect of Contamination

[7] Blodget R.B., Influence of Absorbed Water and Temperature on Tan Delta and Dielectric Constant of Oil-Impregnated Paper Insulation, Trans. AIEE, 1961

[8] Xose M. Lopez-Fernandez, et al., Transformers: Analysis, Design, and Measurement

[9] Stainslaw M. Gubanski, et al., Dielectric Response Methods for Diagnostics of Power Transformers

[10] Roberts Neimanis, Determination of Moisture Content in Mass Impregnated Cable Insulation Using Low Frequency Dielectric Spectroscopy, IEEE Power Engineering Society Summer Meeting, 2000

[11] Chandima Ekanayake, Diagnosis of Moisture in Transformer Insulation - Application of frequency domain spectroscopy, Chalmers University of Technology, 2006

[12] Uno Gafvert, et al., Dielectric Spectroscopy in Time and Frequency Domain Applied to Diagnostics of Power Transformers, Proc. Of the 6th ICPADM, Xi'an, China, 2000

[13] Dag Linhjell, et al., Dielectric Response of Mineral Oil Impregnated Cellulose and the Impact of Aging, IEEE Transactions on Dielectrics and Electrical Insulation, Volume: 14 , Issue: 1, 2007

[14] Issouf Fofana, et al., Low Temperature and Moisture Effects on Oil-Paper Insulation Dielectric Response in Frequency Domain, IEEE Electrical Insulation Conference, 2009

[15] Ohlen, Matz, et al., Best practices for Dielectric Frequency Response measurements and analysis in real-world substation environment, CMD international conference, 2012

[16] Dinesh Chhajer, Electrical Tester, Megger Limited, 2012

17

Proposal of Dynamic Modeling of Distribution System with System Identification

Ryohei Kawagishi[†], Daisuke Yamanaka* and Yasuyuki Shirai**

Abstract – In this study, a new method of building a dynamic load model for dynamic stability analysis was proposed. This method is as follows: First, inject some small electric power disturbance of known pattern to power system and measure the response in distribution system. Next, analyze the measured response using system identification and create the load model. Simulation was performed with PSCAD, and a dynamic load model was build and evaluated. In the simulation, the availability of the proposed method was confirmed. The model could reproduce the response of the stable system precisely. In addition to it, the stability of the system could evaluate with proposed method.

Keywords: Dynamic load characteristics, Distribution system, System identification

1. Introduction

In the past, most of the loads in power systems were static load such as light bulbs and heaters. So the evaluation of stability is done by using static load models. Recently, however, the rate of active and dynamic electric equipment such as energy storage devices, dispersed generators and inverter fed loads in distribution system has increased.

In such situation, an evaluation method is needed to grasp the dynamic characteristic of distribution system appropriately [1]. There are some methods for evaluating the dynamic characteristics of distribution system [2], [3]; however, most of them are based on recording the load responses during accidental disturbance such as some faults. It means that the quality of such evaluation methods is not so satisfying because available data are limited.

Therefore, a new method for evaluating the dynamic characteristics of distribution system from on-line measured data was proposed [4]-[6]. This method is as follows: First, inject some small electric power modulations of known pattern to power systems and measure the responses in the distribution system. Next, analyze the measured responses using system identification.

In this paper, new dynamic load modeling method using this method was proposed, and was tested by numerical simulation to evaluate the validity of the model.

---

†   Corresponding Author: Dept. of Energy Science and Technology, Kyoto University, Japan (kawagishi@pe.energy.kyoto-u.ac.jp)

*   Dept. of Energy Science and Technology, Kyoto University, Japan

**   Dept. of Energy Science and Technology, Kyoto University, Japan (shirai.yasuyuki.7v@kyoto-u.ac.jp)

2. Proposed Method

2.1 Concept of the Method

The proposed method of dynamic modeling of distribution system is described below.

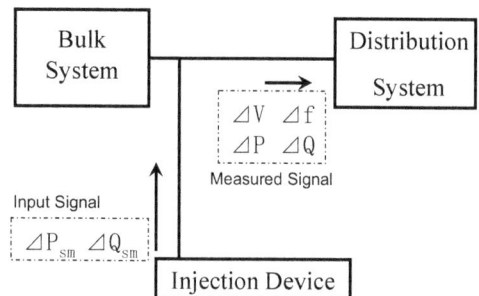

Fig. 1. Conceptual diagram of this study.

Fig.1 shows a distribution system connected to a bulk power system. There is also an injection device connected to a bus between bulk power system and distribution system.

First, generate a small power disturbance (ΔP_{sm}, ΔQ_{sm}) by the injection device. It causes fluctuations of voltage, frequency, active power, and reactive power flow ΔV_L, Δf_L, ΔP_L, and ΔQ_L. Next, calculate transfer functions between ΔV_L or Δf_L and ΔP_L or ΔQ_L with system identification such as Fig.2.

Transfer function includes information of dynamic characteristic of the system, so it is possible to use as dynamic load model. We use the transfer function made by the proposed method as dynamic load model.

2.2 System Identification

In this study, the transfer function of the system is estimated based on Box-Jenkins (BJ) model because it is suitable to remove noise in meaning data.

The order n of input-output transfer function ($B(q)$, $F(q)$ in Fig.3) is set to 3 to 6 each function, and that of noise-output transfer function ($C(q)$, $D(q)$ in Fig.3) is set to 2. In these all cases, transfer functions are calculated. To evaluate them, MDL (Minimum Description Length) shown in Eq.(1) is used .

$$MDL = \left(1 + \frac{2n}{N} \log N\right) \cdot V \tag{1}$$

where N is the number of data, and n is the number of parameters to estimate. In this study, duration of power disturbance was 120s, and sampling interval was 10ms. Therefore, the N was 12,000. Where V is loss function which is given in Eq.(2).

$$V = \sum_{k=1}^{N} \left(Y(k) - Y_{bj}(k)\right)^2 \tag{2}$$

2 sets of 2-input / 1-output BJ model

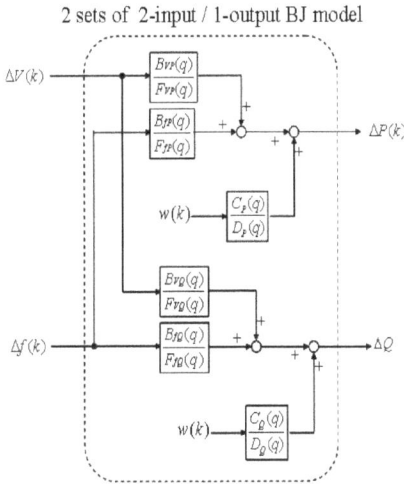

Fig. 2. Transfer function made by the proposed method as dynamic load model.

Y_{bj} is calculated by feeding the input signals into the identified transfer function with a certain combination of orders. Y is the output signal of original system.

The lower the MDL is, the more accurate the calculated model is. Therefore, the transfer function with the lowest MDL is selected as most suitable load model.

BJ model

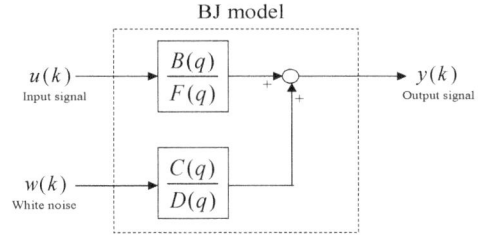

Fig. 3. Box-Jenkins Model.

2.3 Input Signal

In system identification, a chirp signal is used as input signal. A chirp signal is represented as Eq.(3).

$$y = y_0 \sin \frac{2\pi f_{max}}{T} t^2 \tag{3}$$

As shown in Eq.(3), chirp signal is sinusoidal signal of amplitude y_0 and takes time T [s] to change its frequency from 0 [Hz] to f_{max} [Hz]. The wave form of chirp signal is shown in Fig.4. Chirp signal contains all frequency components from 0 [Hz] to f_{max} [Hz]. Therefore, all of the oscillation modes whose natural frequency are between 0[Hz] and fmax [Hz] can be observed. In this paper, those variables are set in Table.1.

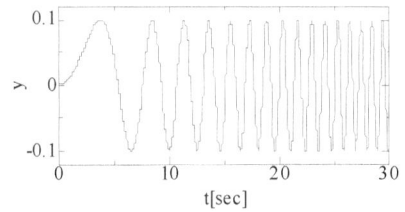

Fig. 4. Form of chirp signal.

Table 1. Configuration of power disturbance

	y_0[pu]	f_{max}[Hz]	T_{start}[s]	T_{end}[s]	Duration[s]
Active Power	0.01	8	0	60	60
Reactive Power	0.01	8	60	120	60

3. Simulation

Simulation was carried out to evaluate the validity of the proposed method. In this study, PSCAD, developed by Manitoba HVDC Research Centre, Inc., was used to simulate power system including dynamic loads.

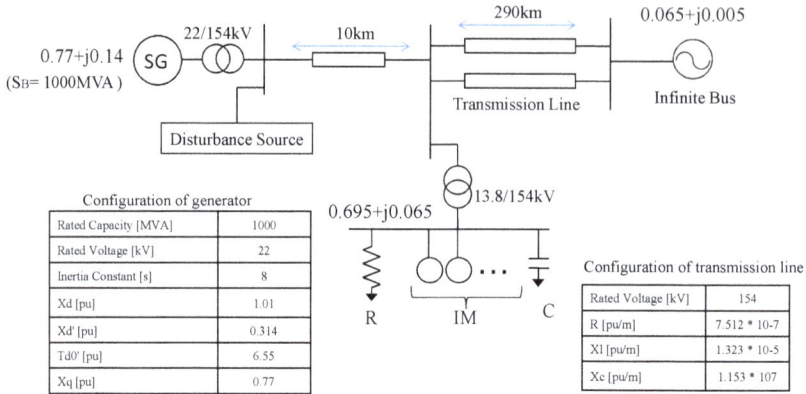

Fig. 5. Simulation Power System.

3.1 Analytical System

In this study, a simple power system shown in Fig.5 was focused on. This system is composed of a generator, infinite bus, distribution system including induction motors and transmission lines between them. Configurations of the generator and transmission lines are shown in Fig.5, respectively.

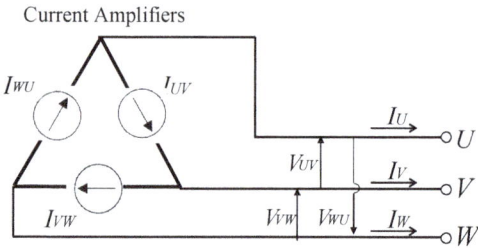

Fig. 6. Disturbance Source.

Between the generator bus and the load bus, the power disturbance source was installed. The power disturbance source was composed of delta-connected ideal current sources, as shown in Fig.6. It is possible to control generation or absorption of active and reactive power simultaneously. Actual devices that can be used as power disturbance source in the real electric power system are energy storage devices such as SMES (Superconducting Magnetic Energy Storage).

The distribution system includes resistive loads, induction motors and capacitors. The stability of whole system is influenced by conditions of distribution system, such as the percentage of motors in the total loads, power flow into Infinite Bus, inertia constant of motors, or load factor of motors, and so on. In this study, percentage of induction motors was changed according to Table.2.

Table 2. Simulation Cases

Case No.	percentage of IM[%]	power flow info Infinite Bus [pu]
1	50	0.8
2	80	0.8

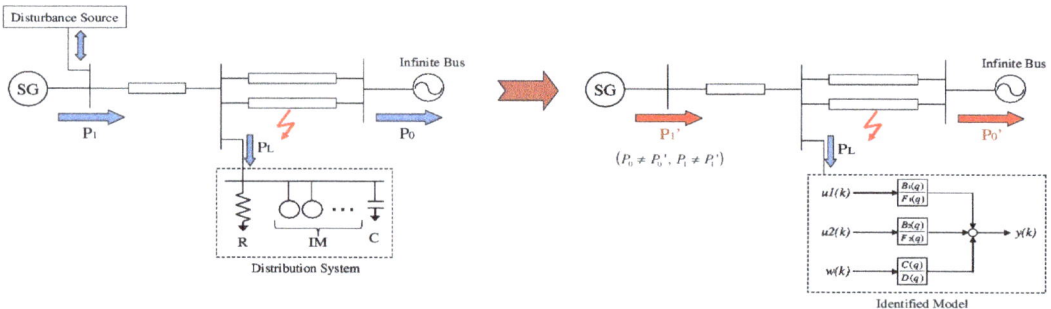

Fig. 7. Dynamic modeling of distribution system.

3.2 Simulation Method

The validity of the model is evaluated with following steps: First, create the transfer function with the proposed method at each case. Second apply simulation of one line to ground fault on the transmission line between infinite bus and distribution system to the default system and the system using dynamic load model built with the proposed method, respectively. The fault point was located 29km off the distribution system, and the fault duration was 100ms. Third, compare those responses from distribution system to evaluate the validity of the model. (Fig.7)

When simulation of one line to ground fault apply to the system using dynamic load model, proposed dynamic load model is composed of variable impedance and identified transfer function as shown in Fig.8.

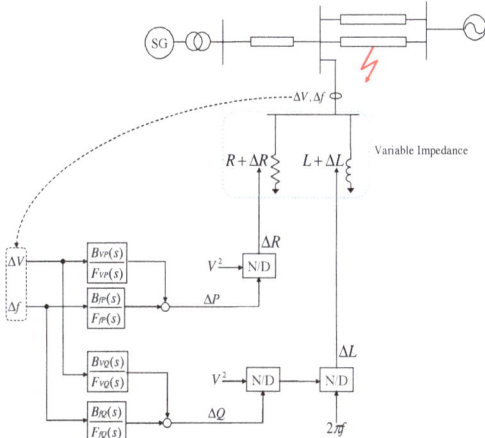

Fig. 8. Dynamic Load Model Consisting of Variable Impedance and Transfer Function.

The identified transfer functions are installed to the PSCAD simulation with following steps: First, voltage and frequency fluctuations are measured at the bus where the model is connected. Next, measured signals are fed into the transfer functions. Finally, variations of resistance or reactance are calculated from the outputs of transfer functions, voltage and frequency according to the equation below:

$$\Delta R = \frac{V^2}{\Delta P} \qquad (4)$$

$$\Delta L = \frac{V^2}{2\pi f \Delta Q} \qquad (5)$$

4. Result and Discussion

4.1 Identified Transfer Function Model

Fig.9 shows the waveforms of generated power disturbance by the disturbance source.

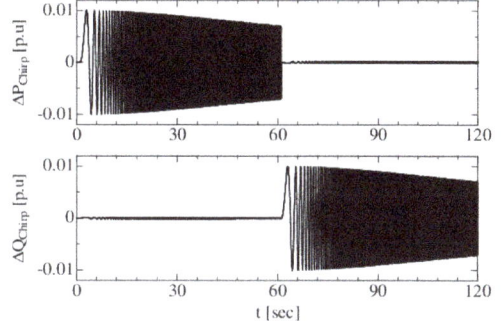

Fig. 9. Active and reactive power disturbance input from disturbance source.

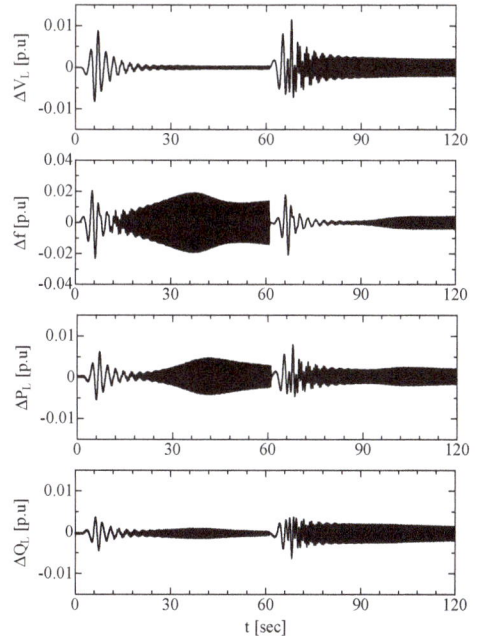

Fig. 10. Measured fluctuations of the distribution system in Case1.

Fig.10 shows the fluctuations of voltage, frequency, active power, and reactive power. The fluctuations of voltage and frequency were used as input signals for system identification. The fluctuations of active and reactive power were used as output signals for system identification. In this figure, these signals were measured in Case1.

Fig.11 shows the Bode diagram of the transfer functions which made by the system identification. These transfer functions were used as dynamic load model.

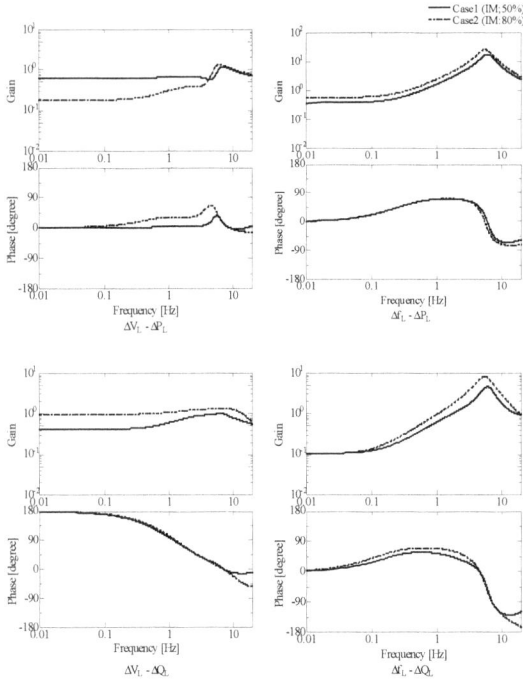

Fig. 11. Bode diagram of transfer function made by system identification.

4.2 Evaluation of the identified load model (Case1)

In this section, it is discussed about the result of Case1. Fig.12 shows the responses of active and reactive power flow into distribution system and fluctuation of voltage and frequency in Case1. In those figures, solid lines depict the responses of original system, and dashed lines depict the responses of the load model.

As shown in figures, the original system is stable, and solid lines and dashed lines are overlapped each other well. For this reason, the load model built with proposed method can reproduce these responses well when the system is stable.

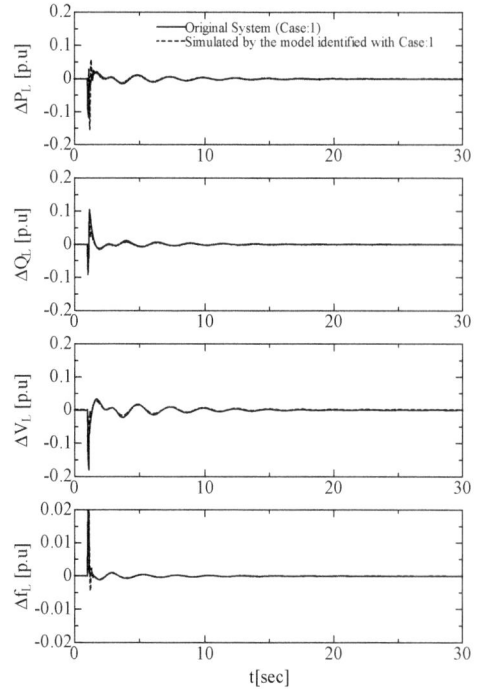

Fig. 12. Load responses of one line to ground fault in Case1. The fault start at 1s and the duration is 100ms. Solid lines depict the responses of original system, and dashed lines depict the responses of the load model.

4.3 Evaluation of the identified load model (Case2)

In this section, it is discussed about the result of Case2. Fig.13 shows the responses of active and reactive power flow into distribution system and fluctuation of voltage and frequency in Case2. In those figures, solid lines depict the responses of original system, and dashed lines depict the responses of the load model.

As shown in figures, the original system is unstable. The system is collapsed within 30 sec. On the other hand, the load model cannot reproduce this. It is considered that, for the large amplitude of voltage fluctuation, nonlinearity of the system is no longer negligible. In the dynamic load modeling method proposed in this study, nonlinearity and feedback loop of the system are neglected. However the trend of becoming unstable can be shown in the figure.

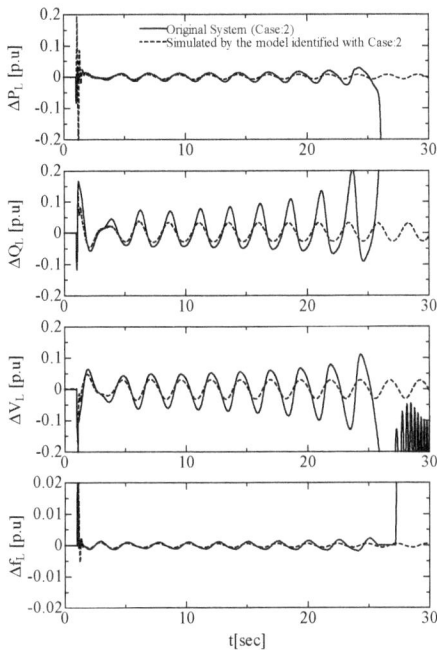

Fig. 13. Load responses of one line to ground fault in Case2. The fault start at 1s and the duration is 100ms. Solid lines depict the responses of original system, and dashed lines depict the responses of the load model.

5. Conclusions

In this study, a new method of building a dynamic load model for dynamic stability analysis was proposed. Simulation was performed with PSCAD, and a dynamic load model was build and evaluated. In the simulation, the availability of the proposed method was confirmed. The model could reproduce the response of the stable system precisely. In addition to it, the stability of the system could evaluate with proposed method.

Even if there is active and dynamic equipment such as induction motors, load fluctuation compensator, and so on in the distribution system, it is possible to build the model which can evaluate the dynamic characteristics of the system with mew method.

References

[1] Tomoyuki Ueda, Shintaro Komami, "Dynamic Load Model in Power System Based on Physical Structure and Measured Data", IEEJ Trans. PE, Vol.126, No.6, p.635-641, 2006.

[2] Kiyoshi Takenaka, Minoru Asada, "Development of novel load dynamic model -Load dynamics model without converging calculation in dynamic of power system simulation-", Technical Report T99018, CRIEPI Report, 2000.

[3] Yoshio Yamagishi, Shintaro Komami, "Power System Dynamic Stability Analysis Considering Dynamic Load and Distributed Generation", IEEJ Trans. PE, Vol.126, No.10, p.977-984, 2006

[4] UDA R, ISHIKAWA H, SHIRAI Y, NITTA T and SHIBATA K, "Fundamental Study on Evaluating Dynamic Load Characteristic in Distribution System by Use of SMES," Proceedings of the International Conference on Electrical Engineering 2008, July 2008.

[5] SHIRAI Y, ISHIKAWA H, UDA R, MIURA H, MIZUTANI K and SHIBATA K, "Evaluating of dynamic characteristic of distribution network including dispersed generator by use of system identification with small power modulation," 17th Power Systems Computation Conference, August 2011

[6] YAMANAKA D, MIZUTANI K, KAWAGISHI R, and SHIRAI Y, "System Identification of Distribution System for Dynamic Stability Analysis" Proceedings of the International Conference on Electrical Engineering 2012, July 2012

18

Analysis of Secondary Arc Extinction Effects according to the Application of Shunt Reactor and High Speed Grounding Switches in Transmission Systems

Seung-Hyun Sohn[†], Gyu-Jung Cho*, Ji-Kyung Park*, Yun-Sik Oh*, Chul-Hwan Kim*, Wan-Jong Kim, Hwa-Jin Oh**, Junh-Jae Yang**, Tomonobu Senjyu*** and Toshihisa Funabashi[§]**

Abstract – When secondary arc is caused by single phase tripping, arc should be extinguished prior to high speed reclosing which can make the system restore. In this paper, we analyzes secondary arc extinction by applying two methods that are shunt reactor and High Speed Grounding Switches (HSGS). To verify the effects of these methods in arc extinction aspects, two factors are discussed. One is secondary arc extinction time, and another is recovery voltage. Especially, secondary arc waveform and amplitude value are analyzed to identify the effects of arc extinction methods. Simulation is conducted by using ElectroMagnetic Transient Program (EMTP) based on actual Korea 765kV transmission system.

Keywords: Secondary Arc Current, Shunt Reactor, High Speed Grounding Switches, Recovery Voltage

1. Introduction

Electricity, produced at nuclear power plant and thermal power plant in Korea, is concentrated in the capital area. Since 765kV transmission lines from these power plants are start points of transmission, improvement of stability and protection in 765kV transmission systems have become growing issues. Thus, the studies about fault and fault clearing on transmission line have been required.

The magnitude of secondary arc current following the primary arc that occurred for circuit breaker operation on single phase to ground fault is proportional to the length of lines, system voltage. Since these factors can affect time of arc extinction, it is need to analyze the secondary arc effect on 765kV transmission line system that has relatively long line length and large system voltage.

When secondary arc current is not extinguished, reclosing causes additional fault, and delayed arc extinction makes dead time increase and system unstable. Thus, various methods of secondary arc extinction like shunt reactor, HSGS and line sectionalizing had been applied on

---
†   Corresponding Author: College of Information and Communication Engineering, Sungkyunkwan University, Korea (sons88@hanmail.net)
*   College of Information and Communication Engineering, Sungkyunkwan University, Korea
**   Korea Power eXchange, KPX, Korea
*** Dept. of Electrical and Electronics Engineering, University of the Ryukyus, Japan
§   Meidensha Corporation, Japan

transmission lines [1]-[2].

In this paper, we explain the secondary arc and arc extinction methods, shunt reactor and HSGS, in section II. In section III, we analyze the secondary arc extinction time and recovery voltage. Also, we compare the arc extinction performance of shunt reactor with that of HSGS based on simulation results. To verify the effects of secondary arc extinction methods, we conduct simulations based on Korea 765kV transmission systems using EMTP.

2. Secondary Arc

2.1 Definition of Secondary Arc

Single pole tripping and reclosing for single phase fault is generally adopted in EHV and UHV transmission lines to improve system stability. When single phase fault occurs, circuit breakers trip to clear the fault. Thus, power is not supplied through fault phase. However, arc current is generated since the capacitive coupling between the fault phase and the two healthy phases [3]. In balanced and transposed single transmission line, secondary arc current is induced by coupling of electrostatic and electromagnetic current that are occurred in two healthy phases [4]. Therefore, secondary arc current can be expressed as equation (1).

$$I_s = I_{sc} + I_{sm} \tag{1}$$

Where,

I_{sc} : electrostatic current induced in healthy phase

I_{sm} : electromagnetic current induced in healthy phase

I_{sc} occurred by electrostatic coupling, is proportional to the length and voltage of transmission line. Also, if compensators are installed on line such as shunt reactor, I_{sm} can be produced by electromagnetic coupling. I_{sm} is produced by mutual induction according to several variables such as line current, secondary arc extinction method and load.

When secondary arc is extinguished, recovery voltage is caused due to phase-to-phase capacitance and mutual inductance in original arc path [5]. When the secondary arc current and recovery voltage in high level, the arc extinction time will be longer. For successful reclosing and recovering transmission line, secondary arc current and recovery voltage must be reduced and eliminated. Thus, effective methods must be used to reduce the secondary arc current and recovery voltage for successful restoration of system by rapid arc extinction [6].

2.2 Methods of Secondary Arc Extinction

2.2.1 Shunt Reactor

When transmission line is energized, large charging current can be generated by shunt capacitance which exists between each phases of line and between phase conductors and ground [7]. Due to charging current, Ferranti phenomena, the voltage of receiving end can be greater than that of sending end, can occur. To prevent Ferranti phenomena and compensate the line, shunt reactor is used on transmission line.

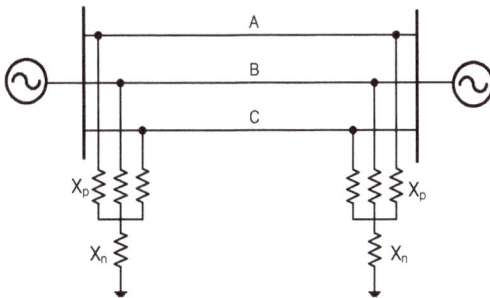

Fig. 1. Transmission Line with Shunt Reactor.

As shown in Fig. 1, shunt reactors are connected to both

ends of transmission line in parallel. When single phase to ground fault occurs, shunt reactors can compensate voltage to line by providing lagging current. The reactor bank extinguishes secondary arc by neutralizing the secondary arc current. In compensation scheme, compensation degree is an important factor because the capacity of reactor is decided by compensation degree.

The relations that calculate the compensation degree and neutral reactor value are shown in equation (2) and (3) [8].

$$F = \frac{1}{X_{eq} \cdot \omega C_1} \tag{2}$$

$$X_n = \frac{\omega(C_1 - C_0)}{3F \cdot \omega C_1 \cdot [\omega C_0 - (1 - F) \cdot \omega C_1]} \tag{3}$$

Where,

ω : frequency

X_n : reactance of neutral reactor

C_1 : positive sequence line capacitance

C_0 : zero sequence line capacitance

F : compensation degree

X_{eq}: equivalent reactance of shunt reactor

2.2.2 High Speed Grounding Switches

HSGS can extinguish the secondary arc rapidly by using the grounding switch that has smaller impedance than that of secondary arc path. However, HSGS has a disadvantage in cost aspects [9]. HSGS are installed at both ends of the UHV transmission line shown in Fig. 2.

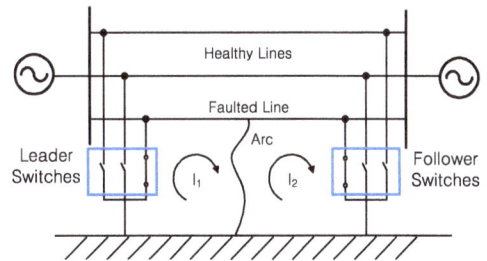

Fig. 2. Transmission Line with HSGS.

When single phase to ground fault occurs, HSGS grounds the fault phase and provides new path that arc current flows. The operating sequence of HSGS is as follow:

1) Single phase to ground fault is occurred and primary

arc is generated.

2) Circuit breakers of fault phase are tripped and secondary arc is generated.

3) Leader switch of HSGS closes, and then closed loop is formed through the arc path and electromagnetic induction current (I_1) flows in that path.

4) Follower switch of HSGS closes, and then another closed loop that electromagnetic induction current (I_2) flows in is formed.

5) HSGS are opened.

6) Circuit breakers are closed to recover the line.

3. Simulation Result and Analysis

3.1 Simulation Conditions

In both case of shunt reactor and HSGS, simulation model system is Korea 765kV transmission system. Also, fault type is single phase to ground fault(A phase) and fault location is middle of the line between Bus 1 and Bus 2. Table 1 shows simulation conditions.   Also, pre-simulation is set as follow:

1) Fault occurred : 0.2s
2) Circuit breaker tripping : 0.2833s
3) Circuit breaker reclosing : 1.2833s

Table 1. Simulation Conditions

Case	Arc Extinction Method
1	Shunt Reactor
2	High Speed Grounding Switches

3.1.1 Case 1 : Shunt Reactor

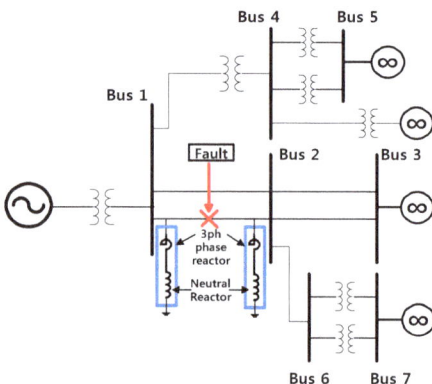

Fig. 3. Korea 765kV Transmission System including Shunt Reactor.

Shunt reactor bank is installed the additional reactor on neutral point, thus 1 bank consists of 4 reactors. Shunt reactor bank is installed on both end of line. Fig. 3 describes 765kV Transmission system in Korea including shunt reactor.

Through the equation (2) and (3), the neutral reactor values, which depends on compensation degree, are calculated. Table 2 shows the neutral reactor values according to compensation degree.

Table 2. Neutral Reactor Values according to Compensation Degree

Compensation degree (%)	Neutral reactor (Ω)
30	∞
35	757.49
40	335.92
45	199.97
50	135.29
55	98.53
60	75.33
65	59.64
70	48.48

3.1.2 Case 2 : High Speed Grounding Switches

Fig. 4 describes Korea 765kV Transmission system including HSGS. Generally, after circuit breaker trips, closing time of HSGS is delayed for 0.167s~0.25s to recognize faulted phase. Also, HSGS is opened 0.167s~0.25s earlier than circuit breaker reclosing to check arc extinction in Korea transmission system. According to HSGS operation scheme in Korea transmission line, after the circuit breaker tripping HSGS's closing time is delayed for 0.17s and HSGS's opening time is delayed for 0.6434s to ensure secondary arc extinction.

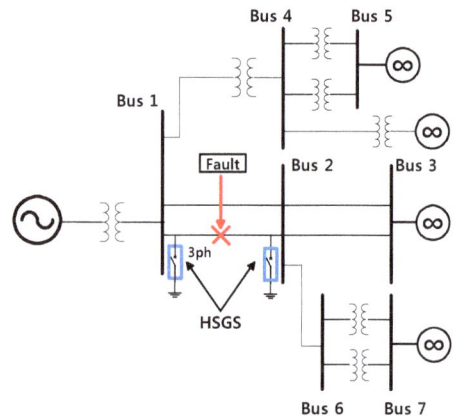

Fig. 4. Korea 765kV Transmission System including HSGS.

Table 3 shows the operation sequence of HSGS for single phase to ground fault.

Table 3. Specific Operation Sequence of HSGS

Operating Status	Operating Time (s)
Fault occur	0.2
Circuit breaker trip	0.2833
Leader HSGS close	0.4533
Follower HSGS close	0.4699
HSGS open	1.1133
Circuit breaker reclose	1.2833

3.2 Simulation Result : Secondary Arc Extinction Time

3.2.1 Case 1

Table 4 and Fig. 5 show simulation results of the arc extinction time when compensation degrees are 40~70%. Numerical simulation results show a trend that arc extinction time decreases as line compensation degree increases.

Table 4. Arc Extinction Time according to Compensation Degree

Compensation degree (%)	Neutral Reactor (Ω)	Arc Extinction Time (s)
0	-	0.978
40	335.92	1.001
45	199.97	0.916
50	135.29	0.849
55	98.53	0.782
60	75.33	0.732
65	59.64	0.682
70	48.48	0.649

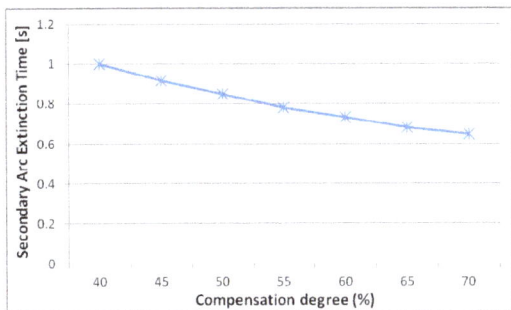

Fig. 5. Arc Extinction Time according to Compensation Degree.

To obtain effective compensation effect, appropriate neutral reactor is necessary to compensate for phase-to-phase capacitance completely. In the case that transmission line has 40% compensation by shunt reactor, arc extinction time is 1.001s. It is longer than the case that line has no compensation. However, neutral reactor value for 40% compensated line is improper to compensate effectively and it causes increased duration of secondary arc current, considerably.

Fig. 6 shows the waveform of secondary arc current when transmission line is compensated by shunt reactor. As shown in Fig. 6, the duration time of secondary arc current is reduced according to application of shunt reactor. Considerable reduction of arc extinction time about 0.329s is obtained by applying shunt reactor which compensates the line 70%.

Therefore in this model system, compensation degree by shunt reactor is meaningful from 45% to 70%. Those compensation degrees have excellent performance in aspects of arc extinction.

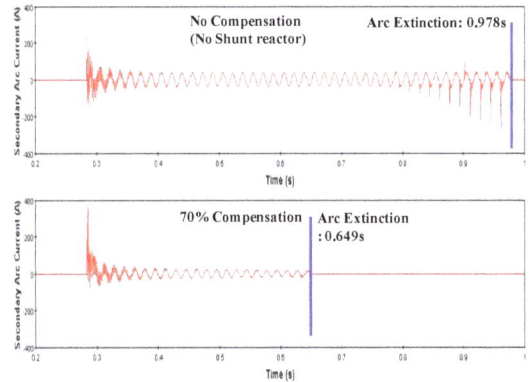

Fig. 6. Secondary Arc Current according to Compensation Degree.

3.2.2 Case 2

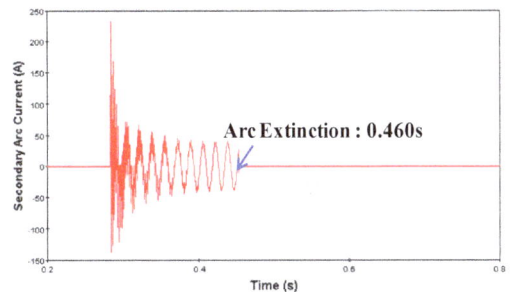

Fig. 7. Secondary Arc Current according to Application of HSGS.

Fig. 7 shows the waveform of secondary arc current when transmission line is equipped with HSGS. Secondary arc current persists until 0.460s, and then it is extinguished.

Comparison of two methods, shunt reactor and HSGS, is shown in Table 5. In aspects of arc extinction time, HSGS makes the arc extinction time shorter than shunt reactor which compensates the line by 70%. It shows that HSGS's performance of arc extinction is better than that of shunt reactor since HSGS can extinguish the arc current quickly by providing low impedance path. The arc extinction speed through the low impedance path is significantly faster than the arc extinction speed through compensation by shunt reactor.

Table 5. Arc Extinction Time according to Arc Extinction Method

Arc Extinction Method	Arc Extinction Time (s)
Shunt Reactor (70% compensation)	0.649
High Speed Grounding Switch	0.460

3.3 Simulation Result : Recovery Voltage

3.3.1 Case 1

Recovery voltage appears across the secondary arc path when the arc is extinguished [10]. Therefore, each case has different recovery voltage duration time because they have different arc extinction time.

Table 6 and Fig. 8 show simulation results of the peak value of recovery voltage when compensation degrees are 40~70%. Numerical simulation result shows a trend that recovery voltage decreases when line compensation degree increases. Trend of peak value of recovery voltage is similar with arc extinction time.

Table 6. Recovery Voltage according to Compensation Degree

Compensation degree (%)	Recovery voltage duration time (s)	Peak Value of Recovery Voltage (kV)
40	1.001 ~ 1.2833	172.992
45	0.916 ~ 1.2833	119.437
50	0.849 ~ 1.2833	86.164
55	0.782 ~ 1.2833	67.403
60	0.732 ~ 1.2833	52.312
65	0.682 ~ 1.2833	42.568
70	0.649 ~ 1.2833	34.639

Fig. 9 shows the waveform of recovery voltage when transmission line is compensated by shunt reactor. As shown in Fig. 9, the magnitude of recovery voltage of

compensated line has smaller than that of not compensated line. Therefore, secondary arc can be extinguished more quickly when shunt reactor is applied.

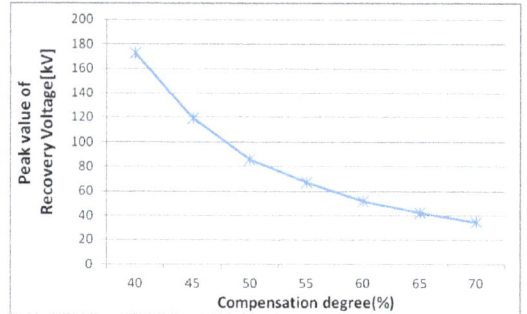

Fig. 8. Peak value of Recovery Voltage according to Compensation Degree.

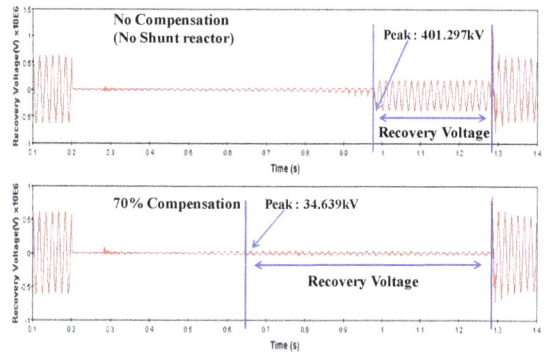

Fig. 9. Recovery Voltage according to Compensation Degree.

3.3.2 Case 2

When HSGS is applied to transmission line, recovery voltage appears in time interval between leader and follower switch closing [11]. Fig. 10 shows the waveform of recovery voltage when transmission line is equipped with HSGS. As shown Fig. 10, recovery voltage occurs from secondary arc extinction time, 0.460s, to HSGS open time, 1.1133s. The voltage that occurs after HSGS open is induced voltage which is caused by other two healthy phases. As shown in Fig. 10, the peak magnitude of recovery voltage is 5666.3V.

Comparison of two methods, shunt reactor and HSGS, is shown in Table 7. HSGS has smaller peak value of recovery voltage than shunt reactor which compensates the line by 70%. This result is also caused by expediting arc extinction of HSGS. Therefore, in both aspects that are arc extinction time and magnitude of recovery voltage, HSGS have appeared higher performance than shunt reactor.

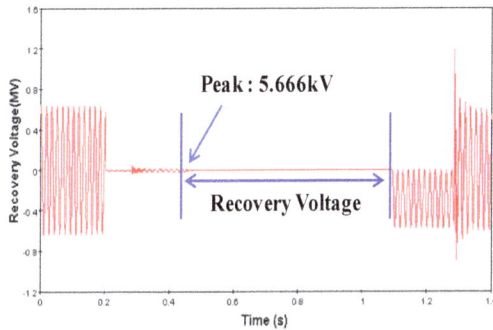

Fig. 10. Recovery Voltage according to Application of HSGS.

Table 7. Recovery Voltage according to Arc Extinction Method

Arc Extinction Method	Peak value of Recovery Voltage (kV)
Shunt Reactor (70% compensation)	34.639
High Speed Grounding Switch	5.666

4. Conclusions

In this paper, secondary arc extinction effects according to application of shunt reactor and HSGS are discussed. For simulation, Korea 765kV transmission line is modeled by EMTP. In case that applying shunt reactor, line compensation degree above 45% are meaningful in arc extinction aspects in modeled system. In addition, according to compensation degree, arc extinction time and magnitude of recovery voltage are changed considerably. Therefore, when system includes shunt reactor, it is important to compute the neutral reactor value which is closely connected with compensation degree properly.

In case that applying HSGS, HSGS show great performance on arc extinction. According to results of simulation, HSGS is regarded as a better method than shunt reactor in arc extinction aspects. However, shunt reactor has an advantage of line compensation but HSGS require specific operation characteristic. Therefore, additional studies that consider various aspects(such as cost) and more simulations are required to obtain practical application effect.

References

[1] KEPCO, "Application of HSGS in 765kV", 1999. 1

[2] Sang-Pil Ahn, "An Alternative Approach to Adaptive Single Pole Auto-Reclosing in High Voltage Transmission Systems Based on Variable Dead Time Control", IEEE Transactions on Power Delivery, Vol. 16, pp. 676-686, Oct. 2001.

[3] E. W. Kimbark, "Suppression of Ground-Fault Arcs on Single-Pole Switched EHV Lines by Shunt Reactors," IEEE Transactions on Power Apparatus and Systems, Vol. 83, Issue 3, pp. 285–290, March 1964

[4] IEEE Committee Report, "Single Phase Tripping and Auto Reclosing of Transmission Lines", Transactions on Power Delivery, Vol. 7, pp. 182-192, Jan. 1992

[5] Yan Xie, Baichao Chen, "Research on Operating Characteristic and Controllability of the Neutral Inductance of Shunt Reactor", Int. Conf. Industrial Mechatronics and Automation, May. 15-16, 2009, pp. 260-263

[6] Z.J Wang, Z.D Yin, M.M Wang, "Influence of Coordinate Controllable Neutral Reactors on UHV Secondary Arc Current", Int. Conf. Energy and Environment Technology, Oct. 16-18, 2009, vol. 2, pp. 156-159

[7] Eithar Nashawati, Normann Fischer, Bin Le, Douglas Taylor, " Impacts of Shunt Reactors on Transmission Line Protection", Oncor Electric Delivery and Schweitzer Engineering Laboratories, 2011

[8] M.R.D Zadeh, M. Sanaye-Pasand, A. Kadivar, "Investigation of Neutral Reactor Performance in Reducing Secondary Arc Current", IEEE Trans. Power Del., vol. 23, pp. 2472-2479, Oct. 2008.

[9] C.H. Kim, S.P. Ahn, "The Simulation of High Speed Grounding Switches for the Rapid Secondary Arc Extinction on 765kV Transmission Lines", Int. Conf. Power System Transients, June. 20-24, 1999

[10] IEEE committee, "Single phase tripping and auto re-closing of transmission lines", IEEE Trans. Power Del., vol. 7, no. 1, pp. 182-192, Jan. 1992.

[11] R.M. Hasibar, A.C. Legate, J.H. Brunke, W.G. Peterson, "The Application of High-Speed Grounding Switches for Single-Pole Reclosing on 500-kV Power Systems", IEEE Trans. Power Apparatus and Systems., vol. PAS-100, pp. 1512-1515, March/April 1981.

19

The Design and Realization of a High Performance Distributed control System for Cascaded H-bridge Converter

Chen Qiurong*, Shi Yu†, Xu Gang*, Liu Shu*, Liu Zhichao* and Shi Shan*

Abstract – A high performance control system for cascaded H-bridge converter is designed and implemented in this paper. The system adopts distributed control to meet the needs of multi-link control and scalability of the Cascaded H-bridge converter. FPGA and CPLD are used as the core controllers: FPGA works as main processor and CPLD works as the execution unit (EU). Different numbers of EUs are added into the system according to different voltage level of the converter system. Optical fiber connections between the EU and FPGA control units ensure its reliability. Finally, several different Voltage level products are presented in this paper. 22kV GTSDC product was established and the device is operating normally in the field to verify the feasibility and reliability of this scheme.

Keywords: Cascaded H-bridge Converter, Distributed control, FPGA, Scalability

1. Introduction

The main task of control system for converter is to produce switching signal of semiconductor power electronic device, thereby obtaining a desired output voltage or current. In addition, the control system should also be able to monitor the working status of the converter, display, record the operating parameters, remote communication and fault handling, etc.

In the past few decades, the hardware circuit of the control system also has developed from the original analog circuit control of discrete elements to a full digital control system which based on microprocessor, microcontroller and DSP (Digital Signal Processor)[1][2].

At the same time, with the capacity and demand of power electronic devices continue to increase; the chain structure can just meet the converter capacity expansion. In recent years, cascaded multilevel converters has gained much attention due to it providing an attractive solution to a transmission application and high power drive system, especially, multilevel converters have been used for STATCOM[3] widely as it can improve the power rating of the compensator to make it suitable for medium or high-voltage high power applications. Cascaded H-bridge Multilevel Inverter (CHMI) is one popular topology of multilevel converters, which uses several low voltage cells, each containing an H-bridge converter. The advantage of this structure is its flexibility. It can meet different voltage

level product through different chain number of the H-bridge converters[4].

In this paper, a high performance control system for this kind of chain converter is designed, implemented and verified. Our products based on this controller are operating normally in many fields.

2. Scheme of the High Performance Control System

2.1 Topology of the Cascaded H-bridge chain converter

Cascaded H-bridge multilevel converter is one popular topology, which uses several low voltage cells, each containing an H-bridge Inverter. Fig.1 shows the topology of the Cascaded H-bridge Multilevel Inverter. Each phase is consist of several cascaded converter units and directly connected to the grid through the filter reactor.

The primary advantage of this structure is its simplicity, the low voltage change rates (dv/dt), and that more H-bridge cells can be cascaded in order to increase the voltage and power level. Also, Multi-level pulse generated can be used to improve the system equivalent switching frequency. After all, A contactor is added in each circuit cell used to bypass the cell when serious failure occurs, and thus to enhance system reliability through module redundancy.

Due to the advantages above, this topology is widely used in chain structure power electronic products. This structure just meets the requirements of scalability of the converter. Different voltage level chain power electronic products consists of different numbers of low voltage cells

†   Corresponding Author: Beijing Sifang automation co.,ltd, China (shiyu@sf-auto.com)

*   Beijing Sifang automation co.,ltd,

added into the converter, such as VSC-HVDC(voltage source converter-high-voltage direct current), SVG(Static Var Generator) and HVFC(high voltage frequency converter).

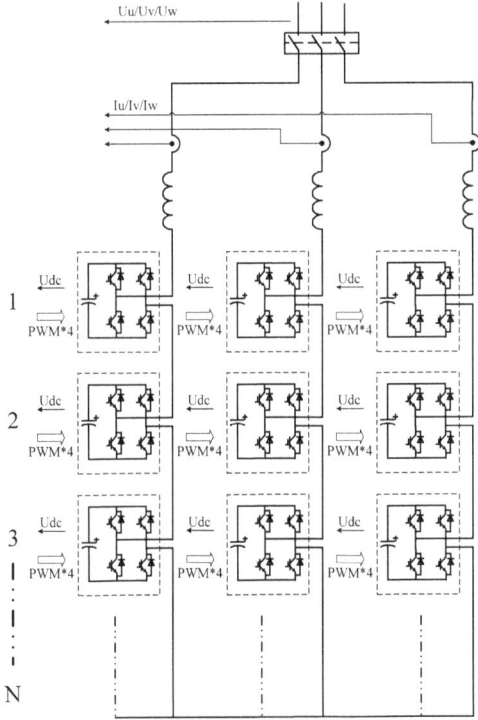

Fig. 1. Topology of the Cascaded H-bridge converter system.

2.2 Control system scheme

With the application of chain structure converter, more and more semiconductor power electronic switching control device were needed, which requires centralized control system to improve the operation and processing capacity.

In order to overcome the existing problems above, we design a distributed converter control system. Fig. 2 shows the hardware structure which contains CPU(Central Processing Unit) +FPGA(Field Programmable Gate Array) as the main controller and CPLDs as the execution units(EUs for short). FPGA is responsible for analog acquisition, data computing processing and producing PWM signal. The relay protection function is realized in the CPU module. Sampling control of each chain is in CPLD module(EUs). We have separated modules to achieve other functions such as HMI and DI/DO.

The most important parts of our design scheme are as follows. Firstly, the distributed structure ensures the flexibility of the system which could easily meet the different voltage level project, such as SVG's and HVFC's. The following section will concretely discuss the benefits of these efficient strategies. Secondly, the FPGA module highly enhances the processing speed of the whole control system which could up to "us" level. FPGA with its powerful parallel processing capabilities, rich I / O pins, and internal logic can easily achieve scalable parallel computing processing of large amounts of data thus to achieve chain extended control of the converter.

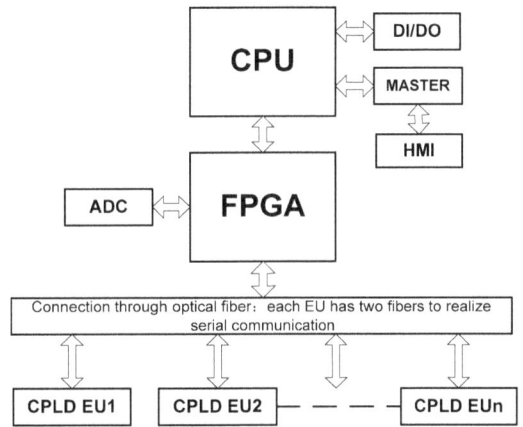

Fig. 2. Hardware structure of the control system for chain converter.

3. Core Technology

In order to optimize the performance of the system, improve the scalability, thus to make the system more flexible to adapt for different engineering applications, advanced FPGA technology and distributed control strategy were adopted.

3.1 Scalable Distributed Control Strategy

The topology graph describes a three-phase circuit, in Fig.1. Each phase loop comprises of three chains. The chain number can be Adjust from 1 to N according to the actual needs. The number of CPLD EUs corresponding configured based on the number of the chain. Therefore, it is much easier to modify the structure to fit for different voltage level project.

Each chain section is composed of the IGBT switching devices, the FPGA control system gathers in total voltage ($U_u U_v U_w$) and current ($I_u I_v I_w$) signals, in fig1. Generate a PWM signal after computing analysis, and then sent the signal through the fiber link to CPLD EUs. Two optical fibers are used between FPGA controller and EU, thereby realizing serial communications. At the same time, CPLD EUs issue commands of FPGA controller and the PWM signal; monitor the operating state of the switching device on its own chain; collect the voltage signal and sent it to FPGAs through fiber-optic serial port.

It's quite flexible to design different control loop for different voltage level project, since the chain number could be determined by the voltage level and the CPLD EUs easily fit to this number.

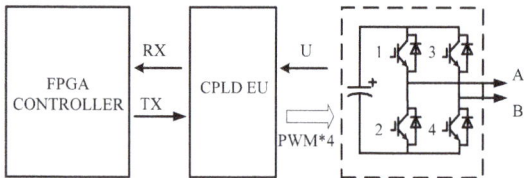

Fig. 3. Communications between FPGA and EUs.

3.2 Advanced FPGA strategy

Why choose FPGA as the critical controller for this system? The main advantage of the digital signal processor within the FPGA is that it can be customized to meet the system requirements. This means that in a multi-channel or high-speed system, users can take full advantage of parallelism within the device, in order to maximize performance, and for low-speed system, more serial ways are used to complete the design[4][5].

The scalability of FPGA dues to it contains a wealth of the external programmable interface. Its internal program runs parallel which has high real-time performance. Take advantage of these, the FPGA controller presented in this paper has achieved the core functions of this high performance converter control system. It includes arithmetic processing of the analog signal acquisition, the extension of the PWM signal modulated output, flexible fiber Channel. For the most part is the management of CPLD EUs.

4. Implementation and Verify

This high performance distributed control system for

Cascaded H-bridge Converter descript in this paper has realized and operating normally as chain power electronic products.

Take GTSDC device for an example: the 22kV/10MW GTSDC(generator terminal sub synchronous damping controller) is one typical device based on the system we discussed. As showed in Fig. 5 and Fig. 6, it is now operating in the Shang Du Power Plant, paralleled in generator stator side directly and realized the sub synchronous current compensation.

Meanwhile, several other products based on this controller presented in this paper have been put into use, such as SVG, HVFC, *etc.* As shown in Table 1, the numbers of CPLD EUs were adjusted according to different voltage level. Therefore, the scalability and reliability could be verified.

Fig. 5. Shang Du Power Plant.

Fig. 6. 22kV/10MW GTSDC Device.

Table 1. Different products based on this controller

	Voltage Level	Chain number	EUs Number
GTSDC	22kV	12	36
SVG	35kV	36	108
HVFC	6kV	6	18
VSC-HVDC	320kV	360	1080

5. Conclusion

This paper describes a high performance distributed control system for Cascaded H-bridge chain converter which can completely satisfy the control function demand of chain converter products, rich in large-scale programmable logic unit. What's more, it is conducive to the expansion of the chain structure of the converter. Several product based on this controller have been operated in field, thus scalability and reliability of the controller could be verified.

References

[1] Zenghuang Qin, 《Electrotechnics》, Higher education press, 2004, Sixth Edition, p177 -p212 ;

[2] Cui Xutao, Yang Rijie, He you, signal processing based on DSP+FPGA experimental system, the development of Chinese Journal of scientific instrument, 2007, volume 28, fifth, p919;

[3] Yang Xingwu ; Jiang Jianguo and Liu Shichao , A Novel Design Approach of DC Voltage Balancing Controller for Cascaded H-Bridge Converter-Based STATCOM, 2009 IEEE 6~(th) International Power Electronics and Motion Control Conference-ECCE Asia Conference Digests;

[4] Zang Yi, Sun Hongge, Cao Yi and Li Jianwei1, Control of Cascaded Inverter with a Novel H-bridge Driver, 2009 IEEE 6~(th) International Power Electronics and Motion Control Conference-ECCE Asia Conference Digests;

[5] Pan Song, CPLD / FPGA application prospects in electronic design, Electronic Technology Applications, 1999.

20

A Study on the Summer and Winter Load Forecasting by using the Characteristics of Temperature Changes in Korean Power System

Jun-Min Cha* and Bon-Hui Ku[†]

Abstract – In Korean power system, the daily load variation is in accordance with the temperature changes in summer and winter time. However, it is not in accordance with temperature changes in spring and autumn time. The cooling load in summer and the heating load in winter contribute to the increasing of the load. Daily temperature changes will cause the load changes. Therefore, the temperature sensitivity is very important in improving the accuracy of the load forecasting during the summer and winter.
This paper proposes the summer and winter load forecasting by using the characteristics of temperature changes in Korean power system. Based on daily peak load of summer and winter, the relationships between characteristics of temperature changes and daily peak load are analyzed in this paper. The characteristics of temperature changes are obtained by using daily peak load and high temperature data in summer and low temperature data in winter during eight years from 2001 to 2008. The sensitivity of daily load in accordance with the increase or decrease in temperature are calculated, respectively. Summer and winter load are forecasted based on the sensitivity between unit temperature changes and daily load variations. Load forecast data with better accuracy are obtained by using the proposed sensitivity than by using the ordinary daily peak load.

Keywords: Load forecasting, Sensitivity of daily load, Temperature changes

1. Introduction

The goal of the electrical power system is to provide high quality power safely and economically, also making effective management possible through load forecast. Power demands change due to industrial development and economic growth, and the growth in power demand, after it reached -3.8% due to foreign exchange crisis, was recovered. After year 2000. the growth rate fell so that after 2008, it was predicted that the growth rate will be very low, about 2.1%. Therefore predicting power demand is needed for stable distribution of power and its accuracy is very important[1-2].

In this study, we have analyzed the relation between properties of temperature change and power demand in summer, which has drastic temperature change effects, finally applying it to power demand prediction in summer and winter.

2. Characteristics of Power Demand

When we look at the characteristics of power demand in Korea, change in amount of load during spring and fall was feeble, while increase in load due to demand in heating during the winter and use of air conditioner during the summer was drastic. Especially during the summer, maximum power is the highest compared to other seasons due to its high temperatures and tropical nights and demand change characteristic is at its biggest, also. In winter, due to the cold weather and supply and use of heating system because of cheap electric charges, heating load occurs, with heating load steadily taking up more proportion among winter maximum load, consequently[1-3].

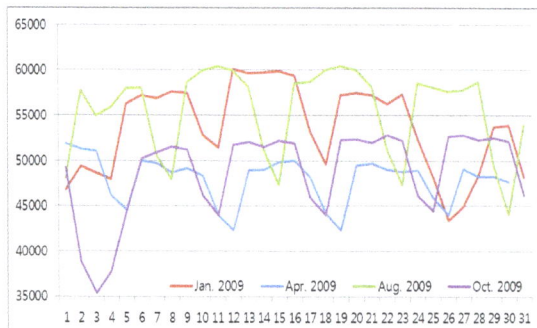

Fig. 1. Seasonal characteristics of power load.

† Corresponding Author: Dept. of Electrical and Electronic Engineering, Daejin University, Korea (kbonhui@nate.com)

* Dept. of Electrical and Electronic Engineering, Daejin University, Korea (chamin@daejin.ac.kr)

Fig. 1 shows seasonal maximum power demand characteristics per hour during January, April, August, and October of year 2009. Demands are similar during the spring(April) and fall(October), while winter(January) and summer(august) power demands are higher.(Excludes October 3rd, or the National Foundation Day of Korea, which is classified as holiday load.)

2.1 Characteristics of the power demand in Spring and Autumn

In spring and fall, change in temperature doesn't effect power demand changes much. This is because demand changes are most effected by air conditioning and heating demands, meaning load changes are not affected by temperature changes as much. Fig.2 and Fig.3 shows temperature and max. load in spring(April) and fall (October).

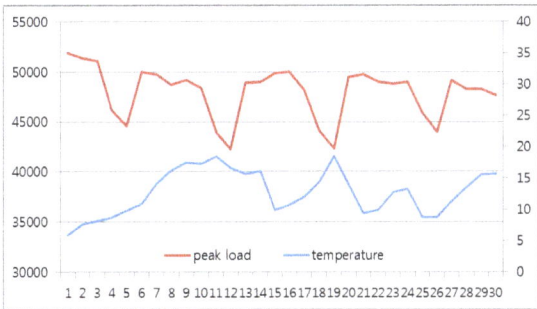

Fig. 2. Temperature and Peak Load(Apr. 2009).

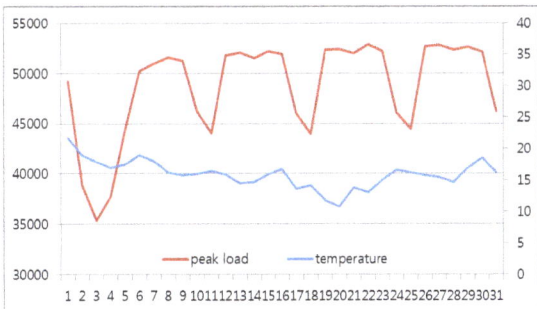

Fig. 3. Temperature and Peak Load(Oct. 2009).

2.2 Characteristics of the power demand in Summer and winter

Fig. 4 shows the relation between temperature and load. The load change in summer, compared with spring and fall,

is more affected by temperature changes. the load is increases, as use of cooling load according to temperature increases.

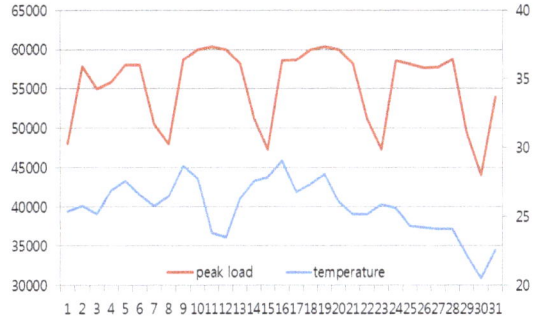

Fig. 4. Temperature and Peak Load(Aug. 2009).

Fig 5 shows the relation between temperature and load in winter. It is noticeable that as temperature goes down, the load increases

Fig. 5. Temperature and Peak Load(Jan. 2009).

3. Case Study

$$I_r = \frac{\Delta L}{L_{real}} \qquad (1)$$

$$R_{pu} = \frac{I_r}{\Delta T} \qquad (2)$$

$$S = \frac{1}{N}\sum R_{pu} \qquad (3)$$

To find the temperature sensitivity :

· **Step 1.** Sort as up-phase and down-phase period by temperature change trends.

· **Step 2.** Using load and temperature data, find the value of temperature relative to the day before(ΔT) and load relative to the day before(ΔL).

· **Step 3.** Find the load intensification factor(I_r) of up-phase and down-phase period.

· **Step 4.** Find the load intensification factor per unit temperature change(R_{pu}) and find the average using the number of data rise and fall of the temperature.

Through these steps, we sort a year's worth of temperature and load data into up-phase and down-phase to find sensitivity. The result is shown on the following Table 1 and Table 2. The temperature sensitivity is the load sensitivity according to rise and fall of per unit temperatures(1 degree).

Table 1. Temperature sensitivity in summer

a) Temperature sensitivity(Increase)	
Upturn	Downturn
0.059	0.034
b) Temperature sensitivity(Decrease)	
Upturn	Downturn
0.059	0.047

Table 2. Temperature sensitivity in winter

a) Temperature sensitivity(Increase)	
Upturn	Downturn
0.001	0.016
b) Temperature sensitivity(Decrease)	
Upturn	Downturn
0.009	0.012

The sensitivity of the temperature change found above is shown differently according to up-phase and down-phase. So, to increase the precision of the demand predictions, we have to apply the sensitivity temperature differently according to change trends[3-5].

3.2 Load Forecasting by using the Characteristics of Temperature

With the sensitivity found above we predict the year 2011's summer and winter power demands. For the demand prediction of summer, we have to use maximum temperatures per day and daily maximum load per day the day and apply sensitivity to the difference of the average temperature of the prediction day and the day before. For the demand prediction of winter, we have to use the lowest temperature of the day. To compare the days for the demand prediction data, we have to use the last 3days' or the last 5days' worth of data, and when the temperature changes drastically, the change rate in demand becomes greater than

it actually is. So, depending on the temperature differences, we add the weighting factor and compare them to the final value for errors.

· **Case 1.** Use max. load (min. load in winter) and max. temperature (min. temperature in winter) of the last 5days for demand predictions.

· **Case 2.** Add weighting factor to the value found in Case 1.
  -Summer (if **temp. diff.**>2 : 0.5) (if **temp. diff.**>4 : 0.3), Winter (if **temp. diff.**>5 : 0.5)

· **Case 3.** Add weighting factor to the value found in Case 1.
  -Sumer (if **temp. diff.**>2 : 0.5) (**temp. diff.**>3 : 0.4) (**temp. diff.**>4 : 0.3), Winter (if **temp. diff.**>5 : 0.5) (**temp. diff.**>6 : 0.4) (**temp. diff.**>7 : 0.3)

· **Case 4.** Use daily max. load (min. load in winter) and daily max. temperature (min. temperature in winter) of the last 3days for the demand prediction.

· **Case 5.** Add weighting factor to the value found in Case 4.
  -Summer (if **temp. diff.**>2 : 0.5) (if **temp. diff.**>4 : 0.3), Winter (if **temp. diff.**>5 : 0.5)

· **Case 6.** Add weighting factor to the value found in Case 4.
  -Sumer (if **temp. diff.**>2 : 0.5) (**temp. diff.**>3 : 0.4) (**temp. diff.**>4 : 0.3), Winter (if **temp. diff.**>5 : 0.5) (**temp. diff.**>6 : 0.4) (**temp. diff.**>7 : 0.3)

3.3 Results Analysis

To predict the year 2011's demand in the summer, we used the year 2009~2010's load and max. temperature data and found the sensitivity depending on temperature changes, and used year 2011 July and August's data separately. The error(%) for each final values are shown in Table 3.

To predict the year 2011's demand in the winter, we used the year 2009~2010's load and min. temperature data and found the sensitivity depending on temperature changes, and used year 2010 December and 2011 February's data separately.

Because the data used in this study was collected for 2 years, 2009 and 2010, it would be possible to even decrease the errors shown in the demand prediction, if we had used more years of data. Especially if the weighting factor is sensitive, or a lot of change in temperature occurs, depending on the range of change in temperature, we can use more years' worth of data for reduction of the errors.

Table 3. Summer-term demand forecast errors

	Jun	July
Case 1	5.24%	8.07%
Case 2	3.11%	4.73%
Case 3	2.69%	4.54%
Case 4	5.72%	6.53%
Case 5	2.89%	4.50%
Case 6	2.72%	4.40%

Table 4 Winter-term demand forecast errors

	January	February
Case 1	8.52%	4.90%
Case 2	5.63%	4.25%
Case 3	4.70%	4.07%
Case 4	8.78%	2.55%
Case 5	5.43%	2.04%
Case 6	4.47%	1.95%

4. Conclusion

This study analyzed the relation between national temperature and the power demand for power demand predictions. The summer and the winter have bigger differences in power demand because of more temperature changes, so the temperature sensitivity should be considered during the demand predictions. with the temperature and load results of 2009 and 2010 to find the sensitivity and finally used it for demand predictions.

1. During the summer and winter, more loads are created, and this is due to the fact that power demands are affected by the temperature. So the application of characteristics of the temperature is needed.
2. The change in the temperature affects the change in the load, so if the sensitivity of the temperature changes was applied, it would be possible to obtain more accurate predictions.
3. Because the demands in the summer and the winter react very sensitively to even small temperature changes, weighting factors can be used for more accurate demand predictions.

Acknowledgements

"This work was supported by the National Research Foundation of Korea(NRF) grant funded by the Korea government(MEST) (No. 20120008367).".

References

[1] Korea Power Exchange, www.kpx.or.kr
[2] KEPCO, "KEPCO in Brief", 2011.4
[3] Sung-Ill Kong, Young-Sik Baek, Kyung-Bin Song, Ji-Ho Park, "The Daily Peak Load Forecasting in Summer with the Sensitivity of Temperature", Trans. KIEE Vol.53 No.6 pp.358-363, 2004. 6
[4] Kyung-Bin Song, Seong-Kwan Ha, "An Algorithm of Short-Term Load Forecasting", Trans. KIEE Vol.53 No.10 pp.529-535, 2004. 10
[5] Bon-hui Ku, Kyoung-Ha Yoon, Jun-Min Cha, Kyung-Bin Song, Kyung-Bin Song, Ung-Ki Baek, "A Study on the Summer Load Forecasting by using the Characteristics of Temperature Change", The Proceedings of Fall Conference of Power Engineering Society of Korean Institute of Electrical Engineers (KIEE), pp.153-155, 2010.11

21

A Voltage Stability Analysis Method of Multi-infeed HVDC Power Systems

Shao Yao[†], Tang Yong[*], Zhang Jian[*] and Li Baiqing[*]

Abstract – According to the Chinese power network planning, the risk of voltage instability at the receiving-end AC power system will be increased significantly, where there will be more HVDC systems feeding into a same AC grid in the future. At present, the effective voltage stability analysis methods of multi-infeed DC (MIDC) system are few, the most widely used method with most research results is the multi-infeed short-circuit ratio (MISCR) index at home and abroad. The deficiency of MISCR index is pointed out. The relationship between multi-infeed interaction factor (MIIF) and the voltage stability of the receiving-end AC system is discussed. Then a new method joint MIIF and MISCR indexes to evaluate the voltage stability of MIDC system is put forward. Finally, simulation tests and analysis results of a large actual grid prove the feasibility and good performance of the proposed new method.

Keywords: Multi-infeed DC power system (MIDC), Voltage stability, Multi-Infeed Interaction Factor (MIIF), Multi-Infeed Short Circuit Ratio (MISCR)

1. Introduction

There With the orderly constructions of China's power grid, many UHV/EHV DC systems will be fed into the East China Receiving-end Power Grid, where several of the DC systems are located in the vicinity of each other. In multi-infeed DC (MIDC) systems, if the electrical couplings between inverter stations are close and the dynamic reactive power supports of the receiving-end AC system are insufficient, when there are serious contingencies happening in the AC system, the risk of the whole system voltage collapse caused by simultaneous DC commutation failure will increase[1]-[10].

At present, many voltage stability analysis methods of MIDC system are derived from the research results of AC system or single-infeed DC system, and the most widely used method is multi-infeed short-circuit ratio (MISCR) method. References [3]–[4] propose the definitions of MISCR respectively. In [3], Node impedance matrix elements are used to describe the effect of DC links, and in [4], the change rate of voltage at converter buses after reactive power disturbance is applied to indicate the interactions between them. Reference [8] proves the consistency of these two definitions based on a decoupled model of MIDC system and gives the index of MISCR for determining the strength of receiving-end AC system.

Reference [5] proposes a novel structure-related index, i.e. the transient voltage supporting index (TVSI) to evaluate the interaction strength of the DC terminals, and it still is a derivative of the concept of short-circuit ratio. However, MISCR index is a relatively macroscopic structural indicator. In some cases, it's insufficient of using MISCR to determine the voltage stability of MIDC systems.

In this paper, the defect of using MISCR method to determine the MIDC system voltage stability is pointed out based on a completely symmetrical two-infeed DC system. The relationship between multi-infeed interaction factor (MIIF) and the voltage stability of MIDC system is studied. Then based on the above analysis, a new method using MIIF and MISCR indexes to determine the voltage stability of MIDC system is put forward. Finally, detailed analysis and simulation tests are carried out in a large actual grid.

2. Evaluation indexes of MIDC system

2.1 Evaluation index for determining the strength of the receiving-end AC system

Historically, an index known as Short Circuit Ratio (SCR) was defined and intended to indicate the ac system strength at the ac/dc interconnection point with respect to the power rating of the single-infeed DC system. And generally think, the higher the value of SCR, the stronger the strength of receiving-end AC system, and the smaller the probability of system voltage instability [1]-[7].

In multi-infeed DC system, considering the interactions between DCs, MISCR is defined as:

---

† Corresponding Author: China Electric Power Research Institute, Haidian District, Beijing 100192, China (yaoshao@epri.sgcc.com)

* China Electric Power Research Institute, Haidian District, Beijing 100192, China

$$M_{ISCRi} = \frac{S_{aci}}{P_{deqi}} \qquad (1)$$

where, i is the number of DC converter station; S_{aci} is the ac short-circuit capacity at the i-th converter ac bus. P_{eqi} is the equivalent DC power considering the impacts of other DC systems.

If the number of DC systems is m, through the Thevenin equivalent method, (1) is extended to:

$$M_{ISCRi} = \frac{S_{aci}}{P_{deqi}} = \frac{U_{Ni}^2 / |\mathbf{Z}_{eqii}|}{P_{dNi} + \sum_{j=1, j\neq i}^{n} |\mathbf{Z}_{eqij}| / |\mathbf{Z}_{eqii}| P_{dNj}}$$
$$= \frac{1}{|\mathbf{Z}_{eqii}| P_{dNi} + \sum_{j=1, j\neq i}^{m} |\mathbf{Z}_{eqij}| P_{dNj}} \qquad (2)$$

where: U_{Ni} is the nominal voltage at the i-th converter bus; P_{dNi}, P_{dNj} are the DC power ratings of the i-th, j-th DC system, respectively; Z_{eq} is the equivalent node impedance matrix of the AC system seen from the ac converter buses; Z_{eqij} is the ith, j^{th} of matrix Z_{eq}; Z_{eqii} is the ith, i^{th} of matrix Z_{eq}.

Based on the value of multi-infeed critical SCR (MCSCR) and the indexes of SCR proposed by IEEE, reference[8] recommends indexes of MSCR as follows: (1) very weak system: $MSCR_i < 2$; (2) weak system: $2 < MSCR_i < 3$; (3) strong system: $MSCR_i > 3$.

2.2 Evaluation index for determining the interaction strength between DC systems

According to [4]–[9], an index known as MIIF, which can be used to provides a first level indication of the degree of the electrical coupling or interaction between two dc systems, has been proposed as the ratio between the voltage changes at inverter ac bus j and i due to a reactive ac power change at inverter ac bus i, and given by:

$$M_{IIFji} = \Delta U_j / \Delta U_i \qquad (3)$$

where, ΔU_j is the observed voltage change at bus j for a small induced voltage change at bus i; ΔU_i is about 1% voltage change at bus i.

Obviously, $0 \leq M_{IIFji} \leq 1$, the larger M_{IIFji} or M_{IIFij} is, the stronger the interaction strength between DC i and DC j will be. Reference [4] indicates that if M_{IIFji} is less than 0.15 when the transmission power of DC i and DC j is similar, the interaction between the two DC systems is

approximately zero, and each DC can be seen as a single-infeed DC system. Simulations in [10] show that, for $M_{IIFji} > 0.6$, the interaction between two DC systems is so strong and the converters can be considered to be on the same bus as regards to commutation failure (CF) behavior.

3. Relationship between MIIF and voltage stability of receiving-end AC system

3.1 Test system model

In order to analyze the relationship between MIIF and voltage stability of multi-infeed DC system, a completely symmetrical test system of two DC infeeds connected through a tie line (z_{12}) is adopted. The circuit is shown in Figure 1. Each DC system uses a standard control strategy of constant current control at the rectifier and constant extinction angle control at the inverter end respectively. The parameters of each DC system are exactly the same. The synthesis load mode, in which an impedance (40%) and an induction motor (60%) are connected in parallel, is used. And the resistances of AC lines are ignored.

3.2 Relationship between MIIF and voltage stability

With the other parameters remaining constant, only changing the value of z_{12}, the relationship between MIIF and MISCR of the above test system are listed in Table 1.

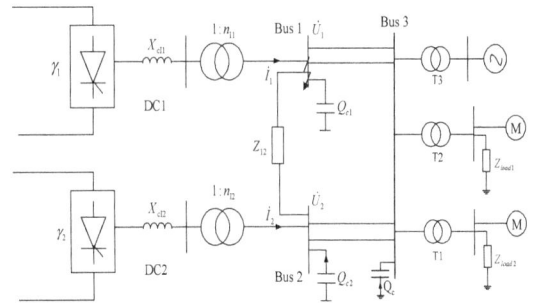

Fig. 1. A schematic diagram of the test system with two DC infeed.

Table 1. MIIF and MISCR of the test system with two DC infeed

No.	z_{12}(p.u.)	M_{IIF21}	$M_{ISCR\,1}$	M_{ISCR2}
1	1.31	0.393	3.23	3.23
2	0.631	0.488	3.23	3.23
3	0.131	0.686	3.23	3.23
4	0.001	0.990	3.23	3.23

Seen from Table 1, with the decrease of z_{12}, $MIIF_{21}$ is increasing, but $MISCR_1$ and $MISCR_2$ are remaining unchanged. This is a special case which will be explained as follows.

The test system shown in Fig. 1 can be equivalent to Fig. 2 under steady-state operation.

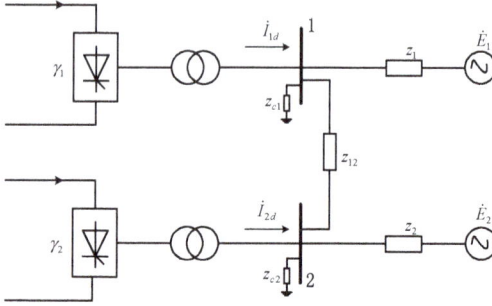

Fig. 2. A simplified model of the test system with two DC infeed.

In the test system, Y and Z are defined as the equivalent node impedance matrix and node admittance matrix seen from the ac converter buses, respectively:

$$Y = \begin{bmatrix} z_1^{-1} + z_{c1}^{-1} + z_{12}^{-1} & -z_{12}^{-1} \\ -z_{12}^{-1} & z_2^{-1} + z_{c2}^{-1} + z_{12}^{-1} \end{bmatrix} \quad (4)$$

$$Z = \frac{1}{|Y|} \begin{bmatrix} z_2^{-1} + z_{c2}^{-1} + z_{12}^{-1} & z_{12}^{-1} \\ z_{12}^{-1} & z_1^{-1} + z_{c1}^{-1} + z_{12}^{-1} \end{bmatrix} \quad (5)$$

where:

$$|Y| = \left(z_1^{-1} + z_{c1}^{-1}\right)\left(z_2^{-1} + z_{c2}^{-1}\right) + \left(z_1^{-1} + z_2^{-1} + z_{c1}^{-1} + z_{c2}^{-1}\right)z_{12}^{-1}$$

Substituting the elements of (5) into (2), MISCR can be rewritten as follows:

$$M_{ISCR1} = \frac{1}{|Z_{11}|P_{dN1} + |Z_{12}|P_{dN2}} \quad (6)$$

$$= \frac{|z_2 z_{c2} z_{12}|}{|z_{c2} z_{12} + z_2 z_{12} + z_2 z_{c2}|P_{dN1} + |z_2 z_{c2}|P_{dN2}}|Y|$$

$$M_{ISCR2} = \frac{1}{|Z_{22}|P_{dN2} + |Z_{21}|P_{dN1}} \quad (7)$$

$$= \frac{|z_1 z_{c1} z_{12}|}{|z_{c1} z_{12} + z_1 z_{12} + z_1 z_{c1}|P_{dN2} + |z_1 z_{c1}|P_{dN1}}|Y|$$

In the actual power system, there is $y_1 \square y_{c1}$, $y_2 \square y_{c2}$. Under the base of DC nominal power and nominal voltage at the converter bus, (6) and (7) can be rewritten by:

$$M_{ISCR1} = \frac{z_1 + z_2 + z_{12}}{z_1|z_{12} + 2z_2|} \quad (8)$$

$$M_{ISCR2} = \frac{z_1 + z_2 + z_{12}}{z_2|z_{12} + 2z_1|} \quad (9)$$

And $z_1 = z_2$, then

$$M_{ISCR1} = \frac{z_1 + z_1 + z_{12}}{z_1|z_{12} + 2z_1|} = \frac{1}{z_1} = M_{ISCR2} \quad (10)$$

As can be seen from (10), as long as the equivalent impedances of AC system remain unchanged, MISCRs will always be constant values.

Assuming there is a 3-phase fault occurring at bus 1 of the AC tie line at 0.2s, and the line jumps at 0.3s after the fault. Curves of voltages and DC powers are as shown in Figure 3.

As can be seen from Fig. 3, with the increase of $MIIF_{21}$, the transient voltage stability of the receiving-end AC system is deteriorated significantly. Even if the strength of AC system is strong ($MISCR_1 = MISCR_2 = 3.23$), when $MIIF_{21}$ reaches a critical value, a serious AC disturbance may cause the whole system to voltage instability.

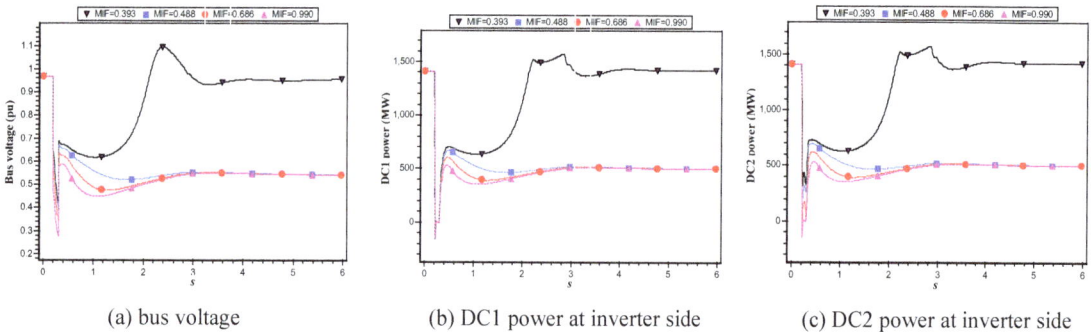

| (a) bus voltage | (b) DC1 power at inverter side | (c) DC2 power at inverter side |

Fig. 3. Simulation results when changing the interaction between DC systems.

4. The voltage stability analysis method joint MIIF and MISCR

According to the analysis results of the previous section, it's not comprehensive enough of simply using MISCR to determine the voltage stability of MIDC system. We should consider both factors of MIIF and MISCR.

Therefore, this paper presents a new voltage stability analysis method with use of MIIF and MISCR: for a MIDC system, if MISCRs at the ac inverter buses are high and MIIFs between DCs are small, the voltage stability of the MIDC system is good; if MISCRs at the ac inverter buses are small and MIIFs between DCs are high, the voltage stability of the MIDC system is poor. For a same MIDC system, the greater the MIIFs are, the higher the probability of system voltage instability is.

And the local power grids with high MIIF between fed DCs are weak areas responsible to voltage collapse in a power system.

5. Simulation analysis

The simulation model which is based on the summer peak load operation mode in the planning of East China power grid in 2015 is shown in Fig. 4. There are a total of 10 DC systems fed into the East China power grid.

By calculating, MIIFs between the 10 DC inverter stations are listed in Table 2.

As seen from Table 2, MIIFs between the inverter stations fed into the same local power grid are greatly significantly greater than MIIFs between the inverter stations fed into different local power grids. And the interaction between Huax and Fengj inverter stations is high; the local power grid with the two DC infeed is voltage stability weak area. MIIFs between DCs fed into Zhejiang power grid between the other DCs are smaller than 0.15. Hence it can be approximated that there are no interactions

between these DCs in addition to Shaox and Zhex inverter stations.

Overall, the interactions between DC inverter stations fed into East China power grid are not great.

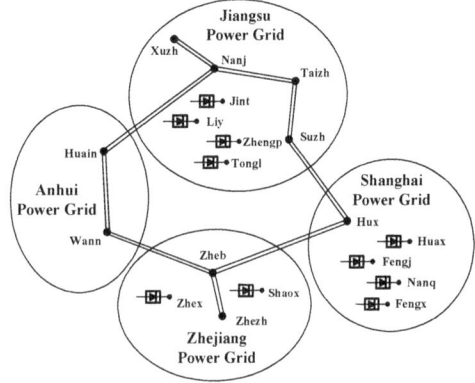

Fig. 4. 1000 kV network structure of East China power grid with 10 DC infeed in 2015.

MISCRs of the DC inverter stations are shown in Table 3. The results are shown that, MISCRs of Jint and Liy inverter stations are greater than 2 and less than 3, MISCRs of other inverter stations are greater than 3. Overall, East China receiving-end power grid is a strong system that its voltage stability supporting capacity to MIDC system is large.

Table 3. MISCRs of DCs in East China Power Grid

DC	M_{ISCRi}	DC	M_{ISCRi}
Zhengp	3.43	Tongl	2.73
Jint	**2.09**	Liy	**2.10**
Nanq	3.64	Huax	3.61
Fengj	3.22	Fengx	3.87
Shaox	3.26	Zhex	3.98

In summary, according to the proposed analysis method, the voltage stability of East China receiving-end power system is good. And normal contingencies will not cause

Table 2. MIIFs between DCs in East China Power Grid

$M_{IIFj,i}$		inverter ac bus j									
		Zhengp	Tongl	Jint	Liy	Nanq	Huax	Fengj	Fengx	Shaox	Zhex
	Zhengp	1.00	0.32	0.31	0.47	0.09	0.16	0.13	0.09	0.13	0.04
	Tongl	0.26	1.00	0.24	0.22	0.13	0.25	0.18	0.11	0.07	0.02
	Jint	0.18	0.17	1.00	0.33	0.06	0.12	0.09	0.05	0.05	0.02
	Liy	0.28	0.16	0.34	1.00	0.05	0.09	0.08	0.05	0.07	0.02
inverter	Nanq	0.03	0.04	0.03	0.02	1.00	0.11	0.14	0.10	0.02	0.01
ac bus i	Huax	0.15	0.30	0.19	0.15	0.38	1.00	**0.52**	0.27	0.09	0.04
	Fengj	0.08	0.14	0.10	0.08	0.31	0.33	1.00	0.24	0.06	0.03
	Fengx	0.09	0.13	0.10	0.09	0.38	0.28	0.39	1.00	0.10	0.04
	Shaox	0.10	0.07	0.07	0.09	0.06	0.08	0.09	0.09	1.00	0.21
	Zhex	0.04	0.03	0.03	0.03	0.03	0.03	0.04	0.03	0.22	1.00

the system to voltage instability. The results are further verified by simulation tool.

Simulation results show that the transient voltage stability of East China power grid in the summer peak load operation mode in 2015 is very good. All unipolar block faults occurred in DC systems, single faults occurred in AC or DC system, three-phase permanent short-circuit fault, which causes the tripout of two circuits, occurred in UHVAC transmission line of receiving-end system and that occurred in the line near commutation buses of inverter stations and majority of DC bipolar block faults occurred in DC systems will not lead to transient voltage instability of East China receiving-end power grid.

The voltage curves of inverter station commutation buses after three phase-to-ground N-2 faults occurring at 1000kV ac lines are shown in Fig. 5.

As seen from Fig. 5, after the faults, bus voltages of the receiving-end AC system drop significantly. But with the fault clearing, bus voltages can restore to new stable values after a period of slight fluctuations. During the fault period, voltage stability at commutation buses in Huax and Fengj inverter stations are worse than the other commutation buses. The local power grid where Huax and Fengj DCs fed into is the voltage stability weak area. The conclusions are

exactly the same with the results concluded from Table 2.

Obviously, the simulation results prove the accuracy of the proposed new method united MIIF and MISCR indexes to evaluate the voltage stability of systems with multiple DC infeed.

6. Conclusion

(1) It's not comprehensive enough of only using MISCR index to determine the voltage stability of systems with multiple DC infeed. Both factors of MIIF and MISCR indexes should be considered.

(2) MIIF, which is an evaluation index of the degree of voltage interaction between two DC links, can reflect the voltage stability weak areas of receiving-end system: the local power grid with high MIIF between fed DC links is weak area responsible to voltage collapse in a power system.

(3) A new method joint MIIF and MISCR to evaluate the voltage stability of systems with multiple DC infeed is proposed: for a MIDC system, if MISCRs at the ac inverter commutation buses are high and MIIFs between DCs are small, the voltage stability of the

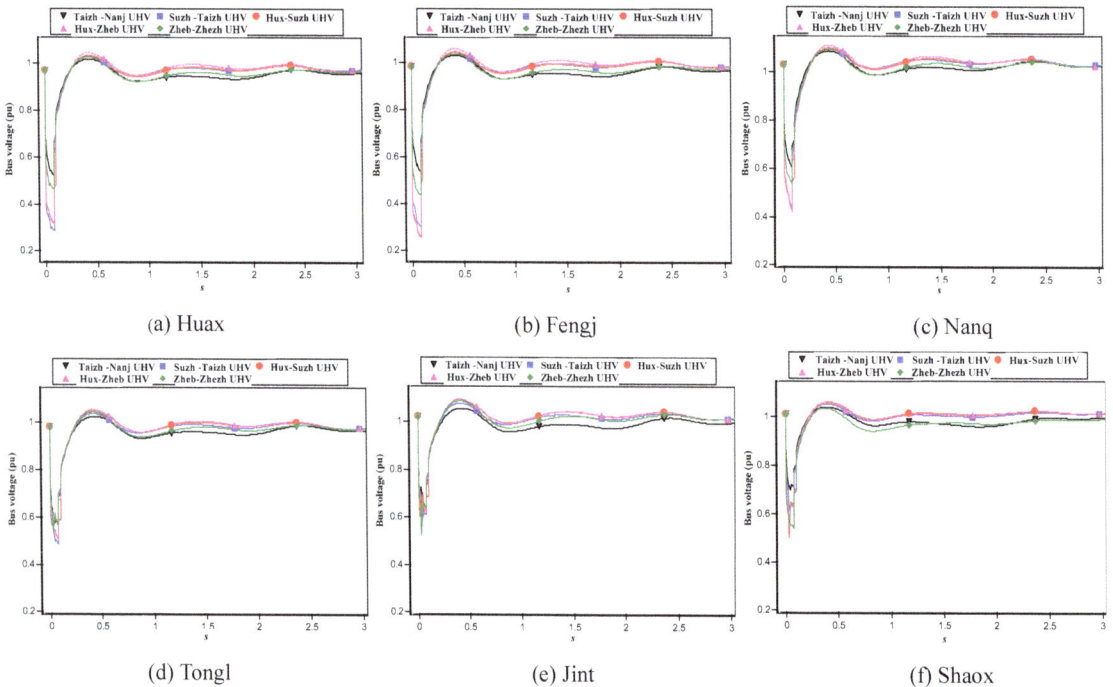

(a) Huax   (b) Fengj   (c) Nanq

(d) Tongl   (e) Jint   (f) Shaox

Fig. 5. Voltage curves of inverter station commutation buses after three phase-to-ground N-2 fault occurring at 1000 kV ac lines in East China power grid.

MIDC system is good; if MISCRs at the ac inverter commutation buses are small and MIIFs between DCs are high, the voltage stability of the MIDC system is poor. For a same MIDC system, the greater the MIIFs are, the higher the probability of system voltage collapse is.

Theoretical analysis and simulation results of a large actual grid prove the accuracy of the proposed new method.

References

[1]   Tang Yong. Power System Voltage Stability Analysis. Beijing: Science Press, 2011: 28.

[2]   Shao Yao, Tang Yong. Current situation of research on multi-infeed AC/DC power systems. Power System Technology, 2009, 33(17): 24-30.

[3]   Paulo F D T, Bernt B, Gunnar A. Multiple infeed short circuit ratio-aspects related to multiple HVDC into one AC network. IEEE PES Transmission and Distribution Conference & Exhibition: Asia and Pacific. Dalian, China: IEEE, 2005: 1-4.

[4]   CIGRE Working Group B4.41. Systems with multiple DC infeed. Paris: CIGRE, 2008.

[5]   Xiaoming Jin, Baorong Zhou, Lin Guang, et a1. HVDC-interaction-strength index for systems with multiple HVDC infeed. CIGRE, Paris, 2010.

[6]   Lin Weifang, Tang Yong, Bu Guangquan. Study on voltage stability of multi-infeed HVDC power transmission system. Power System Technology, 2008, 32(11): 7-12.

[7]   Hong Chao, Rao Hong. The index parameters for analyzing multi-infeed HVDC systems. Southern Power System Technology, 2008, 2(4): 37-41.

[8]   Lin Weifang, Tang Yong, Bu Guangquan. Study on voltage stability of multi-infeed HVDC power transmission system. Proceedings of the CSEE, 2008, 28(31): 1-8.

[9]   Nayak R N, Sasmal R P. AC/DC Interactions in multi-infeed HVDC scheme: a case study. IEEE Power Conference, India, 2006: 5pp.

[10] Ebrahim Rahimi, A. M. Gole, J. B. Davies, et al. Commutation failure analysis in mulit-infeed HVDC systems. IEEE Transactions on Power Delivery, 2011, 26(1): 378-384.

[11] Tang Yong, Bu Guangquan, Hou Junxian, et al. PSD-BPA Transient Stability Program User Manual. Beijing: China Electric Power Research Institute, 2008.

22

Research on Hybrid Energy Storage System of Super-capacitor and Battery Optimal Allocation

Yang Xiu*, Li Cheng[†] and Liu Chunyan**

Abstract – Hybrid super-capacitor and battery energy storage combine the advantages of power-type energy storage element and energy storage components, to avoid the disadvantages of a single energy storage technology, which is one of the important development direction of the energy storage technology. For energy storage applying in high-power, high-capacity and strong volatility applications, the article analyzes the composition of the total cost of the objective function in the hybrid super-capacitor and battery energy storage system entire life cycle, with regard to two decision variables - the number of each energy storage element batch and the batch of energy storage element in the objective function, proposing calculation method of energy storage component number and batch based respectively on Ragone plots and the equivalent cycle life. The example proves the effectiveness of the proposed method.

Keywords: Hybrid storage system, Super-capacitor, Battery, Ragone plots, Equivalent cycle life

1. Introduction

In recent years, with the increasingly serious energy crisis and environmental pollution, solar energy, wind energy and other distributed renewable energy for its abundant resources, little pollution will play an important role in the future energy landscape. Owing to the randomness and volatility of solar and wind, the energy storage system is essential[1].

As the energy storage element, battery have a large energy density to facilitate long-term storage of electrical energy and other characteristics, but it have such disadvantages of small power density, low charge and discharge efficiency, high power, and frequent charging and short discharging cycle life[2]. as a power storage element, Super capacitor have advantages of high power density, fast charge and discharge, high energy storage efficiency, long cycle life, suppressing the short-term energy fluctuation and smoothing the instantaneous energy in the system. However, the energy density is low[3]. It can be seen, the storage battery and super capacitor energy storage are complementtary, Making batteries and super capacitors as energy storage devices at the same time, will improve the performance of the energy storage device greatly[4].

In order to optimize the working process of battery and prolong its service life, the paper[5] proposes the idea of a super capacitor and battery hybrid energy storage and proves theoretically that the hybrid energy storage can take advantage of the complementary characteristics of the batteries and super capacitors, improve the capacity of outputting power, reduce the battery charge and discharge times for extending its life. There are hybrid energy storage application research in the field of electric vehicles[6], engineering machinery[7]. The paper [8-9] study hybrid energy storage applying in independent photovoltaic and other distributed power generation system, the results show that the hybrid energy storage can optimize the battery charge and discharge process, to reduce the number of charge and discharge small cycles, extend battery life. However, the research against the circuit topology and control method of hybrid super capacitor and battery system, lack of theoretical research concerning optimization configuration of hybrid super-capacitor and battery system.

For energy storage applying in high-power, high-capacity and strong volatility applications, the article analyzes the composition of the total cost of the objective function in the hybrid super-capacitor and battery energy storage system entire life cycle, with regard to two decision variables - the number of each energy storage element batch and the batch of energy storage element in the objective function, proposing calculation method of energy storage component number and batch based respectively on Ragone plots and the equivalent cycle life.

[†]   Corresponding Author: Shanghai University of Electric Power, China (lc07130305@163.com)

*   Shanghai University of Electric Power, China (yangxiu721102@126.com)

**   Shenyang University of Chemical Technology, China

2. Ragone Plots and Diagram of the Energy Storage

2.1 Block diagram

Hybrid super-capacitor and battery system consists of the battery, super-capacitor, DC/DC converter and load. Fig. 1 show its structure diagram.

The system configures super-capacitor and battery to transfer energy for load, by controlling the DC/DC converter can actualize energy flow process from super-capacitor to the battery or load, optimize the charge and discharge process , reduce the charge and discharge cycles of battery, prolong its serving life, reduce the cost of hybrid energy storage system and improve its economy.

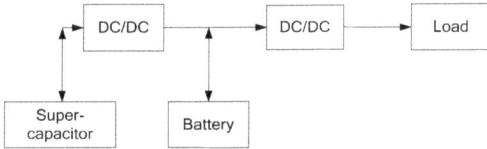

Fig. 1. Hybrid energy storage system diagram.

2.2 Ragone curve of the energy storage element

The abscissa of pe coordinate system is power density (W/kg), the ordinate is energy density (J/kg), the relationship between the discharge power and the energy output of the unit mass storage element can be expressed by Ragone curve [11], denoted by:

$$e = EPD(p) \qquad (1)$$

The Ragone curve of the energy storage element is shown in Fig. 2, which is related to the type of energy storage element, the discharge and working initial conditions of the energy storage element. It can be seen that when the energy storage element discharge by the constant power P_A, the maximum energy output is e_A and the maximum discharge time is T_A, which is also the line slope connecting the point A to the origin O.

For a more complex mathematical model of the energy storage element, such as lead-acid batteries, the Ragone curve can be achieved by inquiring the constant power discharge table in the user manual, according to the power value and the time listed in the table and the curve fitting method. The Ragone curve integrates the state of charge constraints, energy density and power density constraints,

temperature constraints and limits power handling energy constraints of the energy storage element

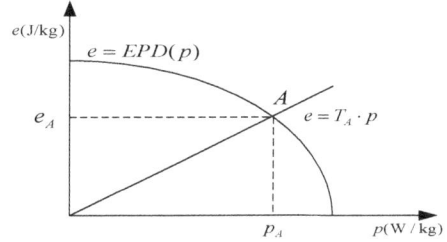

Fig. 2. Ragone plots of the energy storage element.

3. Optimization Model

3.1 The objective function

The costs of hybrid super-capacitor and battery energy storage system in the entire life cycle include the primal investment costs $L1$ causing by the purchase and accessories of the first batch of energy storage components; the secondary investment costs $L2$ causing by the aging and damage of the energy storage component; environmental costs $L3$ generated by recycling the energy storage component failed.

The primal investment costs is determined by the type, the number of the first batch and price of the choosing energy storage element , as follows:

$$L1 = C_b \cdot n_b + C_c \cdot n_c \qquad (2)$$

C_b, C_c – Battery, super-capacitor price;

n_b, n_c – the number of each batch of battery, super-capacitor.

The secondary investment cost is related to the replacement batch, the type, the number of each batch and price of the energy storage element , the relationship as following:

$$L2 = (\kappa_b - 1) \cdot C_b \cdot n_b + (\kappa_c - 1) \cdot C_c \cdot n_c \qquad (3)$$

κ_b, κ_c – the batch of battery ,super-capacitor in the entire life circle .

The environmental cost is related to the replacement batch, the type, the number of each batch and the recycling price of the energy storage element , the relationship as following:

$$L3 = \kappa_b \cdot C_{br} \cdot n_b + \kappa_c \cdot C_{cr} \cdot n_c \qquad (4)$$

C_{br} , C_{cr} – Battery, super-capacitor recycling price.

The optimize design goal is to meet the system performance indicators throughout the life cycle of the energy storage system with minimal investment costs, ie:

$$\min(L1 + L2 + L3) \qquad (5)$$

3.2 Constraints

Hybrid super-capacitor and battery energy storage system need to meet the power balance constraints, as shown in equation 6. The state of charge and power constraints of energy storage component are reflected in its Ragone curve.

$$p_b(t) + p_c(t) = p_L(t) \qquad (6)$$

$p_c(t)$, $p_b(t)$ – super-capacitor and battery output power; $p_L(t)$ – load power. Making the load get through a first-order low-pass filter can obtain the battery load instruction, so that the super capacitor and battery bear respectively small-scale power fluctuation and large-scale power fluctuation, the load curve of energy storage components bearing will affect the decision variables in the optimization objective function that they are the number of each batch and batch within the entire life of the system, the hybrid energy storage system flow chart is shown in Fig. 3.

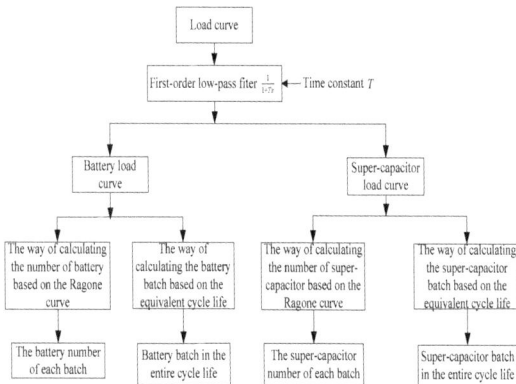

Fig. 3. Hybrid energy storage system flow diagram.

4. Decision Parameters Calculation Way

4.1 The method of calculating the number n of energy storage element

According to the Ragone curve of the energy storage element, the greater the discharge power, the smaller the output energy. If the energy storage element discharge at the constant power $P_{L\max}$, the output energy is $E_{L\max}$, which can meet the parameters $P_{L\max}$ and $E_{L\max}$ of any load.

$P_{L\max}$ represents the peak power of the storage energy load curve, $E_{L\max}$ represents the maximum energy flowing through the energy storage element, i,e :

$$E_{L\max} = \max\left[e_L(t)\right] \qquad (7)$$
$$= \max\left[\int_0^t p_L(\varsigma)d\varsigma\right]$$

Under the initial condition, the Ragone curve is recorded as $e = EPD(p, Q_{u0})$, In order to meet the load , when the energy storage element discharge at the constant power $P_{L\max}$, the output minimum energy should be greater than or equal to $E_{L\max}$. If the mass of energy storage components is m kg:

$$E = m \cdot e = m \cdot PED\left(\frac{P_{L\max}}{m}, \frac{Q_{u0}}{m}\right) \geq E_{L\max} \qquad (8)$$

When the equal sign of formula (8) is established, m is the minimum value to meet the design requirements. Formula (8) deform:

$$\frac{EPD\left(\dfrac{P_{L\max}}{m}, \dfrac{Q_{u0}}{m}\right)}{\dfrac{P_{L\max}}{m}} \geq \frac{E_{L\max}}{P_{L\max}} \qquad (9)$$

The left of the formula (9) is the slope of the line connecting the point in the Ragone curve to the origin, the right is the minimum duration of the energy storage load $T_L = \dfrac{E_{L\max}}{P_{L\max}}$.

In the pe coordinate system, in order to meet the load demand, there is at least a point in the Ragone curve so that the line slope value is greater than or equal to the minimum duration of the storage load connecting the point to the origin.

In conjunction with Fig. 3 can obtain the desired minimum quantity of each batch energy storage element:

$$n = \frac{E_{L\max}}{e_L \cdot m_0} = \frac{P_{L\max}}{p_L \cdot m_0} \quad (10)$$

m_0 – the mass of a single energy storage element.

4.2 The method of calculating the batch κ of energy storage element

The batch κ of energy storage element is determined by the equivalent cycle life N of energy storage components and the designing to operation life DL of the hybrid super- capacitor and battery energy storage system, namely:

$$\kappa = \frac{DL}{N} \quad (11)$$

The serving life of super-capacitor is for decades and the number of cycles up to 500,000 times. So the replacement batch of super-capacitor is constant in a hybrid system design, this article focuses on the solution method of battery batch.

When the hybrid super-capacitor and battery energy storage system is designed, the operating life can be determined, so the batch of the energy storage element is mainly determined by its equivalent cycle life. The battery life is related with its work way, when it work in the cycle way, the cycle life of the battery is mainly related to the depth of discharge, the greater the depth of discharge, the shorter the cycle life of the battery. This approach utilizes predefined state of charge of the battery to charge and discharge ensuring that the battery is neither deeper in the depth of discharge, nor shallower in the depth of charge under normal charging and discharging conditions. From the point of view of energy output, this method is not attractive, because each cycle is only passing part of the available energy, but this cycle way increases the VRLA battery cycles, improves the technology economy of battery energy storage system.

Due to the different discharging depths of the battery in the hybrid system within one cycle, when the battery discharge i th, defined discharge depth:

$$DoD_i = \frac{\int_{t_i}^{t_{i+1}} p_b(\tau) d\tau}{W_b} \quad (12)$$

W_b – Single battery energy

The paper is based on battery life curve, shown in Fig. 4, employs the equivalent charge and discharge cycles to calculate N [12], as shown in equation 13:

$$N = \sum_{DoD_i=10}^{DoD_i=100} \frac{N_{cyc}(DoD_0)}{N_{cyc}(DoD_i)} \quad (13)$$

$Ncyc(DoD_0)$ – The cycle life based on reference depth of discharge DoD_0

$Ncyc(DoD)$ – The cycle life based on depth of discharge DoD

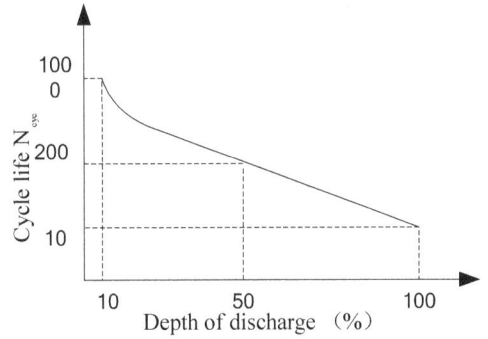

Fig. 4. The relationship between cycle life and depth discharge of a type lead-acid battery.

5. Optimization Algorithm

In the calculation method of the decisive parameters, the mass and batch of energy storage component are based on the load curve assumed by the energy storage element, the time constant T affect greatly the load curve which the energy storage components bear, increasing the time constant T can enhance the time scale of battery load, so that the battery load is more smoother and its proportion is reduced in the energy storage load; reducing the time constant T can lower the battery load time scale, so that the battery load and storage load tends to be consistent, and the battery load proportion is enhanced in the energy storage load.

The equation (5) is a complex function of the time constant T, contain complex non-linear relationship, which is difficult to establish the analytical expression between the time constant and the objective function. In the optimization process, although the time constant is a only variable, by means of the computer simulation listing methods can find the optimal time constant, but it will cause a large amount of calculation and slow convergence.

Matlab optimization toolbox provides a solution for a variety of optimization problems, combining the golden section method and parabolic approximation method can solve the problem of the univariate nonlinear optimal value in fixed interval. So creating a simulation model of the objective function in Matlab and employing optimization toolbox can solve the optimal time constant T^*.

6. Analysis Axample

6.1 Data sources

To verify the rationale and superiority of the proposed hybrid super-capacitor and battery system configuration program, the paper selects the literature[13] with high power, high volatile, obtains the data of load power changing over time after processing, As shown in Fig. 5, and other data parameters are as follows:

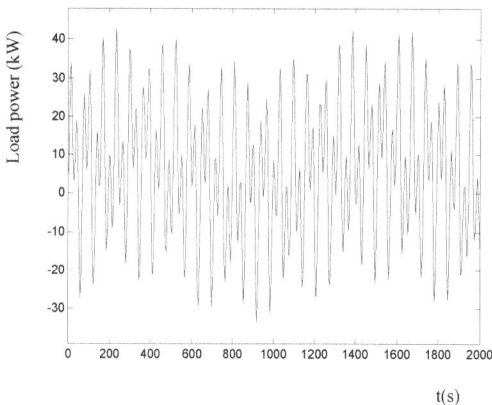

Fig. 5. The power of load.

1) Hybrid energy storage system parameters are shown in Table 1, the price of the energy storage element in the table contains the prices and recycling price.
2) The Ragone curve of the energy storage element
   According to the Ragone curve drawing method

described in this article, the Ragone curve super-capacitor and battery are respectively:

$$e_c = -0.0062p_c + 5.8 \quad 0 < p_c < 935.48 \qquad (14)$$

$$e_b = -0.0015p_b^3 + 0.093p_b^2 - 2.3112p_b + 33.876$$
$$0 < p_b < 36.8 \qquad (15)$$

3) Hybrid system design life is 1000 load curve cycle shown in Fig. 5, and the battery and super-capacitor are charged after each cycle, so that they can be restored to the original level of stored energy.
4) Super-capacitor cycle life is long, so its required batch can be considered 1 in the life-cycle of the hybrid energy storage system.
5) The depth of the reference battery discharge is 50% and its discharging cycle is 200, the relationship between battery cycle life and depth-of-discharge is shown in Fig. 4.

Table 1. The parameters of the hybrid energy storage system

Battery parameters	Value	Super-capacitor parameters	Value
Capacitor /F	2400	Capacity /AH	200
Rated voltage /V	2.7	Rated voltage /V	2
Maximum discharge Current /A	1800	Maximum discharge Current /A	240
Resistance /mΩ	1.5	Resistance /mΩ	0.7
Mass/kg	0.6	Mass/kg	15.5
Price/$ /kg	104.5	Price/$ /kg	4.8

6.2 Results and analysis

Using the optimization algorithm of golden section method and parabolic approximation method in Matlab Optimization Toolbox to configure the hybrid super-capacitor and battery energy storage system, the simulation results are shown in Figs 6,7.

From the simulation results can be seen with time constant increasing over time, the mass of each battery batch is decreasing, the mass of each super-capacitor batch is increasing, the battery batch is decreasing in the entire life of the hybrid system, but the total cost reduces at the first and then increases. Because increasing the time constant will reduce the battery to bear frequent power load fluctuations, reduce the mass of each battery batch and replacing batch in the entire life cycle to lower the installed battery capacity, thus reduce the total cost. However, when the time constant increase to 33 minutes, continue to increase its value, the influence is little on the effect of reducing the battery batch. However, the capacity of the super-capacitor continues to increase, and its price is high, this will lead to increase the cost of hybrid system. When the time optimal constant is 12 minutes in the example, the cost of hybrid systems is lowest. Super-capacitor and battery bearing respectively the load curve are shown in Fig. 8, the optimization results are shown in Table 3; the calculation results of the separate BESS are shown in Table 2. It can be seen that the cost of designing hybrid super-capacitor and battery energy storage system in the paper is 33.8% of the cost of a single battery energy storage when they bear the same load, thus greatly improve the economic of the energy storage.

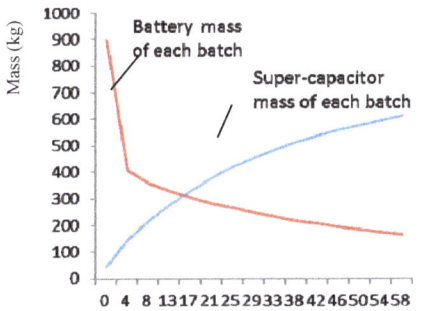

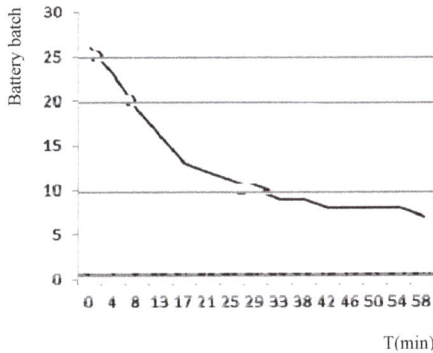

Fig. 8. Load power of energy storage element.

Table 2. Calculation results of single battery energy storage system

Parameters	Value
Battery equivalent discharge cycles over the life	5.4
Battery batch over the life	27
Cost of each battery batch/$	5909.7
Cost of battery/$	159561.9

Table 3. Optimal calculation results of hybrid energy storage system

Parameters	Value
Time constant/min	12
Equivalent battery discharge cycles	3.2
Battery batch over the life	16
Cost of each battery batch/$	1592.6
Cost of super-capacitor/$	28497.1
Cost of battery /$	25481.6
Cost of hybrid system/$	53978.7

Fig. 6. Mass and batch of energy storage element.

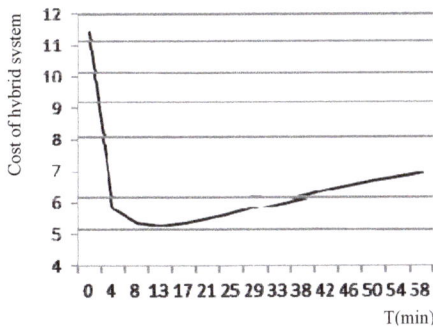

Fig. 7. Cost of the hybrid energy storage system.

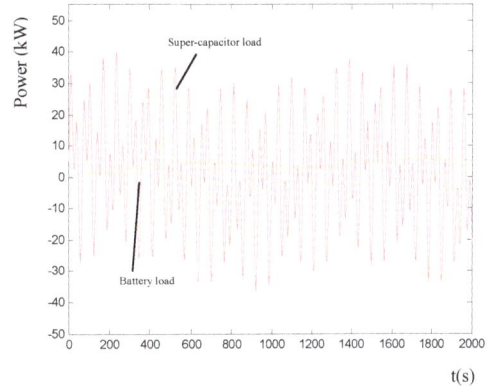

7. Conclusions

For energy storage applying in high-power, high-capacity and strong volatility applications, the article analyzes the composition of the total cost of the objective function in the hybrid super-capacitor and battery energy storage system entire life cycle, with regard to two decision variables - the number of each energy storage element batch and the batch of energy storage element in the objective function, proposing calculation method of energy storage component number and batch based respectively on Ragone plots and the equivalent cycle life. The example proves the effectiveness of the proposed method.

References

[1] WANG Chengshan, WANG Shouxiang. Study on some key problems related to distributed generation systems[J]. Automation of Electric Power Systems, 2008, 32(20): 1-4.

[2] ZHOU Lin, HUANG Yong, GUO Ke, FENG Yu. A survey of energy storage technology for micro grid[J]. Power System Protection and Control, 2011, 39（7）: 147-153.

[3] LU Hongyi, HE Benteng.Application of the super-capacitor in a micro grid[J]. Automation of Electric Power Systems, 2009, 33（2）: 87-91.

[4] GAO Lijun, DOUGAL R A, LIU Shengyi. Power enhancement of an actively controlled battery/ultracapacitor hybrid[J]. IEEE Trans on Power Electronics, 2005, 20（1）: 236-243.

[5] DOUGAL R A, LIU S, WHITE R E. Power and life extension of battery-ultracapacitor hybrids[J]. IEEE Trans on Components and Packaging Technologies, 2002, 25（1）: 120-131.

[6] Sun Liqing, Chen Wei, Wang Renzhen. The application of super-capacitor system in electric vehicles / China Automotive Engineering Society 2003 Annual Conference Proceedings, 14-16 October 2003, Beijing, Beijing: Mechanical Industry Press, 2003.

[7] Zhang Yanting. The energy research of hydraulic excavators based on hybrid motion and energy recovery[D]. Hangzhou : Zhejiang University, 2006.

[8] Tang Xisheng. Research on Energy Management and Stability of Distributed Generation System with EDLC as Energy Storage[D] Beijing : Chinese Academy of Sciences Graduate School (Institute of Electrical Engineering), 2006.

[9] TANG Xisheng, WU Xin, QI Zhiping. Study On a stand-alone system with battery/ultracapacitor hybrid energy storage[J]. Acts Energiae Solaris Sinica, 2007, 28(2) : 178-183.

[10] http://www.mpoweruk.com/performance.htm# Ragone.

[11] Thomas Christen, Martin W. Carlen. Theory of Ragone plots[J]. Journal of Power Sources, 2000, 91(2) : 210-216.

[12] Erik Schaltz, Alireza Khaligh, Peter Omand Rasmussen. Influence of Battery/Ultracapacitor Energy-Storage Sizing on Battery Lifetime in a Fuel Cell Hybrid Electric Vehicle[J]. IEEE TRANSACTIONS ON VEHICULAR TECHNOLOGY, 2009: 3882-3891.

[13] MUYEEN S M, ALI M H, TAKAHASHI R, et al. Wind generator output smoothing and terminal voltage regulation by using STATCOM/ESS [C]// Proceeding of the 2007 IEEE Lausanne Power Tech, July 1-5, 2007, Lausanne, Switzerland: 1232-1237.

[14] Lijun Gao; Dougal, R A; Shengyi Liu. Power enhancement of an actively controlled battery/ultracapacitor hybrid[J]. IEEE Transactions on Power Electronics, 2005, 20(1): 236-243.

23

Average Modeling and Control of Module Multilevel Converter

GuoJu Zhang[†], Yao Chen*, Lisa Qi**, Rongrong Yu* and Jiuping Pan**

Abstract – Modular Multilevel Converter (MMC) has presents great potential in high power quality, low operation loss, scalability and high reliability, which make MMC a good choice for DC transmission, MV drive and other HV or MV applications. This paper describes the average modeling and control of MMC. Firstly, the operation principles of MMC are analyzed，based on which the average MMC model is deduced. Secondly, the control methods for DC voltage control, circulating current suppression, and capacitor voltage balancing, etc are developed. Finally, the effectiveness of the proposed average model is verified by comparing the simulation results of an MVDC distribution system using detailed MMC models. The correctness of the designed control methods also demonstrated through simulation. It is also proven to be feasible of using MMC in MVDC applications.

Keywords: Module Multilevel Converter (MMC), Average model, MVDC system, Simulation analysis

1. Introduction

MMC is first proposed by R.Marquardt in 2002[1], compared with other high-voltage high-power converter topology such as direct series/parallel connection, or multi-level technology, MMC presents great potential in high power quality, low operation loss, high scalability and reliability. These advantages make MMC a good choice for DC transmission, MV drive and other HV or MV applications [2-5].

The current research on MMC focuses on working principles, control strategies, modulation strategies and parameters design, for example direct module method and nearest level modulation[8-9], capacitors voltages balancing scheme[10-11], circulating current control [12-13], etc. However, few papers discuss the average modeling of MMC. Paper [14] proposes a time-domain analytical model which combines switching function and instantaneous power to give the analytical equations for voltage and current of each arm and each sub module. But such time-domain analytical model and other detailed MMC simulation models featured in accurate dynamics have high requirement on both hardware and software simulation platforms, and so that still not satisfactory to support efficient simulation studies.

Average modeling of MMC aims to realize fast, stable, and reliable MMC simulation with proper accuracy. It can reduce the computation load of simulation platforms, and

thus have great engineering application value. This paper describes the average modeling of MMC and control method thereof. In section 2, the operation principles of MMC are analyzed and the average model is deduced. In section 3, the control method including outer loop and inner loop control is developed. In section 4, an MVDC distribution system model is setup with both detailed and average MMC models. The effectiveness of the proposed average model and the control are demonstrated by simulation results.

2. Operation principles

The circuit topology of MMC is shown in Fig. 1, con-sidering the consistency of three phases, only phase a is given in detail. In phase a, there are 2N cascaded sub modules, and 2 valve reactors La which are used to suppress the circulating current in normal operation and to limit the rate of rising of DC short circuit current. The neutral point of phase a is connected to grid through grid reactor Lsa. There are 3 operation states of a half bridge sub module:

1) Blocking state, the pulses of IGBT T1 and T2 are blocked in this state.

2) IN state, T1 is switched off and T2 is switched on. In this state, the output voltage $u_{SM}=u_C$, the current through capacitor $i_C=i_{SM}$, where u_c is capacitor voltage and i_{sm} is output current of the sub module.

3) OUT state, T1 is switched on and T2 is switched off. In this state, $u_{SM}=0$, $i_C=0$.

By controlling the IN and OUT states of each sub

---

†   Corresponding Author: ABB (China) Corporate Research Center, Beijing, China

*   ABB (China) Corporate Research Center, Beijing, China

**   ABB (US) Corporate Research Center, Raleigh, USA

module, the sinusoidal output voltage can be fitted by adding output voltages of sub modules. In steady state operation, DC side current I_d is equally distributed to 3 phases, and AC side current i_a, i_b and i_c are equally distributed in upper and down arms of respective phase.

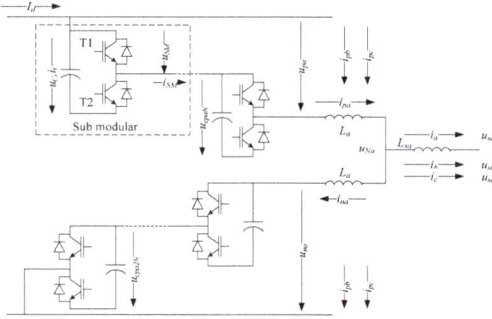

Fig. 1. Topology of MMC.

3. Average model

3.1 Average modelling of sub module

According to the operation states of a sub module, switching function S can be used to represent IN ($S=1$) and OUT ($S=0$) states of a sub module:

$$\begin{cases} u_{SM} = S \cdot u_C \\ i_C = S \cdot i_{SM} \end{cases} \quad (1)$$

In one switching period, assuming the duration when $S=1$ is d, using state space average method, the average model of a sub module is:

$$\begin{cases} u_{SM} = d \cdot u_C \\ i_C = d \cdot i_{SM} \end{cases} \quad (2)$$

3.2 Average modeling of MMC

As shown in Fig. 1, either upper or down arm is composed of N cascaded sub modules. Taking upper arm of phase a as an example, combining with equation(2), average model of this arm can be described by:

$$\begin{cases} u_{pa} = \sum_{i=1}^{N} d_i u_{Ci} \\ i_{Ci} = d_i i_{pa} \end{cases} \quad (3)$$

where u_{pa} and i_{pa} represent upper arm voltage and current of phase a. By replacing the sub-modules with controllable voltage sources in equation(3), the average model of MMC can be obtained as shown in Fig. 2, and the corresponding mathematic model can be described by:

$$\begin{cases} \dfrac{1}{2} V_{dc} = u_{pk} + L_k \dfrac{di_{pk}}{dt} + u_{NO} \\[2mm] \dfrac{1}{2} V_{dc} = u_{nk} + L_k \dfrac{di_{nk}}{dt} - u_{NO} \\[2mm] u_{sk} = u_{Nk} - L_{sk} \dfrac{di_k}{dt} \end{cases} \quad (4)$$

where u_{Nk} is the voltage of neutral point of phase a, and $k=a,b,c$. When the system is symmetrical, $u_{NO}=0$.

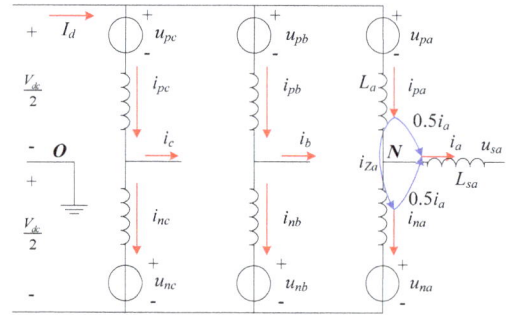

Fig. 2. Average model of MMC.

Comparing with 2-level VSC, one difference is that there exists the circulating current path inside MMC. The circulating current is defined as the current between positive DC line and negative DC line or the current among three phases[15]. The circulating current is used to not only transfer active power between AC side and DC side, but also transfer reactive power among three phases. As mentioned before, the AC side current in each phase is equally distributed in upper and down arms in steady state, then upper and down arms currents i_{pk}, i_{nk} can be described by:

$$\begin{cases} i_{pk} = i_{Zk} + \dfrac{1}{2} i_k \\[2mm] i_{nk} = i_{Zk} - \dfrac{1}{2} i_k \end{cases} \quad (5)$$

where i_{Zk} is the circulating current in phase k. i_{Zk} can be deduced from power conservation law:

$$\begin{cases} i_{Za} = A\left[\cos\varphi + \cos\left(2\omega t - \varphi\right)\right] \\ i_{Zb} = A\left[\cos\varphi + \cos\left(2\omega t + \dfrac{2\pi}{3} - \varphi\right)\right] \\ i_{Zc} = A\left[\cos\varphi + \cos\left(2\omega t - \dfrac{2\pi}{3} - \varphi\right)\right] \end{cases} \quad (6)$$

where $A = \text{M}\cdot\text{I}_{sm}/4$, M is the modulation ratio, I_{sm} is the peak value of AC side current, φ is power factor angle.

In summary, equations (3), (4), (5) and (6) represents the average model of MMC. The developed average model is used in the simulation study in Section 5.

4. Control strategy and modulation method

The mathematical model of MMC in section 3 is analyzed in order to develop appropriate control strategies. Based on equations (4) and (5), u_{pk} and u_{nk} can be described by:

$$\begin{cases} u_{pk} = \dfrac{1}{2}\left(V_{dc} - L_k\dfrac{di_k}{dt}\right) - L_k\dfrac{di_{Zk}}{dt} \\ u_{nk} = \dfrac{1}{2}\left(V_{dc} + L_k\dfrac{di_k}{dt}\right) - L_k\dfrac{di_{Zk}}{dt} \end{cases} \quad (7)$$

As shown in equation(7), there exist two components in u_{pk} and u_{nk}: one is related with AC side current i_k, and the other is related with circulating current i_{Zk}.

There are different control strategies applicable to MMC due to different applications. For examples, DC voltage and unit power factor control can be adopted when a MMC is used as a rectifier; PQ control can be adopted when a MMC is used as an inverter. One common point in these applications is that the output of out loop is the AC side current reference, which means that different control strategies is realized by controlling of fast current inner loop.

According to equation(7), the AC side current control equations in abc phases static coordinates can be described by:

$$\begin{cases} u_{pk1}^{*} = \dfrac{1}{2}V_{dc} - G_{IR}\left(s\right)\left(i_k^{*} - i_k\right) \\ u_{nk1}^{*} = \dfrac{1}{2}V_{dc} + G_{IR}\left(s\right)\left(i_k^{*} - i_k\right) \end{cases} \quad (8)$$

where u_{pk1}^{*} and u_{nk1}^{*} are modulation waves of upper and down arms in phase k. $G_{IR}(s)$ can be transfer function of a PI controller or proportional resonant (PR) controller.

And the circulating current control equations:

$$\begin{cases} u_{pk2}^{*} = -G_{CR}\left(s\right)\left(i_{Zk}^{*} - i_{Zk}\right) \\ u_{nk2}^{*} = -G_{CR}\left(s\right)\left(i_{Zk}^{*} - i_{Zk}\right) \end{cases} \quad (9)$$

where u_{pk2}^{*} and u_{nk2}^{*} are modulation waves of upper and down arms in phase k. $G_{CR}(s)$ can be transfer function of a PI controller or PR controller.

As shown in equation(6), there exist DC component and 2nd harmonic component in the circulating current. The DC component is related to active power transferring through MMC, summation of DC components in three phases is I_d which cannot be eliminated by control method. The 2nd harmonic component, which is used to transfer active power among three phases for energy balancing, is related with capacitor charging and discharging. In steady state, the energy should be distributed equally in three phases. Based on the analysis above, the circulating current reference i_{zk}^{*} can be described by:

$$i_{Zk}^{*} = \dfrac{I_d}{3} + G_{ZR}\left(s\right)\left(u_{cp\_ref} - u_{cpk\_avg}\right) \quad (10)$$

where $u_{cp\_ref} = \dfrac{\sum_{k=1}^{2N}\left(U_{cpak} + U_{cpbk} + U_{cpck}\right)}{3}$ denotes the average energy stored in each phase, $U_{cpk\_avg} = \sum_{k=1}^{2N} U_{cpak}$ is energy stored in phase k, $G_{ZR}(s)$ is transfer function of a PI controller.

For high-power multilevel converters, the most commonly used modulation methods include direct modulation (DM), selective harmonic elimination PWM (SHEPWM), multicarrier PWM, phase-shift carrier PWM (PSCPWM), etc. Among them, PSCPWM can realize high equivalent switching frequency using low physical switching frequency and automatically suppress all low-order harmonics; these advantages make PSCPWM a good choice of modulation strategy for MMC in MVDC applications.

Due to parameters' differences among different sub modules in one arm, the capacitor voltages are different respectively, which adversely influences output voltage quality. In order to eliminate this adverse influence, one method is to sort sub modules according to the capacitor voltage, and then decide IN or OUT state of each sub module according to the direction of arm current. However, this method needs extra resources for comparison and selection, which makes IGBTs switch on and off with high frequency and thus causes higher switching loss [16]. According to the operation principles of sub module, the charging or discharging duration in one switching period can be adjusted by changing IN or OUT duration time. Based on this, a component u_C^* for balancing capacitors voltages is added to the control strategy:

$$u_{Ci}^* = K\left(u_{Cref} - u_{Ci}\right) \cdot \text{sign}\left(i_{SMi}\right) \qquad (11)$$

where u_{Ci}^* is the capacitor voltage of sub module i, $\text{sign}(i_{SMi})$ is sign of current i_{SMi}, K is proportional gain.

Based on control equations above, the control block diagram of MMC is shown in Fig. 3which is also implemented in the simulation study in Section 5.

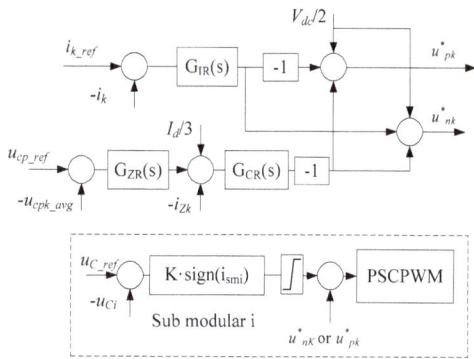

Fig. 3. Control strategy block diagram of MMC.

5. Simulation verification

In order to verify the effectiveness of the average MMC model and the control strategy thereof, both detailed and average MMC models are implemented in a MVDC distribution system model shown in Fig. 4. The simulation environment is Matlab/Simulink/SimPowerSystems.

In Fig. 4, there exists one rectifier, two motor drives and other loads in the MVDC distribution system. DC voltage is controlled by the rectifier and DTC control method is adopted in motor drive. The parameters of MMC converters system are: AC grid line to line voltage 11kV; transformer rated capacity 6MVA and voltage ratio 11kV/6.3kV; rectifier rated capacity 5MVA, DC voltage 10kV; line resistor 0.03 Ohm; inverter rated capacities 2.7MVA and

1.5MVA, motor rated speed 1500rpm. The square torque load representing pump/fan load characteristic is modeled.

The simulation results from both detailed model and average model are shown in Fig. 5 and Fig. 6 for comparison. Sequentially in each figure, they are capacitor voltages (a), DC line voltage (b), active and reactive power of rectifier (c), rotor speed of M1 (d), electrical and load torque of M1 (e). At t=5s, load of M1 is reduced and its rotor speed also reduces from 1500rpm to 1000rpm accordingly. It can be observed the behavior of average model is consistent with that of detailed model under both normal operation and load change scenario, which veried the correctness of the proposed average modeling method developed in Section 3.

Besides, during the load change, the DC line voltage kept unchaged through outler loop control, and the fluctuation of capacitor voltage reduces along with the active power as expected, which demonstrate the effectiveness of the control strategy discussed in Section 4.

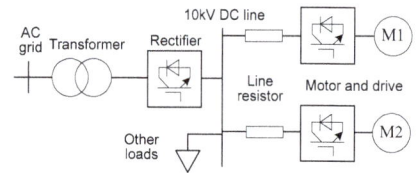

Fig. 4. MVDC distribution system.

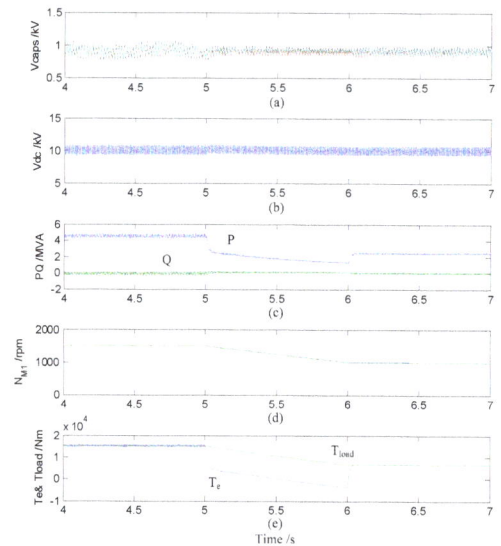

Fig. 5. simulation results of detail model.

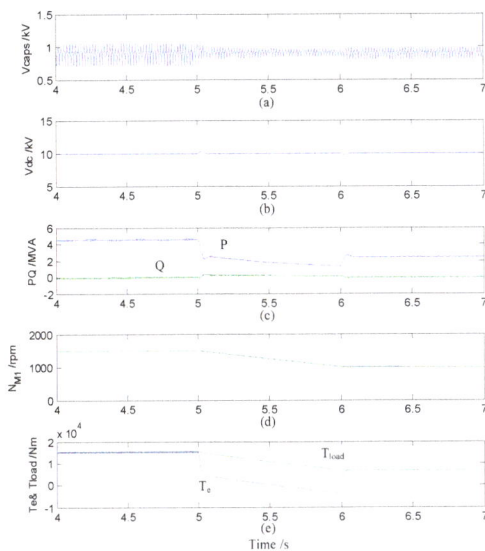

Fig. 6. simulation results of average model.

6. Conclusions

This paper presented the average modeling method of MMC, and based on which elaborated the control mechanism for circulating current suppression and capacity voltage balancing. Taking a MVDC distribution system as an example, the simulation results of the MMC detail and average models are compared and analyzed to verify the effectiveness of the developed average MMC model and control. This example also indicates feasibility of MMC for MVDC applications.

References

[1] A. Lesnicar, R. Marquardt. A new module voltage source inverter topology[C]. European Conference on Power Electronics and Applications (EPE). Toulouse, France, 2003 : 1-10.

[2] S. Allebrod, R. Hamerski, R. Marquardt. New transformerless, scalable Module Multilevel Converters for VDC-transmission[C]. Power Electronics Specialists Conference (PESC). Rhodes, Greece, 2008 : 174-179.

[3] H. Akagi. The state-of-the-art of power electronics in Japan[J]. IEEE Transactions on Power Electronics. 1998. 13(2) : 345-356.

[4] A. Alesina, M. Venturini. Solid-state power conver-

sion : A Fourier analysis approach to generalized transformer synthesis[J]. IEEE Transactions on Circuits and Systems. 1981. 28(4) : 319-330.

[5] M. Glinka, R. Marquardt. A New Single Phase AC/AC-Multilevel Converter For Traction Vehicles Operating On AC Line Voltage[C]. European Conference on Power Electronics and Applications (EPE). Toulouse, France, 2003.

[6] Hagiwara M, Akagi H. PWM control and experiment of module multilevel converters[C]//Proceedings of IEEE Power Electronics Specialists Conference. Rhodes, Greece : IEEE, 2008 : 154-161.

[7] Rohner S, Bernet S, Hiller M, et al. Pulse width modulation scheme for the Module Multilevel Converter [C]//European Power Electronics and Applications Conference. Barcelona, Spain : IEEE, 2009 : 1-10.

[8] Michail V. Analysis, implementation and experimental evaluation of control systems for a module multilevel converter[D]. Stockholm Sweden : Royal Institute of Technology, 2009.

[9] Guan Minyuan, Xu Zheng, Tu Qingrui, et al. Nearest level modulation for module multilevel converters in HVDC transmission[J]. Automation of Electric Power Systems, 2010, 34(2) : 48-52.

[10] Antonopoulos A, Angquist L, Nee H P. On dynamics and voltage control of the module multilevel converter [C]//European Power Electronics and Applications Conference. Barcelona Spain : IEEE, 2009 : 1-10.

[11] Ding Guanjun, Tang Guangfu, Ding Ming, et al. Submodule capacitance parameter and voltage balancing scheme of a new multilevel VSC module [J]. Proceedings of the CSEE, 2009, 29(30) : 1-6.

[12] Rasic A, Krebs U, Leu H, et al. Optimization of the module multilevel converters performance using the second harmonic of the module current[C]//13th European Conference on Power Electronics and Applications. Barcelona, Spain : IEEE, 2009 : 1-10.

[13] Tu Qingrui, Xu Zheng, Zheng Xiang, et al. Mechanism analysis on the circulating current in module multilevel converter based HVDC[J]. High Voltage Engineering, 2010, 36(2) : 547-552.

[14] Wang Shanshan, Zhou Xiaoxin, Tang Guangfu, et al. Modeling of Module Multi-level Voltage Source Converter[J]. Proceedings of the CSEE, 2011, 31(24) : 1-8.

[15] Yang Xiaofeng, Zheng Trillion Q. A Novel Universal Circulating Current Suppressing Strategy Based on the MMC Circulating Current Model[J]. Proceedings of the CSEE, 2011, 32(18) : 59-65.

[16] Zhao Xin, Zhao Chengyong, Li Guangkai, et al. Submodule Capacitance Voltage Balancing of Module Multilevel Converter Based on Carrier Phase Shifted SPWM Technique[J]. Proceedings of the CSEE, 2011, 21(31) : 48-55.

24

A Two-stage CMOS Integrated Highly Efficient Rectifier for Vibration Energy Harvesting Applications

Qiang Li[†], Jing Wang*, Dan Niu** and Yasuaki Inoue*

Abstract – We report, in this paper, a highly efficient CMOS rectifier for low power energy harvesting system. By using two-stage structure and a novel active diode, the rectifier can work at a wide range of input voltage amplitudes of 0.45V up to 1.8V under a standard 0.18um CMOS process. A comparator that consists of common-gate stage and control stage is designed to control the switch of the active diode. The proposed rectifier can achieve a peak voltage conversion efficiency of over 96% and a power efficiency over 90%. Simulated power consumption of the rectifier is 197nW at 450mV, which is about 30% smaller than the best recently published results.

Keywords: Negative voltage converter , Wide range, Rectifier

1. Introduction

The field of energy harvesting technology is growing rapidly in recent years, which is largely stimulated and driven by different research efforts [1]-[3]. It plays a more and more important role in renewable resources and sensor systems. Comparing with conventional battery power systems, energy harvesting device has a long lifetime and is widely used in the system which is difficult to access to. Among different types of energyharvesting technology [4]-[6], vibration energy harvester attracted a great deal of interest and it is well suited for technical environment, such as engines and railroad bridge. In those cases, energy is used to supply low power devices and typical working frequency is between mHz and kHz [7]. Commonly, vibration energy harvesters supply an AC voltage, thus rectifier is needed to change the AC source to DC one which is used to power the loading system. Moreover, the output voltage amplitudes of different micro harvesting generators may vary and output power is usually low, and therefore both efficiency and wide range input amplitude are needed for the rectifier design. Fig. 1 shows the schematic of vibration energy harvesting circuit interface. Based on the previous research, a high efficiency and wide input range rectifier will be presented here.

The paper is organized as follows. In Section II, the topologies and characteristics of passive and active rectifier are given. The rectifier is proposed in section III. Simulation results are given in Section IV, followed by a conclusion in Section V.

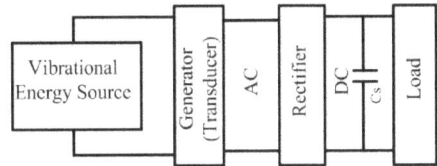

Fig. 1. Schematic of vibrational energy harvester circuit interface.

2. Rectifier Topologies

Conventional rectifiers are achieved by diodes and capacitors. The single-diode half-wave and full-wave rectifier with a diode bridge are commonly used in high voltage applications, since the diode forward voltage drop of 0.7V to 1V cannot be accepted in low voltage systems. Schottky diodes with a low forward voltage drop can replace the common diode to improve the efficiency. However, the high production cost and high reverse leakage current blocks its uses. In CMOS-only implementation circuits, MOS transistors in diode configuration (gate connected to the drain) are usually used. In [8], two diodes of the bridge rectifier are replaced by two cross-couple PMOS transistors. The gates of these PMOS transistors are driven by the input voltage, and the other two diodes are

---

†   Corresponding Author: Graduate School of Information, Production and Systems, Waseda University, Japan (liqiang@fuji.waseda.jp)

*   Graduate School of Information, Production and Systems, Waseda University, Japan

**   Key Laboratory of Measurement and Control of CSE, Southeast University, Nanjing, China

still implemented by diode-connected NMOS transistors, which results in a V_{th} voltage loss.

Although those passive rectifiers' structure are simple, the performances of them are lower. In order to achieve high voltage and power efficiency, active diode was presented and can be found in literatures [9]-[11]. The active diode is composed of a comparator controlled MOS transistor switch, working as an ideal diode with current flowing in only one direction and no voltage drop. By using this concept, the minimum input voltage amplitude is further decreased to around 0.7V [11]. Recently, bulk-input comparator technique is presented in [12], [13] for ultra-low-voltage systems and the rectifier can work well with input voltage smaller than 1V. However, bulk and comparator leakage current increase dramatically when operation voltage is larger than 1V, which results in a bad performance and limits its scope.

In order to overcome these drawbacks, a high efficiency and wide input amplitude range rectifier is presented. The minimum operation voltage is lower than previous published paper and the rectifier can work at a wide range of input voltage amplitudes of 0.45V up to 1.8V compared with 0.5V to 1V in [12] and 0.28V to 0.7V in [13]. This allows the rectifier to work in different types of vibration energy harvesting system.

3. Propose Rectifier

The proposed rectifier is consisted of two stages: Negative voltage converter and comparator controlled active diode. Fig. 2 shows the schematic of the two-stage structure. The requirements for the rectifier are defined as follows: i) Wide working voltage range from 0.45V to 1.8V; ii) Small leakage current and minimum power consumption lower than 200nW; iii) Up to 1kHz frequencies.

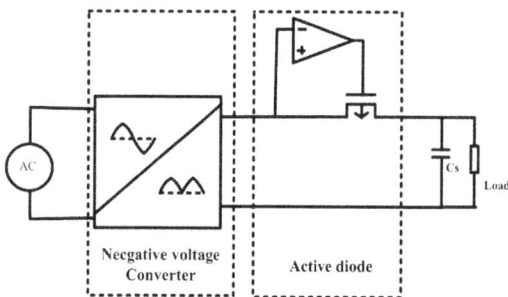

Fig. 2. Schematic of the proposed rectifier.

3.1 Negative voltage converter

The first stage of the rectifier is called negative voltage converter and used to convert the negative half waves of the input sinusoidal wave into positive ones, which is done with two PMOS and two NMOS transistors shown in Fig. 3 [9]. During the whole period of the input, node 1 is always the high potential and node 2 is the low potential. Consequently, the bulk of the PMOS transistor can be directly connected to node 1 and the NMOS to node 2.

In this case, no voltage drop V_{th} occurs. The voltage loss is only $|V_{dsp}| + V_{dsn}$ during each conductive branch, where V_{dsp} and V_{dsn} are caused by the on-resistance of the transistors. Thus, the dropout voltages can be minimized by using large transistor size to decrease the resistance so as to get a small voltage drop. The simulation results show that the voltage drop in this stage can be less than 10mV. Though the negative voltage converter has nearly no voltage drop, it cannot control the current direction, therefore an active diode is necessary to block the reverse current.

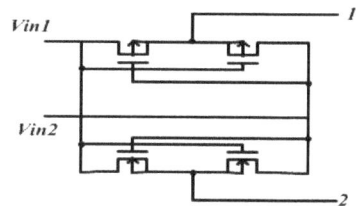

Fig. 3. Circuit diagram of the negative voltage converter.

4. Simulation Results

The main aspects of the rectifier are wide input voltage range, output voltage conversion efficiency, and the power efficiency, which will be discussed respectively.

The circuit was simulated with HSPICE using a standard 0.18um CMOS process. A pure sinusoidal waveformwhose default frequency is 100Hz is applied to the input of the rectifier. Additionally, a capacitance of 10uFand a load of 50KΩ are used.

4.1 Input range and voltage conversion efficiency

The proposed rectifier can work at a wide range of input voltage amplitudes of 0.45V up to 1.8V, which is decided by the threshold voltage and breakdown voltage of the standard 0.18um CMOS process. This allows the rectifier to work in different types of vibration energy harvesting system. The voltage efficiency η_v is defined as the fraction of the output DC voltage V_{out} and the input voltage amplitude $|V_{in}|$, which is shown in Eq. (1). Fig. 5 shows the input voltage range and voltage conversion efficiency versus different ohmic loads. The output voltage efficiency of the rectifier is higher with larger load resistors.

$$\eta_v = \frac{V_{out}}{|V_{inp}|} \cdot 100\%, \qquad (1)$$

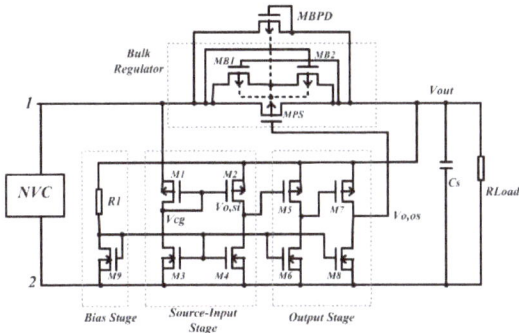

Fig. 4. Schematic of the proposed active diode.

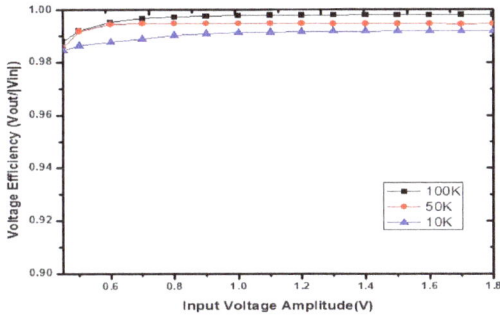

Fig. 5. Voltage conversion efficiency versus input voltage amplitude with 100Hz input.

4.2 Power efficiency

The power efficiency of the rectifier is calculated using

$$\eta_p = \frac{\int_{t_1}^{t_1+T} v_{out}(t) \cdot i_{out}(t) dt}{\int_{t_1}^{t_1+T} v_{in}(t) \cdot i_{in}(t) dt} \cdot 100\%, \qquad (2)$$

where T is one period of the input signal and t_1 is the start time. Fig. 6 shows the power efficiency versus different input voltage amplitude under three types load condition. With increasing R_{load}, the efficiency decreases, because the current through the ohmic load decreases and tends to the current through the comparator.

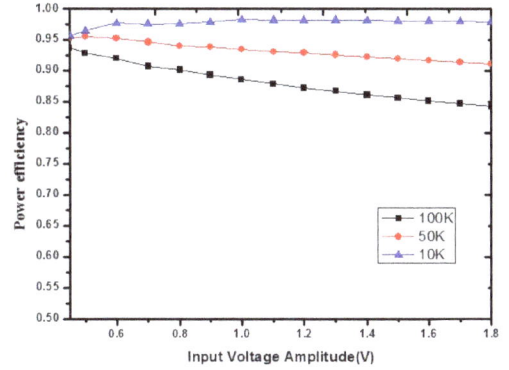

Fig. 6. Power efficiency versus input voltage amplitude with 100Hz input.

4.3 Working frequency

The voltage conversion efficiency versus the input voltage amplitude for different input frequencies is simulated and shown in Fig. 7. The load capacitor is adapted for the applied frequency. The voltage efficiency for the lower frequency is better than 1kHz case. Since higher frequency will results in the delay of the comparator and increasing reverse current. However, at typical energy harvesting frequencies, the working frequency range of this rectifier is sufficient for most applications.

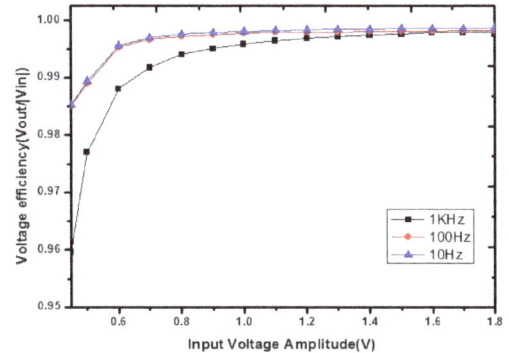

Fig. 7. Voltage efficiency versus input voltage amplitude with RLoad=50KΩ.

Table 1. Performance comparisons between rectifiers

	NOLTA 2011 [13]	TCAS I 2011 [12]	JSSC 2009 [10]	This work
Input Amplitude range	0.28V-0.7V	0.5V-1V	1.2V-2.4V	0.45V-1.8V
Voltage efficiency	76%-97%(Rload=40k)	Over 90%(Rload=50k)	82%-95%(Rload=2k)	95%-99%(Rload=50k)
$|Vin|_{min}$	0.28V	0.5V	1.2V	0.45V
Frequency	10Hz-3KHz	10Hz-10KHz	200KHz-1.5MHz	10Hz-1KHz
Power efficiency	78%-95%	Up to 95%	82%-87%	84%-98%

4.4 Power consumption

The static power consumption of the rectifier is simulated. The results show that the power consumption at 450mV is only 197nW, which is about 30% smaller than the best recently published result [12].

4.5 Performance comparisons

Table 1 shows the performance comparisons of the work with other previously reported rectifiers. The lowest operation voltage of previous rectifiers is 0.5V in [12] with a 500mV low threshold voltage 0.35um CMOS process. However, it cannot work when the input voltage is larger than 1V, which blocks the application field. The rectifier proposed in this paper overcome these drawbacks. It can not only work at a 0.45V input voltage, but also operate at a wide range input voltage amplitudes of 0.45V up to 1.8V. Moreover, peak voltage conversion efficiency and power efficiency are dramatically improved in this work.

5. Conclusion

In this paper, a wide range input amplitude and highly efficient rectifier for energy harvesting system was proposed. The rectifier is well suited for an input amplitude as low as 0.45V and can work at a wide range of input voltage amplitudes of 0.45V up to 1.8V under a standard 0.18um CMOS process. By using a 3-stage comparator controller, the rectifier can achieve a maximum peak voltage conversion efficiency of 99% and a power efficiency of 98%. Moreover, the power consumption of the rectifier is 197nw at 450mv, which is about 30% smaller than the best recently published results.

Acknowledgements

This work is supported by VLSI Design and Education Center (VDEC), University of Tokyo in collaboration with Synopsys, Inc.

References

[1]   Ottman.G.K, Hofmann.H.F, and Lesieutre.G.A, "Optimized piezoelectric energy harvesting circuit using step-down converter in discontinuous conduction mode," IEEE Trans. Power Electron, vol. 18, no.2, pp. 696-703, March 2003.

[2]   Hoang.D.C., Tan.Y.K, Chng.H.B, and Panda.S.K, "Thermal energy harvesting from human warmth for wireless body area network in medical healthcare system," in IEEE PEDS, pp. 1277-1282, November 2009.

[3]   Ramadass.Y.K, Chandrakasan.A.P, "An Efficient Piezoelectric Energy Harvesting Interface Circuit Using a Bias-Flip Rectifier and Shared Inductor," IEEE J.Solid-State Circuits, vol. 45, no. 1, pp. 189-204, January 2010.

[4]   F.Moll, A.Rubio, "An approach to the analysis of wearable body-powered system," Proc. Mixed Design Integr. Crcuits Syst., Gdynia Poland, June 2000.

[5]   E.O.Torres, G.A.Rincon-Mora, "Electrostatic Energy-Harvesting and Battery-Charging CMOS System Prototype," IEEE Trans. Circuits Syst. I, Reg. Papers, vol. 56, no. 9, pp. 1938-1948, September 2009.

[6]   Yang.Guangjun, Bai.Cunru, and Sun.Jing, "Research on Vibration Source and Transfer Path for Power Section of Low Turbulence Wind Tunnel," in IEEE ICMTMA, pp. 231-234, March 2010,

[7]   C.B.Williams and R.B.Yates, "Analysis of a micro-electric generator for microsystems," in Proc. Transducers/Eurosensors, Stockholm, Sweden, vol. 1, pp. 369-372, 1995.

[8]   M.Ghovanloo, K.Najafi, "Fully integrated wideband high-current rectifiers for inductively powered devices," IEEE J.Solid-State Circuits, vol. 39, no 11, pp.1976-1984, November 2004.

[9]   C.Peters, O.Kessling, F.Henrici, M.Ortmanns, and Y.Manoli, "CMOS Integrated Highly Efficient Full Wave Rectifier," Proc. IEEE international Symposium on Circuits and Systems, pp. 2415-2123, 2007.

[10]   Song Guo, Hoi Lee, "An Efficiency-Enhanced CMOS

Rectifier With Unbalanced-Biased Comparators for Transcutaneous-Powered High-Current Implants," IEEE J. Solid-Stae Circuits, vol. 44, pp. 1796-1840, June 2009.

[11] Qiang Li, Renyuan Zhang, Zhangcai Huang, and Y.Inoue, "A low oltage CMOS rectifier for wirelessly powered devices," IEEE international Symposium on Circuits and Systems, pp. 873-876, May 2010.

[12] C.Peters, J.Handwerker, D.Maurath, Y.Manoli, "A Sub-500 mV Highly Efficient Active Rectifier for Energy Harvesting Applications," IEEE Trans. Circuits Syst. I, Reg. Papers, vol. 58, no. 7, pp. 1542-1550, July. 2011.

[13] Dan Niu, Zhangcai Huang, Minglv Jiang, Yasuaki Inoue, "A Sub-0.3V highly efficient CMOS rectifier for energy harvesting applications," IEICE NOLTA, vol. 3, no. 3, pp. 405-416, July 2012.

25

BER Performance of QPSK-Transmitted Signal for Power Line Communication under Nakagami-like Background Noise

Youngsun Kim[†], Soon-Woo Rhee*, Jae-Jo Lee* and Sang Ki Oh*

Abstract – Power line communication (PLC) technology is used in various areas of a smart grid. The extensive use of PLC technology has proved its key role in smart grids throughout the world. Despite PLC's advantages, it suffers from a difficulty in modeling and analysis of its power line channel. Many researchers have experimented and proposed channel models for their own defined environments; as a result, a unified model for the power line has not been presented. To overcome this problem partially, we used a Nakagami-m distribution like background noise and analyzed the performance of the binary transmission system in our previous work. In this paper, we compare the analytical and simulated BER (bit error rate) performances of QPSK (quadrature phase shift keying) transmission system with Gray encoding. We verify that the analyzed and simulated performances are well matched for different values of m.

Keywords: Power Line Communication, Background Noise, Bit Error Rate, Nakagami-m distribution

1. Introduction

To develop a smart grid for electricity services, a reliable and secure communication infrastructure is needed. The communication infrastructure may fully support two-way communication and guarantee inter-network connections using both wired and wireless communication technologies. From various deployment practices and commercial use, power line communication (PLC) has been adopted for automatic meter reading (AMR) and advanced metering infrastructure (AMI) for smart grid last-mile communication [1][2][3]. PLC can cover from the distribution line to the in-home communication network. PLC was also selected as the communication technology for the electric-vehicle charging system by ISO/IEC and its standardization is ongoing [4].

To adopt the PLC to a reliable communication infrastructure of a smart grid, we need to model the channel characteristics of the power line. Though the PLC is in a promising state, it suffers from a difficulty in modeling and analysis of its power line channel. Many researchers have experimented and proposed the channel models for their own defined environments; as a result, a unified model for the power line has not been presented. To overcome this

problem partially, researchers proposed the model for an amplitude distribution of the background noise as a Nakagami-m distribution [5][6]. Nakagami-m distribution is quite useful and widely used as it represents various channel models by changing the m parameter in wireless communication. The amplitude distribution of background noise is represented by the Nakagami-m distribution with $m \approx 1$ at high frequency (25 MHz) while $m <1$ at low frequency (8 MHz) [6].

In our previous work, a closed-form expression for the real part of Nakagami-m distributed background noise was derived, which facilitates a mathematical derivation of the system performance [7]. Using this closed-form expression, we derived the Bit Error Rate (BER) performance of (Binary Phase Shift Keying) BPSK transmission over power line channels and verified its validity by simulations [8].

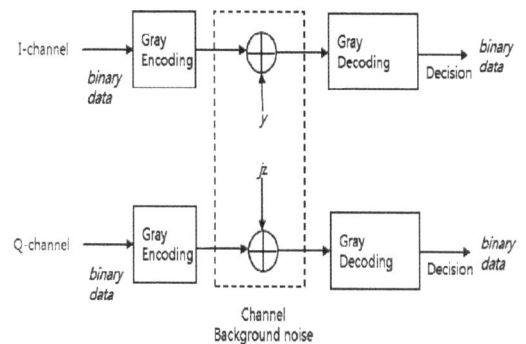

Fig. 1. System model for QPSK transmission.

† Corresponding Author: Power Telecommunication Research Center, Korea Electrotechnology Research Institute (KERI), Korea (yskim@keri.re.kr)

* Power Telecommunication Research Center, Korea Electro-technology Research Institute (KERI), Korea

In this paper, we consider the BER performance of a quadrature-phase shift-keying (QPSK) system with Nakagami distributed background noise. A derivation of SER (Symbol Error Rate) performance was partially introduced in [9]. We apply Gray encoding to the QPSK system for the analysis of BER derivation. The system model is presented in Section 2. In Section 3, we analyze the BER performance of the QPSK system. Simulation results are presented in Section 4. Finally, we conclude in Section 5.

2. System Model

Fig. 1 shows a baseband system model for Gray encoded QPSK transmission. Once 2-bit symbols are selected from a set, $\{\pm 1\}$, in-phase and quadrature binary data pass through the Gray encoder as in Fig. 2. The Gray code gives a characteristic such that two adjacent symbols differ only by one bit. Thus a decision error with an adjacent symbol occurs with only a one-bit error. As a result, we can approximate the BER performance simply by dividing the SER performance by 2. This procedure is widely used for evaluating the performance of an M-PSK transmission system, where the QPSK is a kind of an M-PSK transmission system with M = 2.

Between the transmission and the decision, both channels have real and imaginary background noises, y and z, respectively. In our previous work, we derived the PDF for y and z as follows [8][9],

$$f(y) = \frac{1}{\sqrt{\pi}\,\Gamma(m)} \sqrt{\frac{m}{\Omega}}\, e^{-\frac{my^2}{\Omega}} \left\{ \frac{\Gamma(\frac{1}{2}-m)}{\Gamma(1-m)} \left(\frac{my^2}{\Omega}\right)^{m-\frac{1}{2}} \right.$$

$$\times {}_1F_1\left(\frac{1}{2}, \frac{1}{2}+m, \frac{my^2}{\Omega}\right) + \frac{\Gamma(m-\frac{1}{2})}{\sqrt{\pi}} \qquad (1)$$

$$\left. \times {}_1F_1\left(1-m, \frac{3}{2}-m, \frac{my^2}{\Omega}\right) \right\}$$

and

$$f(z) = \frac{1}{\sqrt{\pi}\,\Gamma(m)} \sqrt{\frac{m}{\Omega}}\, e^{-\frac{mz^2}{\Omega}} \left\{ \frac{\Gamma(\frac{1}{2}-m)}{\Gamma(1-m)} \left(\frac{mz^2}{\Omega}\right)^{m-\frac{1}{2}} \right.$$

$$\times {}_1F_1\left(\frac{1}{2}, \frac{1}{2}+m, \frac{mz^2}{\Omega}\right) + \frac{\Gamma(m-\frac{1}{2})}{\sqrt{\pi}} \qquad (2)$$

$$\left. \times {}_1F_1\left(1-m, \frac{3}{2}-m, \frac{mz^2}{\Omega}\right) \right\}$$

for $0<m<1$ and $m \neq 1/2$. The confluent hypergeometric function of the first kind, $_1F_1$, is defined as [10, Chap. 9.2]

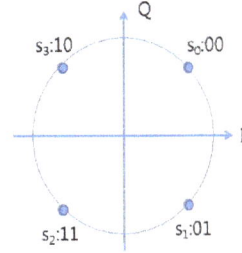

Fig. 2. Gray encoded signal constellation for QPSK.

$$_1F_1(a,b,z) = 1 + \frac{a}{b}\frac{z}{1!} + \frac{a(a+1)}{b(b+1)}\frac{z^2}{2!}$$

$$+ \frac{a(a+1)(a+2)}{b(b+1)(b+2)}\frac{z^3}{3!}\cdots \qquad (3)$$

Then, we concluded that the amplitude noise of the power line shows the same statistical behavior in both the real and imaginary parts of noise [9].

3. Derivation of the BER Performance

In this section, we derive the BER performance of the QPSK transmission system. As shown in Fig. 1, I- and Q-channel binary data are encoded with Gray coding, then, the channel background noise is added. A transmitter sends the complex symbol, $\pm 1 \pm j$, according to the symbol set, $\{s_0, s_1, s_2, s_3\}$ then, the complex decision metric at the receiver, r, is defined as

$$r = \pm 1 \pm j + y + jz \quad, \qquad (4)$$

where the I- and Q-channel data, $\pm 1 \pm j$, are corrupted by a complex Nakagami background noise, $y + jz$. We assume that four symbols are transmitted with equal probability, thus, the optimal threshold for a decision is x- and y-axes. For example, if signal s_2 is transmitted and a noise-corrupted received signal falls in the third quadrant, then the decision is correctly made as s_2; otherwise, it fails. When a symbol s_2 is sent, the probability of symbol s_2 is correctly decided, $p(C\,|\,s_2\,)$, and is represented by

$$p(C \mid s_2) = p(y < 0 \mid s_2)\, p(z < 0 \mid s_2) \qquad (5)$$

which means a noise-corrupted symbol must fall on the third quadrant. We get

$$p(y < 0 \mid s_2) = 1 - p(y > 0 \mid s_2)$$
$$= 1 - \int_1^\infty f(y \mid s_2)\,dy \qquad (6)$$

where $f(y \mid s_2)$ is a conditional density function of y with sending s_2, defined as in Eq. (1);

$$f(y \mid s_2) = \frac{1}{\sqrt{\pi}\,\Gamma(m)} \sqrt{\frac{m}{\Omega}}\, e^{-\frac{m(y+1)^2}{\Omega}} \left\{ \frac{\Gamma(\frac{1}{2}-m)}{\Gamma(1-m)} \left(\frac{m(y+1)^2}{\Omega} \right)^{m-\frac{1}{2}} \right.$$
$$\times {}_1F_1\!\left(\frac{1}{2}, \frac{1}{2}+m, \frac{m(y+1)^2}{\Omega} \right) + \frac{\Gamma(m-\frac{1}{2})}{\sqrt{\pi}} \qquad (7)$$
$$\left. \times {}_1F_1\!\left(1-m, \frac{3}{2}-m, \frac{m(y+1)^2}{\Omega} \right) \right\}.$$

The probability of symbol s_2 being correctly decided, $p(C \mid s_2)$, is

$$p(C \mid s_2) = \left(1 - \int_1^\infty f(y \mid s_2)\,dy \right)^2$$
$$= 1 - 2\int_1^\infty f(y \mid s_2)\,dy + \left(\int_1^\infty f(y \mid s_2)\,dy \right)^2. \qquad (8)$$

Finally, the symbol error probability, P_e, is given as

$$P_e = 1 - p(C \mid s_2)$$
$$= 2\int_1^\infty f(y \mid s_2)\,dy - \left(\int_1^\infty f(y \mid s_0)\,dy \right)^2 \qquad (9)$$
$$= 2P_1 - P_1^2$$

$$P_1 = \frac{\Gamma\left(\frac{1}{2}-m\right)}{2m\sqrt{\pi}\,\Gamma(m)\Gamma(1-m)} \left[x^{2m}{}_2F_2\!\left(m, m, \frac{1}{2}+m, m+1; -x^2 \right) \right]_{x=\sqrt{\frac{m}{\Omega}}}^{\infty}$$
$$+ \frac{\Gamma\left(m-\frac{1}{2}\right)}{\sqrt{\pi}\,\Gamma(m)} \left[x\,{}_2F_2\!\left(\frac{1}{2}, \frac{1}{2}; \frac{3}{2}-m, \frac{3}{2}; -x^2 \right) \right]_{x=\sqrt{\frac{m}{\Omega}}}^{\infty} \qquad (10)$$

where we can evaluate the integral part, P_1, using the result

from [8] as in Eq.(10). For higher values of SNR, P_1^2 becomes very small; then Eq. (9) can be approximated as

$$P_e \approx 2P_1 \qquad (11)$$

To derive the BER performance from the symbol error probability, we can assume that the most probable error occurs only for adjacent signal points due to the Gray encoding. Then the BER can be expressed as

$$P_b = \frac{P_e}{\log 2M} \qquad (12)$$

where M=4 for the QPSK transmission [11]. From Eq. (9) and Eq. (12), we get

$$P_b = P_1 - \frac{P_1^2}{2} \qquad (13)$$

For higher values of SNR (Signal to Noise Ratio), P_1^2 becomes very small; then, Eq. (9) can be approximated as

$$P_b \approx P_1. \qquad (14)$$

Since P_1 is the BER performance of the BPSK-transmission system [8], we can observe that the BER performance of the QPSK-transmission system is approximately the same as that of the BPSK. This is due to the orthogonality of the real and imaginary components of the QPSK transmission system.

4. Simulation Results

We simulated the BER performance of the QPSK-transmission system with the Monte Carlo method. 10^6 symbols are randomly generated at the transmitter. The generated symbols are Gray-encoded and corrupted by complex Nakagami-like background noise; then, the Gray-decoded received symbols are decided by a threshold comparison as presented in Fig. 1. The theoretical SER is presented in Eq. (9) and is approximated in Eq. (11). Also, Eq. (14) is used to calculate the BER performance. For m=0.6 and m=0.9, we analyzed the simulated error rate performances in Fig. 3 and Fig. 4, respectively. The simulation results match well with those of the analysis. The accuracy of the approximated analysis is credible for all SNR values.

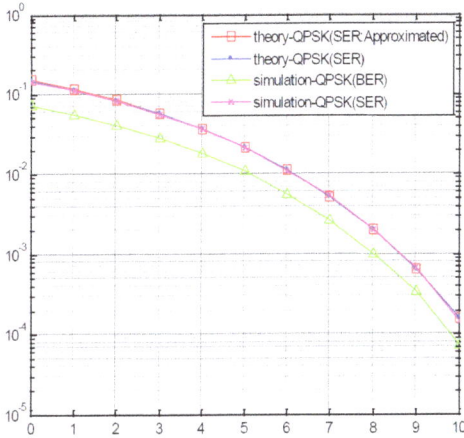

Fig. 3. BER performance with m=0.6 and N=10^6 sample points with Gray encoding.

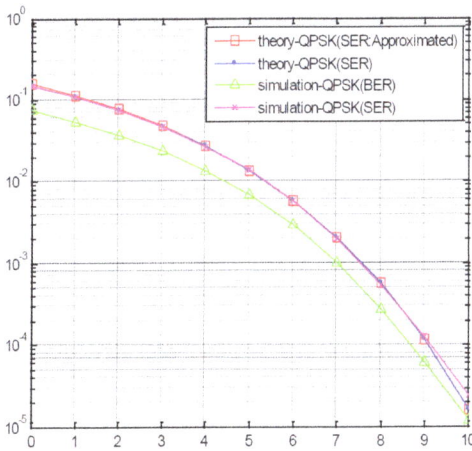

Fig. 4. BER performance with m=0.9 and N=10^6 sample points with Gray encoding.

5. Conclusion

In this paper, we analyzed and simulated the error rate performance of the QPSK-transmitted signal under Nakagami-like background noise. Nakagami distributed background noise is added to both real and imaginary portion of QPSK signal. Gray code is used for the ease of calculation for BER performance. Simulation was done for randomly generated 10^6 symbols through the Monte Carlo method. From simulation, we verified that the analyzed and simulated results are almost same for various range of SNR

values. The BER performance of the QPSK-transmitted signal with imperfect synchronization parameters would be the focus of future work.

Acknowledgements

This paper was fully presented at the ICEE 2013[12] and selected by the editorial board of JICEE for publication.

References

[1] Y. Zhang and S. Cheng, "Power Line Communications", IEEE Potentials, vol. 23, no. 4, pp. 4-8, Oct.-Nov. 2004.

[2] A. Zaballos, A. Vallejo, M. Majoral and J. M. Selga, "Survey and performance comparison of AMR over PLC standards", IEEE Trans. Power Del., vol. 24, no. 2, pp. 604-613, April 2009.

[3] B. Sivaneasan, E. Gunawan and P. L. So, "Modeling and performance analysis of automatic meter-reading systems using PLC under impulsive noise interference", IEEE Trans. Power Del., vol. 25, no. 3, pp. 1465-1475, July 2010.

[4] IEC, "Road vehicles - Vehicle to grid communications interface - Part 1: General information and use-case definition", ISO/IEC 15118-1: 2010.

[5] M. Nakagami, "The m-distribution - A general formula of intensity distribution of rapid fading", in Statistical Methods in Radio Wave Propagation, pp. 3-36, Pergamon Press, Oxford, U.K, 1960.

[6] H. Meng, Y. L. Guan and S. Chen, "Modeling and analysis for noise effects on broadband power-line communications", IEEE Trans. Power Del., vol. 20, no. 2, pp. 630-637, April 2005.

[7] Y. Kim, H-M. Oh and S. Choi, "Closed-form expression of Nakagami-like background noise in power-line channel", IEEE Trans. Power Del., vol. 23, no. 3, pp. 1410-1412, Oct. 2008.

[8] Y. Kim, Y-H. Kim, H-M. Oh and S. Choi, "BER performance of binary transmitted signal for power line communication under Nakagami-like background noise", Proc. Energy 2011 : The First International Conference on Smart Grids, Green Communications and IT Energy-aware Technologies, pp. 126-129, May. 2011.

[9] Y. Kim, H-M. Oh and S. Choi, "Error rate performance of QPSK-transmitted signal for power line communi-cation under Nakagami-like background noise", Proc.

Energy 2012 : The Second International Conference on Smart Grids, Green Communications and IT Energy-aware Technologies, May 2012.

[10] I. S. Gradshteyn and I. M. Ryzhik, Table of Integrals, Series, and Products, 6th ed., San Diego, CA: Academic, 2000.

[11] R. E. Ziemer and W. H. Tranter, Principles of Communications: Systems, Modulation, and Noise, Fourth edition, 1995.

[12] Y. Kim, S. W. Rhee, J. Lee and S. K. Oh, "BER performance of QPSK-transmitted signal for power line communication under Nakagami-like background noise", Proc. of ICEE 2013, July 2013.

26

Dynamic Power System Zone Division Scheme using Sensitivity Analysis

Puming Li[†], Jianing Liu*, Bo Li*, Yuqian Song** and Jin Zhong**

Abstract – A dynamic zone division scheme is presented in this paper based on sensitivity analysis of power system. The concept of this scheme is to divide a large power system into small potential areas, which enclose strong correlated buses and have weak connections with each other. These areas can be considered as different individual zones in system analysis. In N-1 contingency analysis, the computation burden is largely reduced by operating in the level of potential areas, and the system cross-section distribution is determined dynamically. Analysis result can be used as an aid of power system operation and control. In this paper, IEEE-118 bus system is chosen as testing system, and this scheme is proved to be feasible and effective for large scale power system, it will be used to build the functional module of next generation EMS in Guangdong power grid and will serve as a function in power grid operation in the future.

Keywords: N-1 contingency criteria, Potential area, Power system stability, Power system zoning, Sensitivity analysis.

1. Introduction

Continuous development of social economy has led to a significant increase of power demands in power systems. The massive scale and complexity of modern system network cannot afford a large-scale blackout. On the other hand, the rising share of renewable energy sources (RES) such as wind power and solar power have changed the stability boundary of power system. All of these adverse factors require more resourceful and precise response of system control and operation. The traditional and widely served operating patterns, however, are based on the operators' personal experiences or large scale computing and processing of EMS system, neither of which is capable enough for fast response of modern power system control. Therefore, the system operator would prefer a more stable and robust network structure.

The goal of accurate operating activities can be achieved by simplifying power system network structure and highlighting the area that is sensitive to system operating activities. Hence, a dynamic zoning scheme using system sensitivity analysis is proposed for power system stability analysis and assistant decision making.

System sensitivity has always been widely used as a

powerful tool for system stability analysis. In [1] and [2], sensitivity is used to determine the optimal placement of devices for system stability maintaining, such as power system stabilizer (PSS), as well as the performance evaluation of these devices. A modeling approach for macro-scale power system was presented in [3]. In addition, sensitivity factor was applied as a very important concept for optimization of nonlinear models of power system [4].

However, for an actual large scale power system like the network in Guangdong, China, it is quite complex and consists of thousands of parameters. The result of sensitivity analysis might provide suggestions to system operation, but the amount of data can be still quite large and may not be able to improve the efficiency of human operation. Therefore, system zoning might be capable for simplification of system operation.

Power system zoning has been widely discussed. For system analysis and control, different principle and scheme of power system islanding and zonal analysis were presented in [5]-[8]. A functional focused method was introduced and estimated in [8].

In this paper, the concept of system sensitivity is considered as the principle of system zoning. Using this dynamic zoning scheme and sensitivity analysis, power system can be divided into a series of small potential areas, and these areas can be considered as relatively individuals. System buses allocated within one area are strongly linked with each other. Different areas, on the other hand, do not correlated substantially. Consequently, power system analysis can be operated in a level of potential areas rather than buses.

† Corresponding Author: Guangdong Power Dispatch Center of Guangdong Power Grid Co., China (lipuming@126.com)

* Guangdong Power Dispatch Center of Guangdong Power Grid Co., China (libo@gddd.csg.cn)

** Department of Electrical and Electronic Engineering, the University of Hong Kong, Hong Kong (yqsong@eee.hku.hk)

With potential areas determined by sensitivity analysis, N-1 contingency analysis can be simplified greatly. Besides, by allocating a series of potential areas and N-1 contingency analysis, a dynamic distribution of power system cross-section can be determined in a very short period of time. The result will be provided to the system operator by showing the cross-section distribution and highlighting the sensitive areas that need to be aware of.

The dynamic zoning scheme is proposed to be applied as a functional module in the project of next generation EMS construction in Guangdong power Grid, China. The rest parts of this paper are organized as follows: Section 2 introduces the principle and detail techniques of sensitivity analysis and system zoning. Standard IEEE-118 bus system is used as test system to study the effect of the zoning scheme in Section 3. The conclusion is drawn in Section 4 to summarize the work of this paper.

2. Sensitivity Analysis in Power Network

Sensitivity analysis is often used to calculate the influence of variations of one parameter on other parameters. Although this problem can be solved by calculating the power flow in the system, the amount of the computation is extremely high especially when the network is complex.

In power system analysis, power flow equations could be generally expressed as:

$$\begin{cases} f(x,u) = 0 \\ y = y(x,u) \end{cases}$$

where x is state variables such as bus voltage magnitude and phase angles; u is the variables we want to control such as the power output of generators and y is dependent variables such as power flow through the line. Thus, the first equation $f(x,u)=0$ is the power balance equation and the second equation $y=y(x,u)$ is the correlation between line flow and bus voltage.

The system is assumed to operate at (x_0, u_0) and the system steady-state operating point has moved to $(x_0 + \Delta x, u_0 + \Delta u)$. After first-order Taylor expansion at (x_0, u_0), the power flow equations could be expressed as:

$$\begin{cases} f(x_0 + \Delta x, u_0 + \Delta u) = f(x_0, u_0) + \dfrac{\partial f}{\partial x}\Delta x + \dfrac{\partial f}{\partial u}\Delta u = 0 \\ y_0 + \Delta y = y(x_0, u_0) + \dfrac{\partial y}{\partial x}\Delta x + \dfrac{\partial y}{\partial u}\Delta u \end{cases} \quad (1)$$

As (x_0, u_0) is the steady-state operation point of the system, we have:

$$\begin{cases} f(x_0, u_0) = 0 \\ y_0 = y(x_0, u_0) \end{cases} \quad (2)$$

After substitution, Equation (1) can be expressed as:

$$\begin{cases} \dfrac{\partial f}{\partial x}\Delta x + \dfrac{\partial f}{\partial u}\Delta u = 0 \\ \Delta y = \dfrac{\partial y}{\partial x}\Delta x + \dfrac{\partial y}{\partial u}\Delta u \end{cases} \quad (3)$$

$$\begin{cases} \Delta x = -\left(\dfrac{\partial f}{\partial x}\right)^{-1}\dfrac{\partial f}{\partial u}\Delta u = S_{xu}\Delta u \\ \Delta y = \dfrac{\partial y}{\partial x}\Delta x + \dfrac{\partial y}{\partial u}\Delta u = (\dfrac{\partial y}{\partial x}S_{xu} + \dfrac{\partial y}{\partial u})\Delta u = S_{yu}\Delta u \end{cases} \quad (4)$$

where,

$$\begin{cases} S_{xu} = -\left(\dfrac{\partial f}{\partial x}\right)^{-1}\dfrac{\partial f}{\partial u} \\ S_{yu} = (\dfrac{\partial y}{\partial x}S_{xu} + \dfrac{\partial y}{\partial u}) \end{cases} \quad (5)$$

S_{xu} and S_{yu} are sensitivity matrices of x and y depending on u, respectively.

In Guangdong Power Grid, the Generation Shift Distribution Factor (GSDF) is used as the sensitivity matrix to divide the system into zones. The element $S(i,j)$ in the GSDF matrix represents the power flow variation in branch j if the output of generator i increases by one unit. If $S(i,j)$ is large enough (e.g. larger than 0.4), it means that generator i and branch j are so strongly correlated that they should be zoned into the same area. By searching along the indices of generators, we can get all the branches and buses that are strongly correlated to a certain generator. Thus the potential zoning results of the system are obtained.

After we get the potential zoning results, the whole grid could be divided into several potential areas. Next step is to do N-1 contingency analysis to determine the stability boundary between different areas. All the buses and branches in the same potential area can be regarded as one bus, decreasing the computation amount of N-1 contingency analysis. After N-1 contingency analysis, potential areas which satisfy N-1 contingency criterion could be considered as one zone and those tie lines which do not satisfy N-1 criterion are regarded as zone boundaries and should be taken into account. The flow chart of the zoning scheme is shown in Fig. 1.

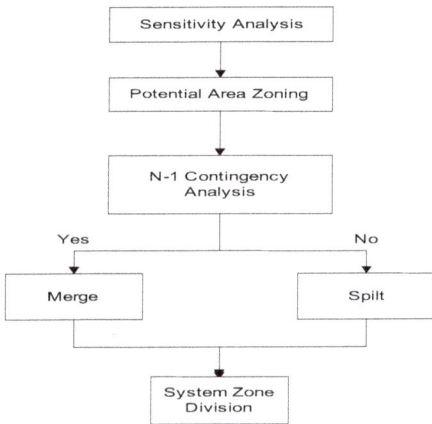

Fig. 1. Flow chart of zoning scheme.

3. Case Study

In this section, the sensitivity analysis algorithm proposed in Section 2 is tested with the IEEE-118 bus system on the software platform Matpower 4.2. The diagram and bus parameters of the system are given in [9].

3.1 Potential Grid Zoning

Part of the sensitivity matrix is shown in Table 1. The result of potential zone division is given in Fig. 2. It is

indicated in Fig. 2 that the whole system could be divided into six potential areas. Buses and branches in the same potential area are strongly geographically correlated as expected. Although some buses are geographically close to each other, they should not be zoned into the same area due to the direction of power flow, such as bus 24 and bus 72. The major role of the sensitivity analysis is to identify the boundaries between these buses, which are not easy to identify intuitively.

3.2 N-1 Contingency Analysis

After the system is divided into several potential areas, all the buses and branches in the same area could be regarded as one bus. The N-1 contingency analysis mainly focuses on the tie lies between different potential areas.

It is illustrated in Fig. 3 that if one of the tie lines between Area 2 and Area 3 (from Bus 30 to Bus 38) is out of service, the N-1 contingency analysis shows that the power flow in the following branches varies considerably: branch from Bus 15 to Bus 33; branch from Bus19 to Bus 34; branch from Bus 23 to Bus 24 and branch from Bus 37 to Bus 38. Thus, the potential areas we got previously reassemble and the system is re-zoned into three parts, which is given in Fig. 3. The system operator only needs to monitor the power flows in the tie lines between different zones. Consequently, the efficiency of power dispatch could be improved.

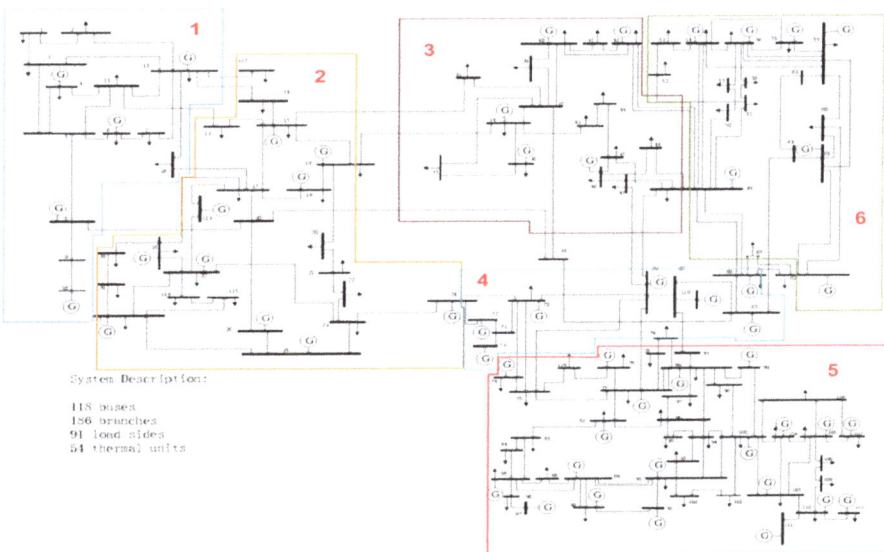

Fig. 2. Potential zone divisions on IEEE-118 bus system.

Table 1. GSDF at Bus 4

Generation Shift Distribution Factor at Bus 4							
Branch	GSDF	Branch	GSDF	Branch	GSDF	Branch	GSDF
1	0.020	51	0.006	101	0.002	151	0.004
2	0.020	52	0.072	102	0.032	152	0.002
3	0.819	53	0.072	103	0.002	153	0.002
4	0.045	54	0.527	104	0.569	154	0.001
5	0.081	55	0.072	105	0.090	155	0.002
6	0.081	56	0.072	106	0.088	156	0.001
7	0.000	57	0.072	107	0.482	157	0.004
8	0.606	58	0.072	108	0.150	158	0.002
9	0.000	59	0.082	109	0.126	159	0.002
…	…	…	…	…	…	…	…
36	0.025	86	0.003	136	0.001	186	0.005
37	0.606	87	0.003	137	0.001		
38	0.104	88	0.002	138	0.000		
…	…	…	…	…	…		
45	0.105	95	0.007	145	0.000		
46	0.002	96	0.521	146	0.000		
47	0.002	97	0.016	147	0.001		
48	0.115	98	0.017	148	0.004		
49	0.002	99	0.017	149	0.008		
50	0.021	100	0.002	150	0.001		

4. Conclusions

The dynamic power system zoning scheme with sensitivity analysis is studied in this paper. The basic algorithm of sensitivity analysis is introduced. The effectiveness and validity of the scheme are tested with standard IEEE-118 bus system on the software platform Matpower. It can be concluded that zoning scheme proposed in this paper is helpful for system operators especially when the system scale is very large and this scheme is being applied as a functional module in the project of next generation EMS construction in Guangdong Power Grid.

References

[1] Sil A, Gangopadhyay T K, Paul S, et al. Design of robust power system stabilizer using H∞ mixed sensitivity technique[C]//Power Systems, 2009. ICPS'09. International Conference on. IEEE, 2009: 1-4. [2] Alstom supplies integrated solar/CC project in Morocco pp.8-10 Vol.152 No.1 2008 POWER

[2] Chureemart J, Churueang P. Sensitivity analysis and its applications in power system improvements[C]// Electrical Engineering/ Electronics, Computer, Telecommunications and Information Technology, 2008. ECTI-CON 2008. 5th International Conference on. IEEE, 2008, 2: 945-948.

[3] Chen Z, Roy K. A power macromodeling technique based on power sensitivity[C] //Proceedings of the 35th annual Design Automation Conference. ACM, 1998: 678-683.

[4] Daniels A R, Lee Y B, Pal M K. Nonlinear power-system optimisation using dynamic sensitivity analysis[J]. Electrical Engineers, Proceedings of the Institution of, 1976, 123(4): 365-370.

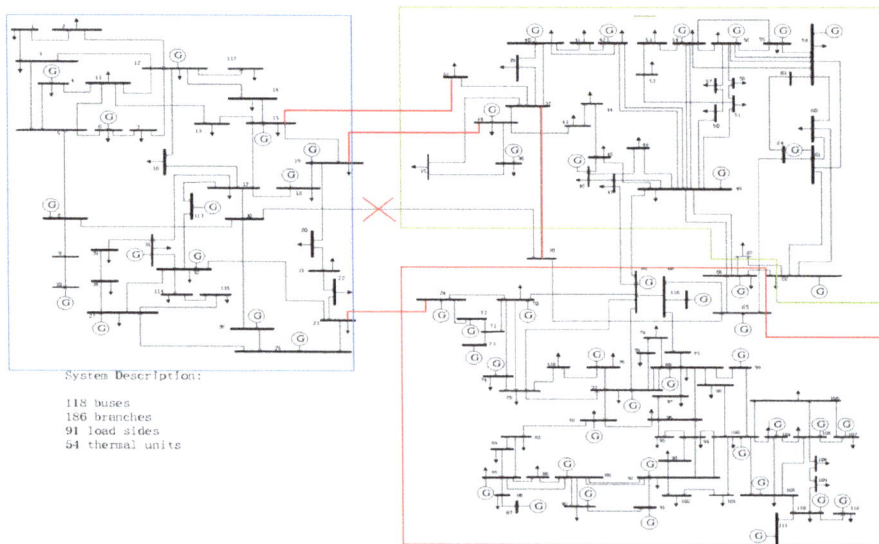

Fig. 3. Zoning results after N-1 contingency analysis.

[5] Jouybari-Moghaddam H, Hosseinian S H, Vahidi B. Active distribution networks islanding issues: An introduction [C]// Environment and Electrical Engineering (EEEIC), 2012 11th International Conference on. IEEE, 2012: 719-724.

[6] Sun R, Wu Z, Centeno V A. Power system islanding detection & identification using topology approach and decision tree[C]//Power and Energy Society General Meeting, 2011 IEEE. IEEE, 2011: 1-6.

[7] Mulhausen J, Schaefer J, Mynam M, et al. Anti-islanding today, successful islanding in the future[C]// Protective Relay Engineers, 2010 63rd Annual Conference for. IEEE, 2010: 1-8.

[8] Chenggen W, Baohui Z, Zhiguo H, et al. Study on power system self-adaptive islanding [C]//Advanced Power System Automation and Protection (APAP), 2011 International Conference on. IEEE, 2011, 1: 270-274.

[9] Power Systems Test Case Archive. [Online]. Available: http://www.ee.washington.edu/ research/pstca/

27

Research on the Security Threats and Strategy in Smart Grid Application

Guanjun Ding†, Bangkui Fan*, Teng Long, Haibin Lan*, Jing Wang* and Zhiyong Chen****

Abstract – The smart grid is evolving rapidly from a relatively isolated environment to an opened one. The adoption of information and communication technologies can make greater connectivity and interoperability between components. However, the increased connectivity also brings the challenge to security. First of all, in allusion to the current study on security threats to smart grid simple and deficient, the paper induces and analyzes it thoroughly, which can also correct some misunderstanding in a certain degree, from the view of system level, services, confidentiality and so on. Secondly, based on the analysis of security threats, the paper constructs the security strategy for smart grid. It consists of four steps, which are discussed detailedly one by one. It can provide the protection effectively. Finally, relevant conclusions are made.

Keywords: Security threats, System level, Confidentiality, Security strategy, Risk assessment, Smart grid

1. Introduction

In a world where protecting the environment is a major concern, it is important to find cost-effective ways of reducing power usage and increasing energy independence [1]. The smart grid concept is a next generation power grid in which the electric power flow is controlled flexibly by fully utilizing the latest information and communication technologies (ICT) [2], can intelligently integrate the actions of all users connected to it-generators, consumers and those that do both, as shown in Fig. 1.

Fig. 1. Schematic diagram of smart grid framework.

By enabling both new and existing electric grid components to communicate with each other, power utilities can better monitor conditions, collect information and remotely control devices. Consequently, the smart grid has the capability to optimize power resources, reduce costs, increase reliability and enhance electric power efficiency [3].

However, the increased connectivity also presents challenges, especially in security [4]. Because of the inherent nature of the technology and the services it supplies, the unfortunate reality is that the smart grid becomes a prime target for acts of vandalism and terrorism. In other words, the smart grid faces the security threats [5]. Therefore, the transformation of traditional power grid to smart grid requires an intrinsic security strategy to safeguard this critical infrastructure.

In view of the existing study on security threats to smart grid which are simple, superficial and deficient, this paper induces and analyzes the security threats to smart grid thoroughly, in section 2, from the aspect of system level, services, confidentiality and so on. Furthermore, based on the research of security threats, the paper puts forth and designs the related security strategy for smart grid. It comprises four steps, and each step is discussed and explained in detail, in section 3. And finally, the conclusions follow in section 4.

2. Induction and analysis of the security threats to smart grid

It's often assumed that security threats exclusively come from hackers and other outside individuals, groups with malicious intent [6]. However, utility staff and other insiders also could pose a risk, because they have

† Corresponding Author: Beijing Information Technology Institute, Beijing, China (guanjunding@163.com)
* Beijing Information Technology Institute, Beijing, China {(bangkuifan, haibinlan)@163.com}
** Beijing Institute of Technology, Beijing, China (longteng@bit.edu.cn)
*** National University of Defense Technology, Changsha, China

authorized access to one or more parts of the power system. Insiders know sensitive pieces of information, such as access to a secure perimeter, passwords stored in system database, cryptographic keys and other security mechanisms which are targets of compromise. Also, not all security breaches are malicious, some result from accidental misconfigurations, failure to follow procedures and other oversights.

So combined with the abovementioned considering, and from view of system level, services and confidentiality, the security threats to smart grid can be categorized into three groups, i.e., system level threats which attempt to take down the grid, theft of services which attempt to steal electrical services and threats to confidentiality/privacy which attempt to compromise the confidentiality of data on the system.

2.1 System level threats

The intent of system level threats is to take down part or all of smart grid, by denying operators access to the radio field, RF spectrum, individual radios or communications modules within meters. For instance, attackers with malicious intent could attempt to change alarm thresholds or issue unauthorized commands to meters, change programmed instructions in the meter or other control device on the grid. Such actions can result in premature shutdown of power or processes, damage to equipment or even disabling of control equipment. To be specific, the following parts compose system level threats.

(1) Radio disturbed or takeover

This threat is characterized by attempts to take over one or more radios or RF communications modules in meters so they belong to attackers [7]. The most common situation of this threat is firmware replacement. Attackers try to insert modified firmware into a device and/or attempt to spread the fakes to numerous devices, as shown in Fig. 2.

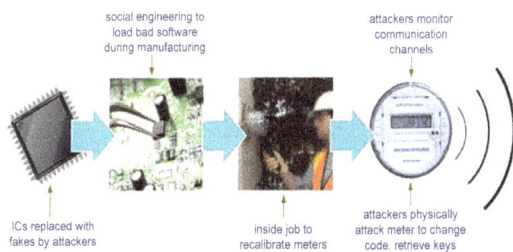

Fig. 2. An example of the threat of radio disturbed or takeover.

(2) Network barge in by attackers

This threat results from stranger radios which attempt to join the RF network or prevent the communications modules from communicating properly. For instance, attackers may intend to use the communications modules to piggyback unauthorized traffic on the network, and try to prevent the communications modules from sending or receiving traffic or use stranger radios to intercept/relay traffic. Also, attackers may attempt to modify the radio or communications modules' credentials to assume a different role.

(3) Services denied

This type of threat could lead all or part of the network becoming unusable. Concretely speaking, it comprises the following cases:

- RF spectrum jamming prevents signal from being received [8].
- Routing black holes, whereby a node is hacked so that it is advertised as the shortest path to everywhere, result in all traffic getting directed to it.
- Protocol-level and RF level jabbering lead a legitimate node co-opted to send so much traffic that the other nodes can't communicate.
- The method of subverting or crashing a device's operating system or applications by overloading memory buffers, i.e., stack smashing, leads data exposed, lost and corrupted [9].
- Kill packets, which are protocol packets, cause radios to crash or to become unreachable via the RF field.
- Attacks on the cryptographic system or protocols lead penetration and degradation of the system [10].
- Environmental attacks, due to physical damage, severe weather or natural disasters, lead service disrupted.

(4) Compromise of back office

Sometimes unauthorized individuals get access to the smart grid management database, so they could bring down the entire grid. Similarly, attackers could change the credentials to which radios respond and potentially bring down the grid, with access to the database which stores privilege data. Also, unauthorized access to billing and other back office systems would open the gate for theft of services and compromise customer privacy.

(5) Compromise of credentials

Credentials prove the identity of an entity on the system and grant that entity access to the communications network, involving access points, communications modules/meters, operations and management systems. Compromise of credentials could make attackers access the communications systems for any purpose, such as denial or theft of services.

2.2 Theft of services

Besides potential attacks on the smart grid itself, utilities face threats which cause theft of services and prevent the operator from collecting revenue. For instance, individual meters or groups of meters can be subverted to misreport, shifting from a higher priced tier to lower priced one. The type of this threat consists of as follows.

(1) Cloning

By cloning, an attacker can replace a meter or radio ID with a duplicate designed to report zero usage.

(2) Migration

An attacker can swap a meter/communications module from a location reporting high usage with a meter/ communications module from a location reporting low usage, to reduce reported usage and associated bills.

(3) Meter/communications module interface intrusion

The communications module inside each meter is connected to meter via a serial port, which can be disconnected so that the meter doesn't report usage. Alternately, an attacker could try to break into the communications module to change usage information.

2.3 Threats to onfidentiality/privacy

Some attacks can result in personally identifiable information disclosed. These threats to confidentiality/ privacy are as follows.

(1) RF interception

Passive eavesdropping on the radio network can enable attackers to capture packets.

(2) Backhaul IP network interception

Fig. 3 shows this type of threat. It could lead information intercepted as it traverses the backhaul network.

(3) Compromise of forwarding point

If a node on the network compromised forwards traffic to an unauthorized individual or entity, confidential information could be exposed [11].

(4) Compromise of meter

This threat consists of any privacy threat related to

physical compromise of a meter.

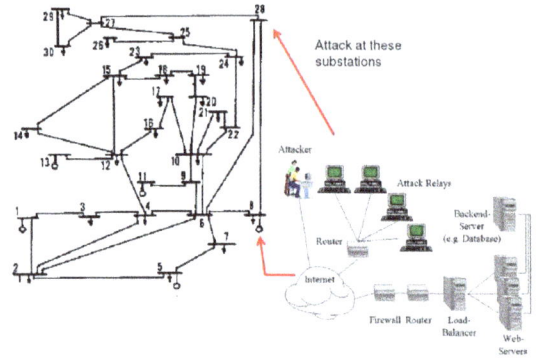

Fig. 3. Schematic diagram of backhaul IP network interception.

3. Security strategy for smart grid

The security strategy for smart grid examines both domain specific and common requirements when developing a mitigation strategy to ensure interoperability of solutions across different parts of the infrastructure. The primary goal of the security strategy should be on prevention. However, it also requires that a response and recovery strategy be developed in the event of a network attack on the electric system.

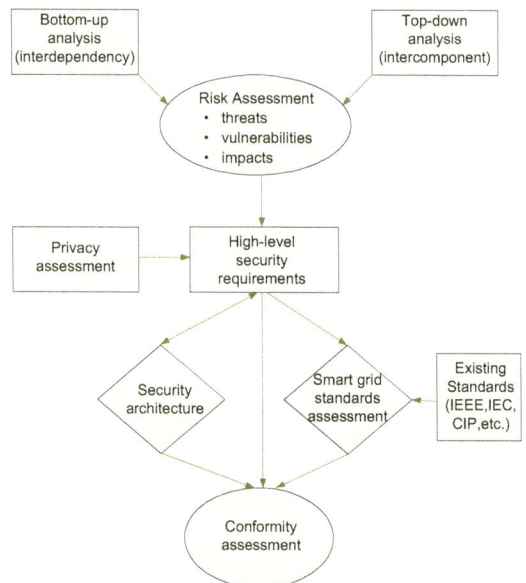

Fig. 4. Flow diagram of the security strategy.

Fulfilment of the security strategy requires the definition and implementation of an overall security risk assessment process for the smart grid. Following the risk assessment, the next step is to select and tailor the security requirements. And then security architecture and smart grid standards assessment are needed. Final step is conformity assessment. Fig. 4 shows the flow diagram of the proposed security strategy. Furthermore, following Fig. 4, each step of the security strategy is discussed detailedly.

(1) Risk assessment

It is the first step of security strategy. It includes threats, vulnerabilities and specifying impacts undertaken from a high-level, overall functional perspective, as shown in Equation (a).

$$\text{Risk} = \text{Threats} \times \text{Vulnerabilities} \times \text{Impacts} \quad (a)$$

Its output is the basis for the selection of security requirements and the identification of gaps in guidance and standards related to the security requirements. The process of risk assessment is based on existing risk assessment approaches developed by both the private and public sectors to produce an assessment of risk to smart grid and its domains and sub-domains, such as homes and businesses. In addition, because smart grid includes system from the information technology, telecommunications and electric sectors, the process of risk assessment is also applied to all three sectors as they interact in smart grid.

Both the bottom-up and the top-down approaches are considered as overall analysis in implementing the risk assessment.

The bottom-up approach focuses on well-understood problems that need to be addressed, such as key management for meters, authenticating and authorizing users to substation intelligent electronic devices [12], and intrusion detection for power equipment. An incident in one infrastructure can potentially cascade to failures in other systems/domains, so interdependencies among smart grid systems/domains are also considered when evaluating the impacts of a security incident.

The top-down approach develops the logical interface diagrams. Some examples of the logic interface categories include business-to-business connections, interfaces to the customer site and interfaces between sensor networks and control systems. A set of attributes, e.g., wireless media, integrity requirements and inter-organizational interactions, are defined and the attributes allocated to the interface categories, as appropriate. This logical interface category/attributes matrix is utilized in assessing the impact

of a security compromise on confidentiality, integrity and availability.

(2) High-level security requirements

The next step is to select and modify the security requirements. On the one hand, system level security requirements are developed for smart grid applications, such as advanced metering, distribution automation, third-party access for customer usage data, home area networks, etc. And on the other hand, a compensating security requirement is implemented in place of a recommended security requirement, to provide a comparable level of protection for the information/control system. More than one compensating requirement may be required to offer the comparable protection for a particular security requirement. For instance, a utility with significant staff limitations may compensate for the recommended separation of duty security requirement by strengthening the audit, accountability and personnel security requirements within the information/control system.

In addition, because the evolving smart grid presents potential privacy risks, a privacy impact assessment should be performed. Several general privacy principles are used to assess the smart grid, findings and recommendations are also developed. The privacy recommendations offer a set of privacy requirements which should be considered when utilities fulfil smart grid information systems.

(3) Security architecture and smart grid standards assessment

Then, the following step is to build a logic reference model to develop the security architecture. The logical reference model consolidates the individual diagrams into a single diagram and expands upon the conceptual model. It identifies logical communication interfaces between actors. The additional functionality of smart grid is also included in this logical reference model.

Additionally, standards that have been identified should be assessed to determine relevancy to smart grid security. In this process, gaps in security requirements are identified and recommendations are made for addressing these gaps. The conflicting standards are also identified with recommendations.

(4) Conformity assessment

The final step is to develop a conformity assessment program for security. The conformity assessment program should assure interoperability, network security and other relevant characteristics. On the one hand, developing a conformity testing framework is to perform an analysis of existing smart grid standards conformity testing programs. And on the other hand, feedback from relevant bodies is another important aspect. Errors, clarifications and enhancements are typically identified to existing standards throughout the normal conformity testing process.

4. Conclusions

The paper induces and analyzes the security threats to smart grid thoroughly, which makes up the corresponding research insufficiency and deeply enhances the understanding in this area. Additionally, it corrects some misunderstanding about threats to smart grid in some degree.

The proposed security strategy can effectively provide the protection required to ensure the effective operation of the smart grid, supports both reliability of the grid and confidentiality of the information transmitted, also ensure confidentiality, integrity and availability of the smart grid infrastructure, including, for instance, control systems, sensors and actuators etc.

References

[1] U.S. Department of Energy, The smart grid: an introduction, 2008.

[2] V. C. Gungor, D. Sahin, T. Kocak etc., Smart grid technologies: communication technologies and standards, pp.529-539, Vol.7, No.4, 2011, IEEE Transactions on Industrial Informatics.

[3] S. Galli, A. Scaglione, Z. Wang, For the grid and through the grid: The role of power line communications in the smart grid, pp.998-1027, Vol.99, No.6, 2011, Proc. of IEEE.

[4] IEC TS 62351-7, Power systems management and associated information exchange-Data and communications security, International Electrotechnical Commission, 2010.

[5] ABB Corporation, Security in the smart grid, 2009.

[6] KEMA and ENA, UK Smart Grid Cyber Security Report, http://sec.jrc.ec.europa.eu/, 2011.

[7] Carpenter, Matthew, Wright etc., Advanced metering infrastructure attack methodology, 2009.

[8] Brodsy, Jacob, McConnell etc., Jamming and Interference Induced Denial-of-Service Attacks, IEEE 802.15.4-Based Wireless Networks, 2009.

[9] Ebinger, Charles, Massy etc., Software and hard targets: enhancing Smart Grid cyber security in the age of information warfare, 2011.

[10] IEEE WGC6, Trial Use Standard for a Cryptographic Protocol for Cyber Security of Substation Serial Links, 2009.

[11] Industrial Defender, Smart Grid Safety vs Confidentiality, 2011.

[12] IEEE, IEEE Standard for Substation Intelligent Electronic Devices (IEDs) Cyber Security Capabilities, 2007.

28

Optimal Sizing of BESS for Customer Demand Management

Kyeong-hee Cho*, Seul-ki Kim* and Eung-sang Kim†

Abstract – The paper proposes an optimal sizing method of a customer's Battery Energy Storage System(BESS) which aims at managing the electricity demand of the customer to minimize electricity cost under the time of use(TOU) pricing. Peak load limit of the customer and charging and discharging schedules of the BESS are optimized on annual basis to minimize annual electricity cost, which consists of peak load related basic cost and actual usage cost(energy cost). The optimal scheduling is used to assess the maximum cost savings for all sets of candidate capacities of BESS. An optimal size of BESS is determined from the cost reduction rate(hereinafter reduction rate) curves via capacity of BESS. Case study uses real data from industry customer and shows how the proposed method can be employed to optimally design the size of BESS for customer demand management.

Keywords: Battery energy storage system, Optimal sizing, Charging and discharging schedule, Peak load limit, Demand management, Electric cost, Reduction rate

1. Introduction

Due to the importance of demand management and the energy storage device dissemination policy, the time of use pricing that had been applied to consumers with contract demand of 1,000[kW] or more has been extended as the consumers with contract demand of 300[kW] or more and planning to be extended to the consumers of about 160,000 houses with 100[kW] or greater in the future [1, 2]. In case of the consumers getting applied with the time of use pricing, about 30~40% rate increase is expected. As the gap of time of use pricing is planning to be extended, it seems that the need of Battery Energy Storage System(BESS) for the demand management of consumers would increase rapidly.

As BESS has high investment costs, it is very important to design proper capacity by analyzing the cost saving effect followed by demand management operation precisely. In relation to this, variable researches have been performed [3~7].

This paper presents a method on optimal sizing of BESS for customer demand management in order to reduce the electricity cost under the time of use pricing. The maximum reduction rate of electricity cost by presenting a model to establish a detailed annual optimum charging and discharging schedule on the rated output and capacity

candidate group to propose a method of selecting the optimum capacity from the relationship of maximum reduction rate on the rated output and capacity gained from the calculation result. Calculating the maximum reduction rate according to the capacity is the point in selecting the optimum capacity while this depends on the comprehendsiveness and accuracy on setting up optimum charging and discharging schedule of BESS. In the case study, the optimum charging and discharging schedule was set up to estimate the maximum reduction rate and showed the process of selecting the optimum capacity based on the actual data of one consumer using the proposed method.

2. Calculation of Electricity Cost

The electric charge system being applied by power companies are composed of base rate and energy rate while the amount of charge is determined by including the power industry infrastructure fund (3.7%) and VAT (10%) to the sum base rate and energy rate. Therefore, the monthly electricity cost is calculated as the formula (1) ~ (3).

$$Energy\ Cost[Won] = [\sum_{i=1}^{24}(P_i \times C_i)] \times n \times 1.137 \tag{1}$$

$$Base\ Cost[Won] = Contract\ demand[kW] \times \\ Demand\ charge[Won\ kW] \times 1.137 \tag{2}$$

$$The\ monthly\ electricity\ cost[Won] = Energy\ cost[Won] + \\ Base\ cost[Won] \tag{3}$$

Here, i : Time[hour]
n: The number of days in the month

† Corresponding Author: Smart Distribution Research Center, Korea Electrotechnology Research Institute, Korea (eskim@keri.re.kr)
* Smart Distribution Research Center, Korea Electrotechnology Research Institute, Korea (kx1004xh@keri.re.kr)

P_i: Energy in i hour[kWh]

C_i: Energy charge in i hour[Won/kWh]

The unit price of base rate and energy rate vary depending on the power supply method (high voltage or lower voltage), by contracted type (for resident, for general purposes, for industries, for education or for agriculture, etc.) while the application of base rate on all contract types except for the power for residential is based on the contract demand. However, the base rate may be estimated using peak load among peak load of December, January, February, July, August, September or this month's portion from the last 12 months including the month of checking the meter for the customers who have installed the digital power meter [8].

3. Optimal Sizing Method of BESS

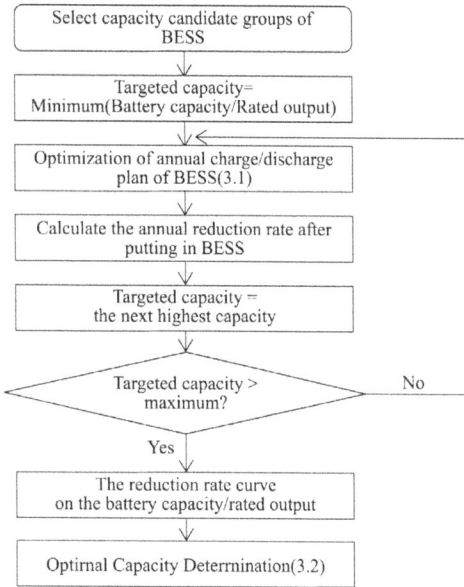

Fig. 1. Optimal Capacity Determination Method of BESS.

Fig. 1 is the overall flow chart on the optimal capacity determination method of consumers.

First, many capacity candidate groups formed by the combination of BESS battery capacity and rated output of charge/discharge device (hereinafter battery capacity/rated output) are selected. Among these, the optimum annual charging and discharging schedule is set up using the combination with minimum battery capacity/rated output as the target capacity while calculating the reduction rate after

putting in the BESS. As the same calculation is performed repeatedly by setting the next highest capacity as the target capacity, the target capacity becomes the maximum capacity combination of the capacity candidate group.

The annual maximum reduction rate according to all capacity candidate groups is calculated. The maximum reduction rate curve on the battery capacity/rated output can be drawn. Then the optimum battery capacity/rated output can be estimated based on the results.

3.1 Optimization of Annual Charging and Discharging Schedule of BESS

The lowest limit value of peak load is estimated on the battery capacity/rated output of target BESS in Figure 1 to optimize the daily charging and discharging schedule of BESS.

The demand by hour of each day according to the optimized daily charging and discharging schedule of the representative date of each month is called the load so that the monthly demand and annual demand can be calculated as formula (4) ~ (5) by multiplying the number of days by each month while the annual electric cost can be calculated as formula (6) using formula (1) ~ (3).

$$The\ monthly\ demand\ [kWh] = (\sum_{i=1}^{24} P_i[kW]) \times n \quad (4)$$

$$The\ annual\ demand\ [kWh] = [\sum_{j=1}^{12}\sum_{i=1}^{24} P_i[kW]) \times n_j] \quad (5)$$

$$The\ annual\ electric\ cost\ [Won] = \sum_{j=1}^{12}\sum_{i=1}^{24} \begin{pmatrix} P_{ij}[kWh]) \times C_{ij}[Won/kWh] \times n_j \\ +Contract\ demand[kW] \\ \times Demand\ charge[Won/kW] \end{pmatrix} (6)$$

Here, j : The number of month

n_j : The number of days in j month

P_i : Energy in i hour in j month [kWh]

C_i : Energy charge in i hour [Won/kWh]

Perform same operation until the lowest limit of peak load becomes identical as the maximum load of the consumer before installing BESS by repeating the optimization of daily charging and discharging schedule as the annual electric cost is calculated and stored to consistently increasing the lowest limit of peak load in Fig. 2. If the annual electric cost according to the increase of lowest limit of peak load is compared, while the base rate gets reduced as the maximum load gets lower, the energy rate increases or converges as the charge/discharge range of battery gets more restriction. Therefore, there is a nature of being in conflict by the maximum load although the reduction of base rate and reduction of energy rate have difference in their degree. The minimum annual electric cost is discriminated from the result of annual electric cost on the various setup of maximum load gained by repeated performance and this is selected as the optimum annual charging and discharging schedule.

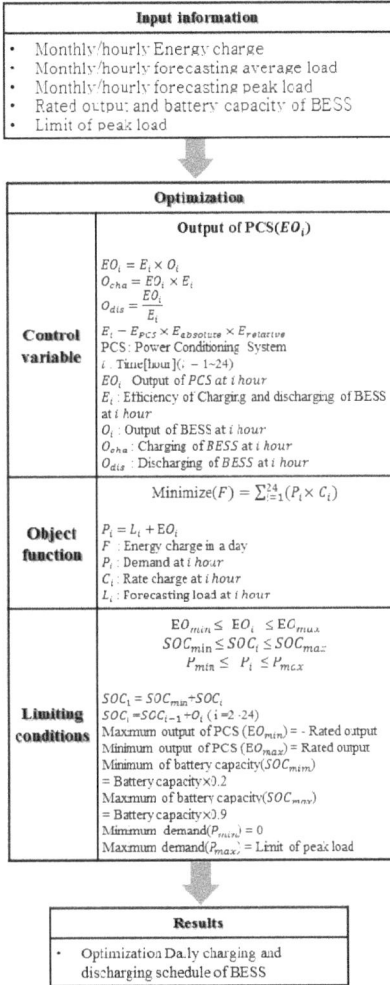

Input information
• Monthly/hourly Energy charge • Monthly/hourly forecasting average load • Monthly/hourly forecasting peak load • Rated output and battery capacity of BESS • Limit of peak load

Optimization	
Control variable	**Output of PCS(EO_i)** $EO_i = E_i \times O_i$ $O_{cha} = EO_i \times E_i$ $O_{dis} = \dfrac{EO_i}{E_i}$ $E_i = E_{PCS} \times E_{absolute} \times E_{relative}$ PCS : Power Conditioning System i : Time[hour]($i = 1 \sim 24$) EO_i : Output of PCS at i hour E_i : Efficiency of Charging and discharging of BESS at i hour O_i : Output of BESS at i hour O_{cha} : Charging of BESS at i hour O_{dis} : Discharging of BESS at i hour
Object function	$\text{Minimize}(F) = \sum_{i=1}^{24} (P_i \times C_i)$ $P_i = L_i + EO_i$ F : Energy charge in a day P_i : Demand at i hour C_i : Rate charge at i hour L_i : Forecasting load at i hour
Limiting conditions	$EO_{min} \le EO_i \le EO_{max}$ $SOC_{min} \le SOC_i \le SOC_{max}$ $P_{min} \le P_i \le P_{max}$ $SOC_1 = SOC_{min} + SOC_i$ $SOC_i = SOC_{i-1} + O_i$ ($i = 2 \sim 24$) Maximum output of PCS (EO_{min}) = - Rated output Minimum output of PCS (EO_{max}) = Rated output Minimum of battery capacity(SOC_{min}) = Battery capacity×0.2 Maximum of battery capacity(SOC_{max}) = Battery capacity×0.9 Minimum demand(P_{min}) = 0 Maximum demand(P_{max}) = Limit of peak load

Results
• Optimization Daily charging and   discharging schedule of BESS

Fig. 2. Optimization Daily Charging and Discharging Schedule of BESS.

3.1.1. Calculating the Lowest Limit of Peak Load

The method of estimating the lowest limit of peak load to reduce the peak load as much as possible while satisfying the restricting conditions using BESS is followed.

First, the value that has subtracted the rated output of storage system from the peak load of consumer is set as the initial value on the lowest limit of peak load and performed the optimizing operation on the maximum load date of each month from January to December. If the solution does not exist, calculate the lowest limit by repeating performance until the optimum solution exists by increasing the lowest limit. Thus, the base rate can be reduced as much as possible by limiting the peak load using the lowest limit of peak load.

3.1.2. Optimization Daily Charging and Discharging Schedule

The process on the optimization of daily charging and discharging schedule of BESS can be shown as Fig. 2.

In order to create the optimization model, four factors including the input information of model, control variable, object function and limiting conditions must be entered. This paper proposes the optimization function to build up the daily charging and discharging schedule of the storage system with the minimized energy rate as object function by setting up the output of PCS as the control variable. A nonlinear optimization method was used to optimize by setting the minimization of annual electric cost by multiplying the monthly number of days to the daily energy rate according to the daily charging and discharging schedule of the storage system as final object function.

Efficiency of charging and discharging of BESS of control variable can be calculated that the commonly used PCS efficiency was entered as 95% and the absolute efficiency was entered as 100% here.

3.2 Optimal Capacity Determination

The process of estimating the maximum capacity of BESS using the maximum reduction rate result of the annual electric cost by each capacity candidate group of BESS is followed.

First of all, calculate the reduction rate curve on the candidates rated output of BESS and its base capacity combination. Here, the base capacity is defined by the battery capacity that has multiplied the parameter to the rated output. In the reduction rate curve on the candidates rated output, the interval with the greatest slant is calculated. The optimum rated output is estimated from this interval, then the rate reduction rate curve by varying the battery capacity is illustrated, the interval with greatest slant is found by calculating the slant of reduction rate on the battery capacity and the greatest value among these intervals is selected as the optimum battery capacity.

4. Case Study

4.1 Analysis on the Load Pattern of Target Consumers

The monthly load pattern of business day, legal holiday and peak day of the target consumer was shown as Fig. 3~5.

The target consumer of this paper is one of industry, the contracted power is 8,500[kW] and the peak load of surveyed load data in 2012 was 38,707.2[kW].

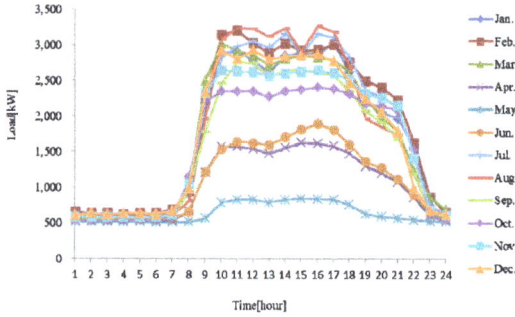

Fig. 3. Monthly Load pattern curve during business days.

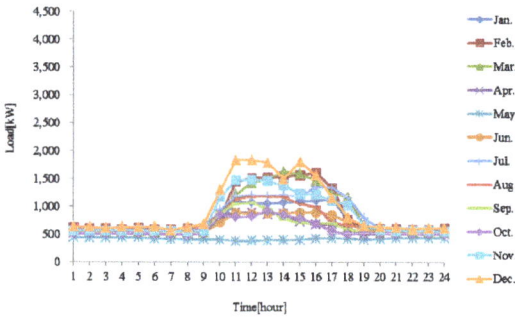

Fig. 4. Monthly Load pattern curve during legal holidays.

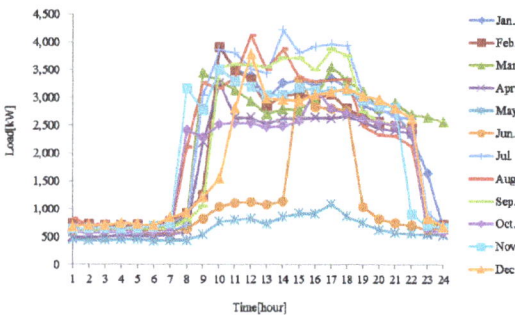

Fig. 5. Monthly Load pattern curve during a peak day.

The monthly one day load pattern from January to December has organized the survey data of target consumers as monthly and day of week to be classified based on business day and legal holiday to be averaged by each hour.

Using the monthly peak load date on the survey load data of the target consumer, the pattern of monthly peak load date was calculated as shown in Fig. 5.

4.2 Optimization of Charging and Discharging Schedule of BESS

The standard electricity cost before putting in BESS of 2012 has been calculated using formula (1) ~ (3) after multiplying respectively load data of Fig. 3~5 and a number of days of each month.

Table 1. The Reduction Rate according to Case #

Case#	Rated output [kW]	Battery capacity [kWh]	Limit of peak load [kW]	Reduction rate [%]
1	400	3,200	3,735.2	6.4
2	600	4,800	3,535.2	9.3
3	800	6,400	3,335.2	12.1
4	1,000	8,000	3,135.2	14.8
5	1,200	9,600	2,935.2	17.2
6	1,400	11,200	2,744.4	19.2
7	1,600	12,800	2,638.9	20.9
8	1,800	14,400	2,533.6	22.4
9	2,000	16,000	2,435.0	23.8
10	2,200	17,600	2,375.4	25.1
11	2,400	19,200	2,374.6	25.9
12	2,600	20,800	2,374.6	25.9
13	2,800	22,400	2,374.6	26.1
14	3,000	24,000	2,374.6	26.2

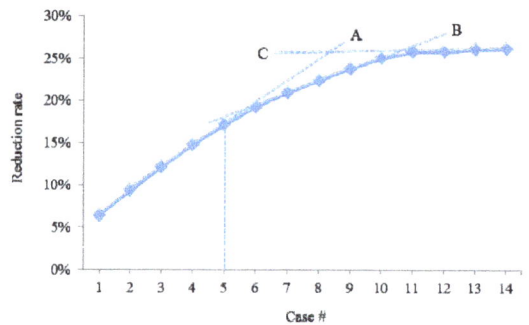

Fig. 6. The Reduction Rate Curve according to case of BESS.

The capacity candidate group of BESS on the target consumer has fixed the rated output range as 400~3,000[kW] and the parameter of the battery capacity as 8. The annual charging and discharging schedule of BESS has been optimized to calculate the electric cost using the algorithm of optimum capacity selection method of BESS mentioned before by increasing the rated output 200[kW] at a time. The result of rate reduction rate according to the capacity candidate group of BESS in Table 1 has been shown as the graph of Fig. 6.

The reduction rate curve in Fig. 6 can be divided into the interval with slant of A, B and C. As slant A is the greatest, the interval which is the rated output 400~1,200[kW] becomes the optimum rated output range with the greatest reduction rate effect. Therefore, as the highest efficiency compared to the capacity is produced when the output of BESS is 1,200[kW] and the capacity is 9,600[kWh], 1,200[kW] could be selected as the optimum rated output.

The annual charging and discharging schedule has been optimized to show the result as table 2 by fixing the optimum rated output as 1,200[kW] according to chapter 3.2 and changing the battery capacity as 1~15 time rate.

The rate reduction rate graph on each capacity of table 2 was shown as Fig. 7.

Table 2. The Reduction Rate according to Case #

Case#	Rated output [kW]	Battery capacity [kWh]	Limit of peak load [kW]	Reduction rate [%]
1	1,200	1,200	3,653.0	4.7
2	1,200	2,400	3,411.0	6.6
3	1,200	3,600	3,263.5	9.6
4	1,200	4,800	3,172.4	11.5
5	1,200	6,000	3,090.9	13.1
6	1,200	7,200	3,010.1	14.7
7	1,200	8,400	2,935.2	16.2
8	1,200	9,600	2,935.2	17.2
9	1,200	10,800	2,935.2	18.1
10	1,200	12,000	2,935.2	18.7
11	1,200	13,200	2,935.2	19.0
12	1,200	14,400	2,935.2	19.0
13	1,200	15,600	2,935.2	18.9
14	1,200	16,800	2,935.2	18.9
15	1,200	18,000	2,935.2	19.0

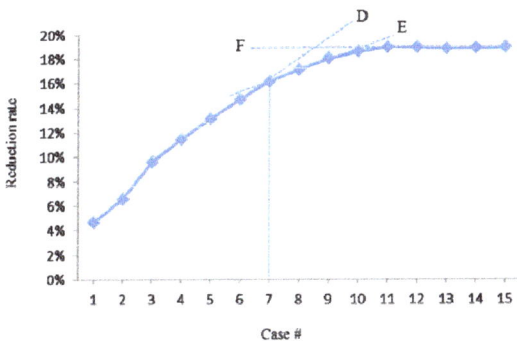

Fig. 7. The Reduction Rate Curve according to Cases.

In Fig. 7, the reduction rate curve can be divided into the interval with slant of D, E and F. As slant D is the greatest, case 7 becomes the optimum battery capacity. Therefore, it is reasonable to select the rated output of 1,200[kW] and battery capacity of 8,400[kWh] as the optimum capacity of BESS.

4.3 Results using Optimal Capacity BESS

Through the case study, the BESS with rated output of 1,200[kW] and battery capacity of 8,400[kWh] could be selected as the optimum capacity of target consumer.

According to charging and discharging schedule of optimal capacity BESS, the fact that load graph (before putting in BESS) gets changed as demand graph (after putting in BESS) and the demand graph gets restricted according to the demand limit could be seen.

The charging and discharging scheduling of BESS is shown as Fig. 9. This is the optimum charging and discharging schedule of BESS compared to the capacity that has optimized both base rate and energy rate while the result on the peak load has been shown in Fig. 9.

The change of electric cost followed according to BESS input is shown as Fig. 10.

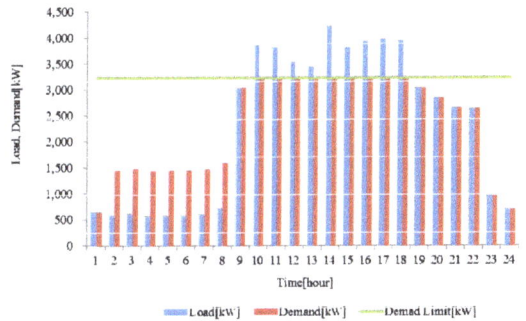

Fig. 8. Load and Power during a Peak Day.

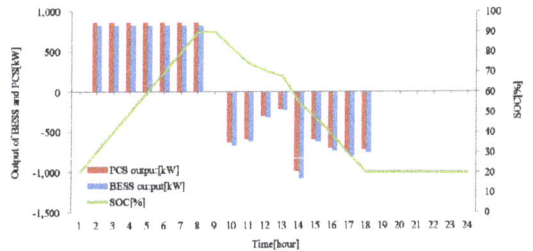

Fig. 9. BESS and PCS Output during a Peak Day.

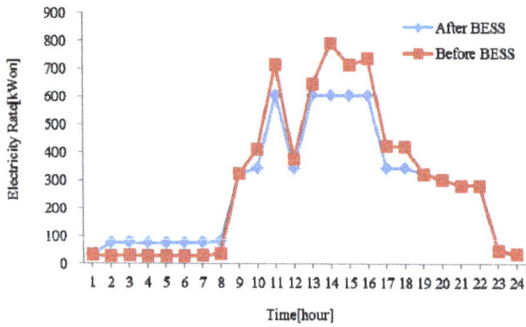

Fig. 10. Change of the Energy Cost by Optimal BESS under the Time of Use Pricing during a Peak Day.

After the BESS input, the load gets reduced by the discharge of peak time slot and the load by the charge of off-peak time slot has increased so that the reduction on the area of graph which is the total rate according to time-of-use pricing could be seen.

5. Conclusion

In this paper we proposed the method of optimal sizing of Battery Energy Storage System (BESS) for consumer demand management. In order to gain the annual maximum reduction rate, the established model on the annual charging and discharging schedule of BESS that has optimized the annual peak load limit value and hourly charging and discharging schedule of battery system have been presented to suggest the method of selecting the optimum capacity based on the maximum reduction rate of electric cost toward the rated output and battery capacity derived from these.

This study was focused on presenting a method of optimizing BESS that can be used on the site by adequately reflecting the characteristics of consumer load, considering the efficiency of battery according to the charge/discharge speed as well as presenting a model to optimize the maximum demand limit value and hourly charging and discharging schedule of the consumer having direct impact on base rate and the energy rate after analyzing the load pattern of monthly business days, legal holidays and peak load days of the target consumer in detail. A design method that can be more generally used is planning to be developed through the development of an algorithm that has adjusted to the economic efficiency model including future investment costs.

References

[1] Ministry of Knowledge Economy, *First Basic Plan of Smart Grid*, Korea, June, 2012

[2] Ministry of Knowledge Economy, *Promotion Plan of the high-capacity power storage*, Korea, 2012. 7. 27.

[3] Jong-Seok Hong, Jae-Chul Kim, Joon-ho Choi, "Determination of Optimal sizes of Battery Energy Storage System Considering Rate-Of-Return for Customers-side", *KIEE Fall Conference for Power Engineering Society*, PP. 146-148, 2001.

[4] S. X. Chen, H. B. Gooi, and M. Q. Wang, "Sizing of Energy Storage for Microgrids", *IEEE Trans. Smart Grid*, Vol. 3, No. 1, pp. 142-151, 2012

[5] A. Oudalov, D. Chartouni, and C. Ohler, "Optimizing a Battery Energy Storage System for Primary Frequency Control", *IEEE Trans. Power Systems*, Vol. 22, No. 3, pp. 1259-1266, 2007.

[6] P. Mercier, R. Cherkaoui, and A. Oudalov, "Optimizing a Battery Storage Energy System for Frequency Control Application in an Isolated Power System", *IEEE Trans. Power Systems*, Vol. 24, No. 3, pp. 1469-1477, 2009

[7] Kyeong-hee Cho, Seul-Ki Kim, Eung-sang Kim, "Optimal Capacity Determination Method of Battery Energy Storage System for Demand Management of Electricity Customer", *The transactions of the Korea Institute of Electrical Engineers*, Vol. 62, No.1, pp.21~28, 2013.

[8] http://cyber.kepco.co.kr

29

EMTP Studies and Simulation Tests on Hongkong Electric's 275kV Transmission System Protection Devices

Kin-Wah Yeung[†] and Man-Kong Yeung, Jonathan*

Abstract – Hongkong Electric has been powering up the city of Hong Kong for more than 120 years. The supply reliability has been consistently rated above 99.999% since 1997 and its electrical protection systems have also achieved excellent performance over the years. This paper examines the impacts of Hongkong Electric's all cable 275kV transmission system on the proper functioning of the protection system and how the adoption of EMTP studies for various fault scenarios and the associated simulation tests on protection devices help to tackle these impacts. Examples of Protection system changes, the recommended modifications to relay design and CT requirement as a result of the simulation test results which confirm the effectiveness of 275kV simulation tests are elaborated.

Keywords: EMTP study, Protection simulation test

1. Introduction

Hongkong Electric has been powering up the city of Hong Kong for more than 120 years. The supply reliability has been consistently rated above 99.999% since 1997 and its electrical protection systems have also achieved excellent performance over the years. This paper examines the impacts of Hongkong Electric's all cable 275kV transmission system on the proper functioning of the protection system and how the adoption of EMTP studies for various fault scenarios and the associated simulation tests on protection devices help to tackle these impacts. Examples of Protection system changes, recommended modifications to relay design and CT requirement as a result of the simulation test results which confirm the effectiveness of 275kV simulation tests are elaborated.

2. Characteristic of Hongkong Electric's 275kV system

Hongkong Electric introduced its 275kV transmission system superimposing on the 132kV system in year 1981. The network mainly comprises of underground & submarine cables and large capacity supergrid transformers directly connected to 275kV cables. The use of all cable network ensures supply continuity in inclement weather, such as typhoon, and is ideal for a densely populated cosmopolitan city like Hong Kong that demand good supply reliability and quality. Furthermore, HEC's 275kV transmission system is solidly earthed through neutrals of 275kV transformers and the system is characterized by:

- High capacitance due to all cable design
- High X/R ratio
- Earth fault current greater than three phase fault current

Consequently, when there is a fault in the 275kV cable system, the fault currents will be rich in DC component, low order harmonics and high frequency transients. Similarly, during the energization of a supergrid transformer through a cable circuit, the inrush currents will also possess these high DC component, low order harmonics and high frequency transients in current. Protection relays subjected to such fault/inrush currents may:

- Be damaged or mal-operate causing unnecessary tripping
- Fail-to-operate or operate slowly causing system stability problem

In the early stage of 275kV operation, relay damage and mal-operation had occurred and subsequent investigation revealed that they were caused by the large high-frequency surge and high DC component in the energization inrush current of the supergrid transformer circuit. After investigation it was recommended that certain modification

†   Corresponding Author: Technical Services Department, Transmission & Distribution Division, The Hongkong Electric Co. Ltd., Hong Kong (kwyeung@hkelectric.com)

*   Technical Services Department, Transmission & Distribution Division, The Hongkong Electric Co. Ltd., Hong Kong (jmkyeung@hkelectric.com)

and improvements be applied on the protective relays in order to achieve protection system security and reliability. Modifications included:

a. Enhancement of low pass filter to reject high frequency components
b. Addition of surge absorbers to the relay CT input and CT secondary circuits
c. Addition of 'Fault Detector Control' to lock relay operation for the initial 10ms to allow DC components to decay
d. Additional distance line protection was introduced to work in parallel in order to enhance the overall line protection scheme

3. EMTP System Studies and Relay Simulation Tests

3.1 First 275kV System Study

To meet the rapid growth of Hong Kong's economy and the increasing electricity demand in 1980's, the company expanded its 275kV network with the installation of four more 275/132kV supergrid transformer circuits. EMTP computer study on the severity of surge/harmonic currents was then performed to assess the suitability of the protection relays/schemes as the system changed. The study concluded that the surge and harmonic currents would lead to mal-operation of the current differential relay and distance relay being adopted at that time. Based on the study result, the relay manufacturer modified the relay designs and input circuits as remedy solution. In order to demonstrate the reliability, stability and operating speed of the modified relays/schemes, simulation tests were carried out by the relay manufacturer using digital computer simulator to covert the computer study results into analogue waveforms for relay secondary injection tests. The modified relays/schemes and other main protection relays were tested with satisfactory results except that one type of distance relay was found with slow response at low generation condition.

Based on the above experience, EMTP study and simulation test on 275kV system were considered necessary to ensure effectiveness of our protection system under different system configurations, generation mix and fault conditions.

EMTP study aims to identify the following major behavior of the network:

- Surge and Harmonic contents

- System overvoltage caused by switching and ferro-resonance
- Review of insulation coordination
- Fault breaking capability of 275kV Circuit Breakers and
- Maximum X/R ratio and DC decay time

Existing 275kV protection relays (and other new relays to be introduced) were then put to simulation tests to confirm that they could cope with the future network behaviours. Modification of the existing protection relays/schemes/settings would be proposed if necessary. Only those relay types which passed the simulation tests with satisfactory results could be employed in the 275kV system.

3.2 275KV System Studies and Simulation Tests

Five 275kV system studies and associated protection system simulation tests were carried out between 1985 and 2013.

Based on the results of simulation tests, modifications in the 275kV protection system were recommended and carried out, such as the modification of relay internal circuits, settings and/or scheme design. Some of the key findings are elaborated below:

3.2.1 Requirements of Protection CT

To ensure no saturation of CT core throughout the fault period, the calculation of knee point voltage of protection class PX CT should follow the following formula:

$$V_k > \frac{I_f}{N}(\frac{X}{R}+1)\times(R_{CT}+R_L+R_R)$$

where V_k :   CT knee point voltage
           I_f :   Maximum system fault current
           N :   CT ratio
           X/R : System X/R at point of fault
           R_{CT} : CT resistance
           R_L :   CT connection lead resistance
           R_R :   Relay burden in ohm

According to the system study results, the X/R ratio at 275kV busbar was calculated to be as high as 80. With such a high X/R ratio, the required CT knee point voltage will be in the order of tens of thousand volts and it is technically impossible to fabricate such CT's.

The protection CT for 275kV line protection is specified in a way that it shall not saturate during the maximum relay operating time which is specified to be 50ms. The formula to calculate knee point voltage is modified as follows:

$$V_k > \frac{I_f}{N}\left[\frac{X}{R}\times(1-e^{-\frac{t}{tp}})+1\right]\times(R_{CT}+R_L+R_R)$$

where   t: Maximum relay operating time
              tp: System time constant at point of fault

Therefore, a CT knee point of 1800V is adequate for the 275kV line protection and subsequently proved by simulation tests.

3.2.2 Relay CT input circuit

During the 275kV simulation test in 1995, one of the numerical line differential relays mal-operated in three external fault cases. The reason for the mal-operation was found due to the saturation of the input transformers at the CT input of the relay. This was an interesting result as protection engineers normally focus on the integrity of the protection CT rather than the input transformer inside the relay. This mal-operation case indicated that DC component might also affect the relay even it was correctly conveyed from the primary by the CT. The design of the analogue input circuit of the relay therefore required improvement.

After analysis and review of the relay scheme with the relay supplier, the internal input transformers of all in-service relays of that kind were subsequently replaced as a remedy action to improve the reliability and stability of the protection.

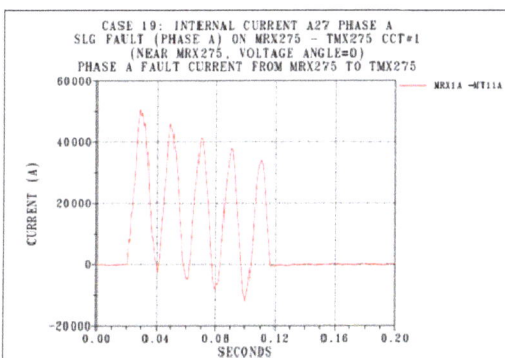

Fig. 1. Typical 275kV fault current with DC offset.

3.2.3 Application design modification

Another finding that affected our relay application was

the slow operation (and even non-operations in some trials) and chattering of contact output for the pilot wire line protection relay during test cases of cable fault on generator circuits. In these test cases, the 275kV station end relay which sensed the main fault current was restrained by high order harmonic components and operated at a lower speed. The relay at the generator end, which sensed the current contributed from the generator, operated at high speed less than 20ms by the high differential quantity as expected.

The above result was undesirable as the overall operation time would exceed the system stability limit. It was important to maintain the designed fault clearance time in order to avoid system stability and minimize damage to equipment. Furthermore, chattering of output contact would damage the relay contacts and create voltage surges onto the DC system that were harmful to other control/protective devices.

The recommended solution by the relay manufacturer included the addition of direct transfer tripping from generator end to remote end to mitigate the problem of slow/non operation at the strong in-feed end. Moreover, fast pick-up/slow drop-off auxiliary relay was introduced as repeat relay for the pilot wire relay to solve the contact chattering problem. The associated protection site modifications were completed by summer 2000.

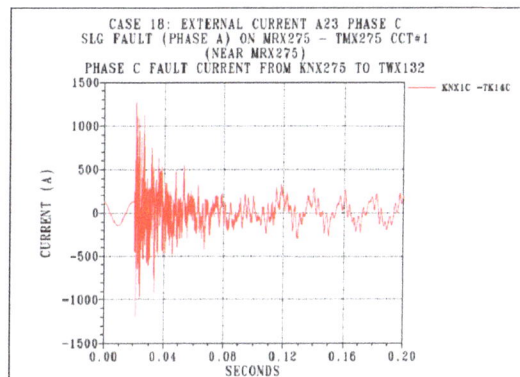

Fig. 2. Typical 275kV fault current with high frequency components.

3.2.4 Filtering circuit and Sampling frequency

During the 275kV simulation test in year 2002, one of the numerical current differential relays was found mal-operated during several external fault test cases.

The simulation test current waveforms were again characterized by large amount of high frequency components. Specifically, 950Hz and 1050Hz components were the dominant frequencies other than the power system fundamental 50Hz in these cases. Such high frequency components posed an aliasing problem to the current differential relay, which has a sampling frequency of1000Hz.

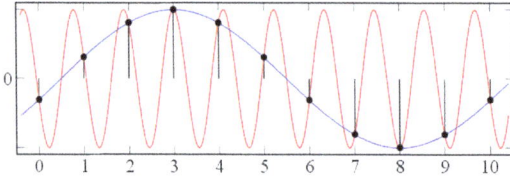

Fig. 3. Two different sinusoids that fit the same set of samples.

Furthermore, it was revealed that its low-pass analogue filtering was not capable of filtering out the high frequency components around 1000Hz. Besides, the local and remote relays were not taking the current samples at the "exact" same time instant. All the factors together made the current differential relays misinterpreted the external fault current as an internal fault with fundamental frequency of 50Hz in some of the test cases. That particular relay was therefore not recommended to be applied in our 275kV system.

3.2.5 Transformer energization inrush

Transient transformer inrush currents can exceed the nominal current and may achieve the rated value of the short-circuit current of the power transformer. The amplitude is decaying very slowly and reaches its steady-state magnetizing current after a few seconds.

Initial magnetizing due to switching in a transformer is considered the most severe case of an inrush. When a transformer is de-energized (switched-off), the magnetizing voltage is taken away, the magnetizing current goes to zero while the flux follows the hysteresis loop of the core. This results in certain remnant flux left in the core. When the transformer is re-energized by an alternating sinusoidal voltage, the flux becomes also sinusoidal but biased by the residual flux. The residual flux may be as high as 80% of the rated flux, and therefore, it may shift the flux-current trajectories far above the knee-point of the characteristic resulting in both large peak values and heavy distortions of the magnetizing current.

Transient inrush currents having a high DC-component and being rich in 1st and 2nd harmonics affect the power quality and can trip protective relays. The 275kV system

study calculated the energization inrush current assuming a high remnant flux level up to 80% and got the maximum current by switching at random voltage angles and using the transformer hysteresis loop data from supplier. The resulted waveforms were employed to test the stability of the transformer biased differential protection relays.

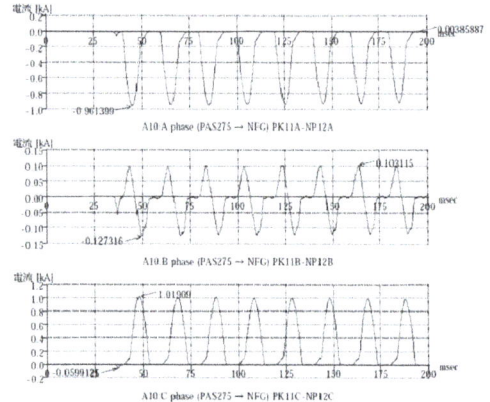

Fig. 4. Transformer energization inrush calculated by EMTP study.

Transformer biased differential relay is featured with 2nd harmonic detector for blocking unwanted operation due to energizing inrush. The 2nd harmonic ratio 2f/1f is normally recommended at 10% ~15%. During the simulation tests, the transformer biased differential protection relays mal-operated in some of the inrush test cases. After subsequent investigation, it was discovered that different relays sensed different levels of 2nd harmonic component of the inrush current due to different filtering characteristics. The 2nd harmonic component seen by some relays could be as low as 6-10%. However, the 2nd harmonic detection setting could not be further lowered to 5% because this would easily block the relay operation during genuine internal fault. Modern numerical transformer protection relays introduced other inrush detection methods like "pattern recognition", "4th harmonic detection", "DC level blocking" and "cross phase restrain/blocking". These methods were verified through the simulation tests but so far a "perfect" solution has not been proved yet.   Advice from the power transformer supplier was sought to avoid heavy energization inrush current due to high remnant flux left in the core. It was recommended that after DC injection test (e.g. dc resistance measurement) on the transformer windings, the transformer should be demagnetized by special AC voltage pattern injection.

3.2.6 High frequency fault current and harmonic analysis

Another case of relay mal-operation observed during the recent simulation test was an "external fault" applied to a pair of analogue type line current differential relay. The test waveform applied was rich in high frequency harmonics above 1kHz as shown in Fig. 5.

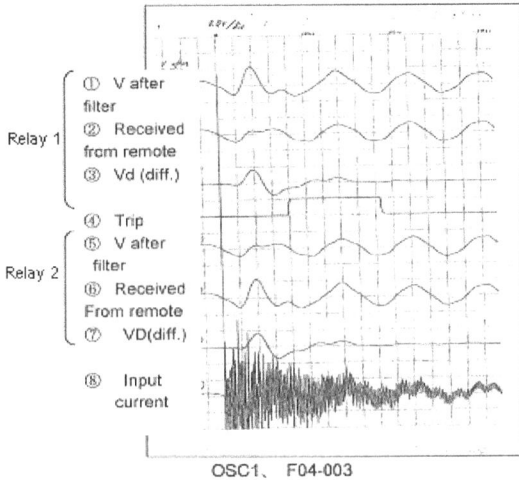

Fig. 5. Simulated fault current and voltage measured at relay input circuits.

The mal operation from the pair of relays was due to the slight mismatch of the CT input analogue filtering characteristics between the relays of both ends. In particular, the output voltage from the input filter circuit of one of the relay was significantly different from that produced from the other ones for the high frequency region above 1kHz. Such mismatch resulted in an abnormal differential quantity that reached the relay operation criteria. However, such mismatch was not reviewed by fundamental (50Hz) injection before the simulation test.

The relay supplier investigated the case and explained that the input circuit of the "aged" relay was calibrated at factory within designed error limit 30 years ago. The reason for the drifted filter characteristic was likely due to component aging. The simulation test was later repeated on selected in-service relays of different manufacturing years and confirmed all relays stable in the tests. The mal-operation case was considered an isolated case, but it gave us important insight on how our routine maintenance test could be improved to reveal such kind of component deterioration that might affect protection reliability.

Based on this result, our maintenance and commissioning strategy was subsequent reviewed and it was recommended to add frequency response test for the 275kV analog line protection relays to keep track of the relay performances at high frequencies instead of only relying on existing practice of adopting simple fundamental injection test.

3.2.7 Improved test method for high impedance differential protection

Testing of high impedance differential relays normally involves generating of CT output voltage waveforms in order to test and verify the through fault stability and in zone fault sensitivity of the scheme. The original test method for high impedance differential relays in previous simulation tests adopted protection class auxiliary current transformers to create a "simulated" waveform that resembles the current waveform from a saturated CT when subjected to large fault current and DC offset. Such test arrangement is shown in figure 6.

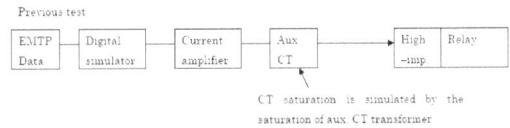

Fig. 6. Previous test set up for High Impedance Relay.

One problem from this test arrangement was that the generated voltage waveform was sometimes too high and might damage the auxiliary CT during tests during repeated tests. Besides, due to knee point limitation of the auxiliary CT (~400V), the test could not fully simulate the in zone fault as the actual 275kV high differential protection CT's were with knee point around 1000V. The test result by using auxiliary CT was therefore questionable in some cases.

A new approach of test arrangement was proposed in recent round simulation test and is shown below:

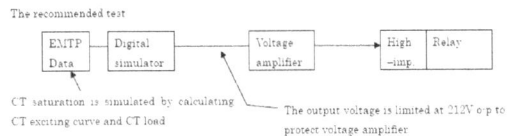

Fig. 7. New recommended test set up for High Impedance Relay in 2012.

The main idea of the new approach is to simulate the "ideal" CT secondary output voltage waveform by software calculation taking into account the CT excitation curve and CT secondary load including the relay, setting resistor & metrosil The CT circuit parameters were entered into the TACS (Transient Analysis of Control System) model in EMTP for calculation, and the test output voltage was injected directly to the relay from a digital controlled voltage test set.

The two waveforms shown in Figs 8 & 9 were examples of test waveforms that show the subtle difference between traditional method and the new method.

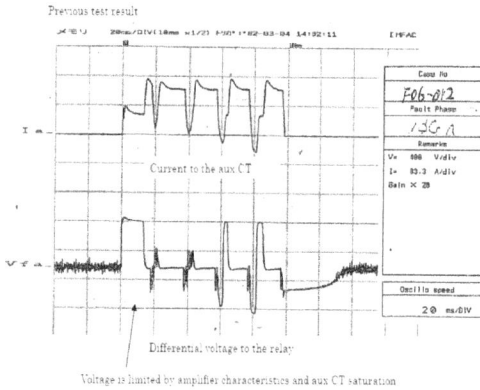

Fig. 8. Test voltage output from traditional set up using auxiliary CT.

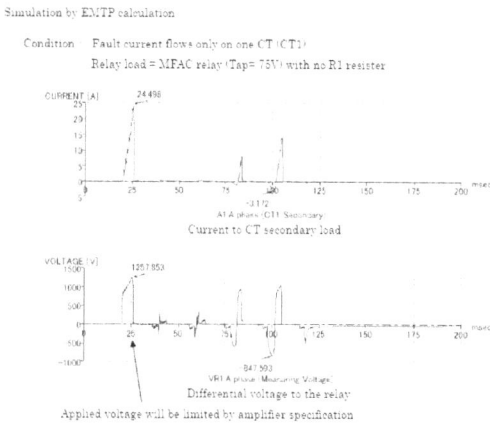

Fig. 9. Test voltage output from new set up.

The new test method only required direct voltage injection to the test relay, which made the test simpler and improved the accuracy because limitation of auxiliary CT was no longer a problem. Some uncertain relay operations

encountered in previous simulation tests were proved satisfactory in the recent round by using the new simulation test method.

4. Conclusion

The authors have shared how the 275kV EMTP system studies and simulation tests revealed weaknesses in the protection system and scenarios that were normally difficult to anticipate and encounter in regular protection design. It provided information that helped protection engineers to continuously improve protection system by relay design & setting review, relay component modification / improvement, relay selection / substitution, protection scheme review and routine maintenance test review. From system management point of view, the 275kV simulation test for protection devices plays an irreplaceable role in ensuring power network reliability. The simulation test is therefore considered a valuable risk identification process that should be adopted when system changes.

Acknowledgements

The author is grateful to the Management of The Hongkong Electric Company, Limited for the kind permission to publish this paper.

References

[1] HK Electric 275kV System Study and Simulation Test Reports (1985-2013)
[2] P.K. Chan, "Management of Protection Systems to Support Superb Electricity Supply Reliability in Hongkong Electric" in *The International Conference of Electrical Engineer 2009*
[3] Andreas Ebner, "Transient Transformer Inrush Currents due to Closing Time and Residual Flux Measurement Deviations if Controlled Switching is used", unpublished.

30

Solid-State Fault Current Limiter based on Magnetic Turn off Principle

Ji-Seong Kang[†] and Young-Hyun Moon*

Abstract – In this paper, a new type of Solid-State Fault Current Limiter(FCL) is proposed. Magnetic Turn off(MTO) FCL is composed of transformer cores and windings. And it utilizes thyristor controller to implement limiting operation. The primary winding is connected to the power grid. The secondary & tertiary windings are wound opposite direction to the primary winding. During the 'Turn On' state, magnetic fluxes produced in the primary winding are canceled out by the secondary winding. The secondary & tertiary windings are controlled by thyristors, and this state makes no impedance throughout the FCL. When fault occurs, MTO FCL will be immediately switched to 'Turn Off' state and this makes the secondary & tertiary windings open. After this operation high impedance will be induced throughout the FCL. MTO FCL is expected to be more practical to implement, because it does not use any superconducting materials or additional complex cooling systems.

Keywords: Fault Current Limiter, Magnetic Turn Off, Fault Current

1. Introduction

The power system capacity in South Korea has rapidly increased during the last decade and it has grown over 80[GW]. As the network capacity increases, the fault current level has also grown. It definitely threats the power system security. Unfortunately, the fault current level exceeded the interrupting capability in several dense network areas in Korea. If we fail to cut off the fault current immediately, the fault will be expanded to a large blackout and expensive facilities can be damaged.

The main purpose of Fault Current Limiter is to reduce the fault currents so that the Circuit Breaker(CB) can easily terminate the fault condition. By using FCLs, we can get the effect of upgrading the breaker capacity, which means that the breaker can terminate relatively large fault current than its own capacity. It helps to save the investment cost for the utility companies. In climate change point of view, the current limiter contributes to reduce SF6 gas demand which is normally needed to make a new Gas Circuit Breaker(GCB).

There are many researches going on various types of FCLs[1][2][3]. We know that the majority of the researches are focusing on developing the Superconducting FCL. But the SFCLs are expensive because superconducting material itself is expensive. And it is needs to be cooled continuously. SFCLs cannot maintain normal state without secure cooling system, which is another disadvantage of SFCLs[4][5][6].

This paper presents a new research activity on developing a new type of FCL based on the MTO principle. MTO FCL can effectively limit the fault current while not using any superconducting materials. Main features and experimental model of MTO FCL will be explained throughout this paper.

2. Magnetic Turn off Principle

We have primary ring and secondary ring sharing same core in the Fig. 1. Assume that both rings are superconducting rings for easier understanding. The primary ring has its original current I1. Magnetic flux Φ1 arises due to I1. Now we insert the secondary ring inside the core, then we can see the current I2 flows opposite direction to I1. And the magnetic flux Φ2 cancels out the primary flux Φ1.

Based on this state, let's consider we take the secondary ring out of the core vertically. We can expect that the I1 will remain the same, while I_2 will be changed to I'. And if I' has certain value, it will create spark at the last moment of taking out. By calculating I' we can know the electrical transients from taking out the secondary ring.

---

†   Corresponding Author: Power Grid Protection Team, Korea Power Exchange, Korea (toasty@kpx.or.kr)

*   Dept. of Electrical and Electronic Engineering, Yonsei University, Korea (moon@yonsei.ac.kr)

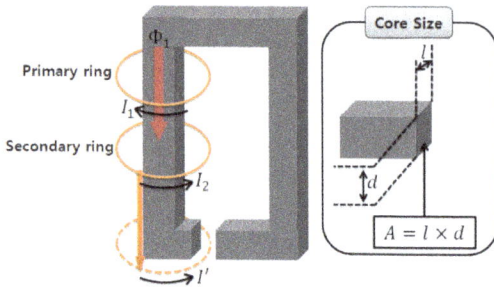

Fig. 1. Magnetic Turn off Principle.

The relation between secondary flux Φ_2 and its inductance L_2 will be described as,

$$L_2=(N_2\Phi_2)/I_2=(N_2/I_2)(N_2I_2/R_2)=(N_2)^2/R_2 \qquad (1)$$

And according to Fleming's right-hand rule, we can contain the induced voltage V_e,

$$V_e=L_2(dI_2/dt)=Blv=Bl(dx/dt) \qquad (2)$$

Now let Δt the time duration for secondary ring to pass through the air gap, and Δx the distance of movement. Then we can Fig. out the value of ΔI, the current change when secondary ring has taken out.

$$L_2(dI_2/dt)=Bl(dx/dt) \qquad (3\text{-}1)$$
$$L_2\Delta I_2=Bl\Delta x \qquad (3\text{-}2)$$
$$\Delta I=(Bld)/L_2=(R_2Bld)/(N_2)^2=I_2/N_2 \qquad (3\text{-}3)$$

Because the size of ΔI is the same amount of I_1 and it flows opposite direction, we can know that the current I_2 will be changed to zero at the taking out moment. And there arises no electrical spark as well.

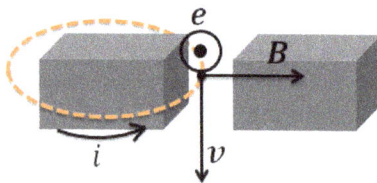

Fig. 2. Direction of the eliminating current.

By doing this vertically we can get rid of the transients. If we take out the secondary ring horizontally instead of vertically, spark will arise.

Current Limiters need to be changed from 'no impedance state' to 'high impedance state' when fault occurs. And it is necessary to make sure no electrical reactions happen. The only way to do so is changing mechanical flux route same direction to the magnetic flux flow. We can define this effect as 'Magnetic Turn off', meaning it turns off the electrical transient reactions by using magnetic switching.

3. Fundamentals of MTO FCL

Basically the structure of MTO FCL is similar to the transformers. MTO FCLs normally use the same transformer windings. Old models usually utilized mechanical movement of the transformer core to controls the magnetic fluxes.

During the normal state, the two windings are arranged as cancelling position. Fluxes made from each winding are all cancelled because the direction is opposite. The impedance of the circuit is nearly zero.

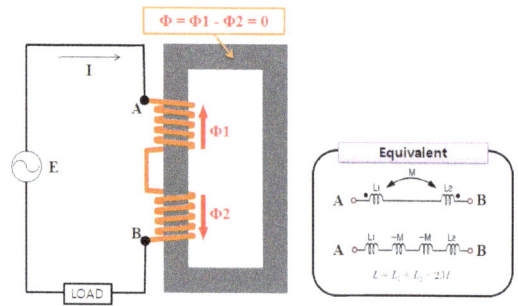

Fig. 3. Equivalent circuit during Normal State.

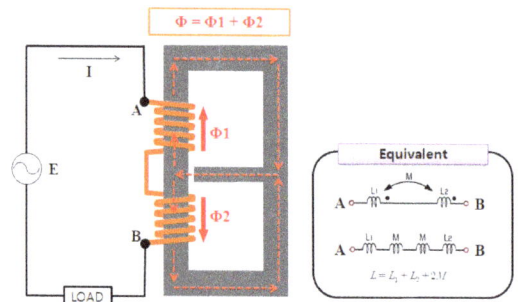

Fig. 4. Equivalent circuit during Fault State.

When fault occur, transformer core changes its mechanical structure. And then the core shows additional flux route, which is not have been before. According to this change, the arrangement of the two windings is also changed from 'differential' to 'additional'. So the flux flows through the core, and the FCL makes high impedance as well.

But in practical, this type of FCL is difficult to implement for several reasons. The most challenging problem is supplying mechanical force. Mechanical force is required to connect and disconnect the flux route automatically. So it is necessary to secure reliable actuation all the time. To compensate this problem, the rotating type model was developed.

3.1 Rotating Type Model

Rotating type is more advanced MTO FCL model to settle disadvantages written before. This type of FCLs can alternate the normal state and fault state continuously. The main characteristic of this model is the rapid and stable state transition by its rotation.

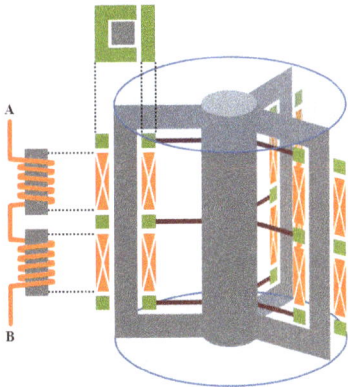

Fig. 5. Equivalent circuit during Normal State.

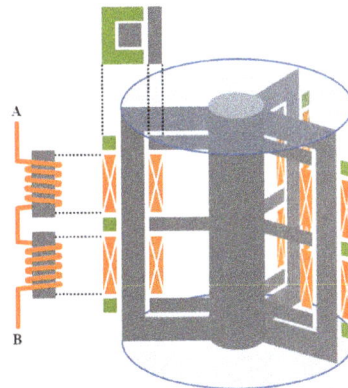

Fig. 6. Equivalent circuit during Fault State.

The main frame is similar to the three phase transformer. Each core leg is divided by several stages. Some stages rotate whereas other stages are fixed. The rotate stage is composed of two blades. One is core blade and the other is copper blade. Two blades always pass through the same stage with 60 degree displacement.

This structure enables a rapid and stable action. But in terms of mechanical force, there should be a large actuating torque to make heavy metal blades rotate. It may need an oil pressure system to operate, and it may not operate within sufficient time to coordinate with the CBs.

3.2 Experimental MTO FCL Model

An experimental model was made to test and implement current limiting effect of the MTO FCL. The structure details are shown in Fig. 7.

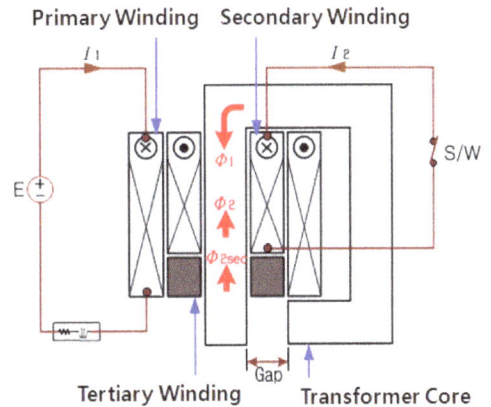

Fig. 7. MTO FCL Experimental Model(1).

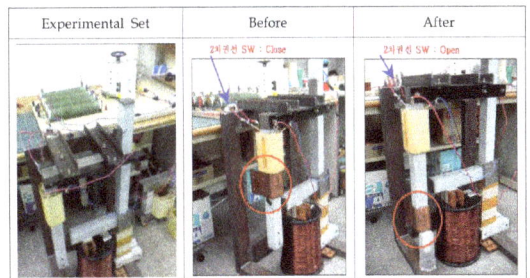

Fig. 8. MTO FCL Experimental Model(2).

In the normal state, the secondary winding switch is closed so the overall magnetic flux flows becomes nearly zero[7]. When fault occur, the secondary winding switch operates and make secondary winding open. But the flux still remains zero until the tertiary winding is taken out. As soon as the secondary winding switch is opened, mechanical force is applied on the tertiary winding so that the winding can be removed out of the core. After this process, no cancelling fluxes are induced against the primary winding flux. This makes high impedance on this experimental model.

The parameters of the experiment are shown in Table 1, and the experimental results are shown in Table 2.

Table 1. Experimental Model Parameters

Parameter	Value	Parameter	Value
AC Source(V)	150	Load resistance RD(Ω)	4.0
Primary winding R1(Ω)	0.40	Secondary winding R2(Ω)	0.35
Primary winding inductance L1(mH)	59.88	Secondary winding inductance L2(mH)	68.32

Table 2. Experimental Model Results

Applied Voltage	50.5V	100.1V	147.6V
Voltage (Before)	7.0V	14.2V	21.7V
Voltage (After)	47.3V	93.3V	140.5V
Current (Before)	9.77A	19.35A	27.70A
Current (After)	2.27A	4.51A	6.93A

We could verify the current limiting effect from this experiment, but due to air gap it is difficult to minimize leakage flux. Good current limiters should have little impedance during the normal state. But we need at least one air gap in this model, because air gaps act as an important role delivering route to implement the mechanical actuation.

4. Solid-State MTO FCL

In chapter 2 we saw how to take off one of the winding without spark. This spark can be told as 'arc' if we talk about the transmission voltage level. So it is important to prevent electrical transients when the FCL transforms to the current limiting stage. Using the mechanical actuation is simple and easy to apply the MTO concept on the conventional FCLs. But it showed us the force supplying and flux leakage problems.

After a research we found out a new method to change magnetic flux route. Instead of moving cores or windings mechanically, we can also achieve the same effect by using electrical switching.

4.1 Structure and Operation

If we can implement 'taking out the winding' electrically instead of implementing it mechanically, we can operate the FCL without any mechanical power. Moreover the FCL would be more reliable than before. By simply applying sequential thyristor switching, the extraction of secondary and tertiary winding can be implemented.

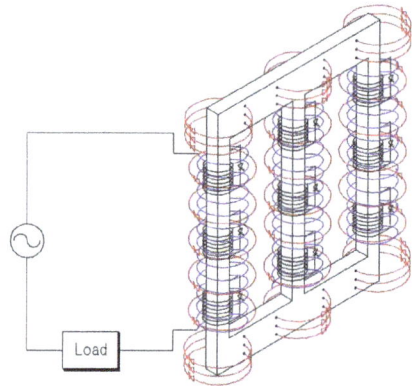

Fig. 9. Structure of Solid-State MTO FCL.

Solid-State MTO FCL operates every operation by thyristor switching. Therefore the operation time is very short (scope of micro seconds), and it enables to limit the first cycle of fault current wave right after the detection of protective relays. Considering the existing HVDC technologies, this model can also be implemented on the transmission voltage level, higher than 154kV network.

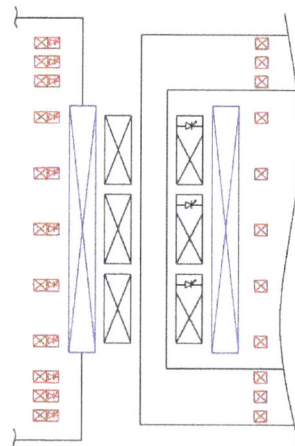

Fig. 10. Switching elements of Solid-State MTO FCL.

During the normal state all thyristors are closed. The primary winding is connected in series to the network. It is likely to produce flux, but the flux can't flow because it is blocked by other windings. The secondary and tertiary windings are short circuited, and they produce opposite direction flux flow. Therefore the sum of three winding fluxes cancels out one another.

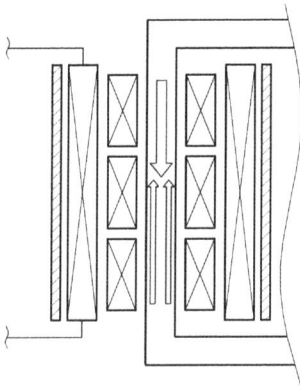

Fig. 11. Solid-State MTO FCL during Normal State.

When fault occur, the very first step is thyristor switching. All thyristors of the secondary winding are opened at the same time. The important thing in this step is, opening time of all thyristors can't be ideally same. There should be one thyristor which is opened last sequence. And this last-opening-thyristor must cut the whole current induced by canceling flux, if there is not the tertiary winding.

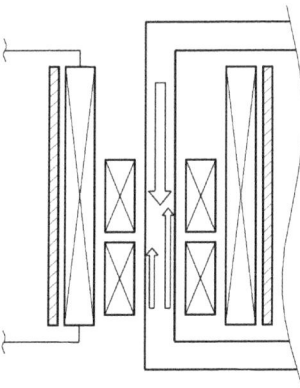

Fig. 12. Solid-State MTO FCL during Fault State.

Tertiary winding is also controlled by thyristor and it does not operate yet. It waits until all secondary winding thyristors become open. Due to this tertiary winding, the last-opening-thyristor of the secondary winding can be safely opened. These sequences are illustrated in Fig. 12 through 14.

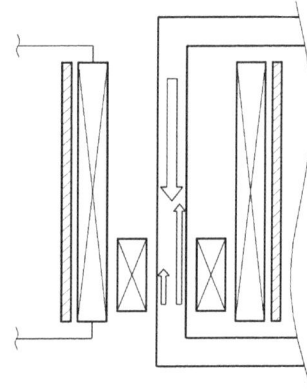

Fig. 13. Solid-State MTO FCL during Fault State.

As secondary winding is opened, the only source of the canceling flux against primary winding is the tertiary winding. The overall reactance still remains zero in this step.

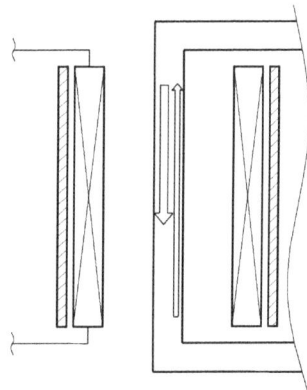

Fig. 14 Solid-State MTO FCL during Fault State

After this stage, tertiary winding thyristors start to open. But unlike secondary winding they operate sequentially. The thyristor located at the top of the FCL cuts first, and the thyristor located at the bottom of the FCL cuts last. As we learned in Chapter 2, the open sequence direction must be same as the flux direction of the primary winding. That's the only way to protect the last-opening-thyristor.

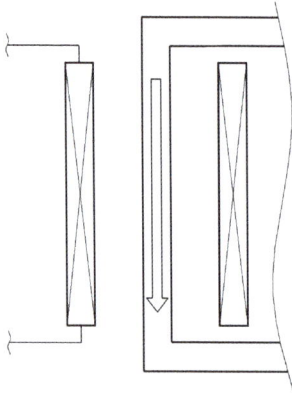

Fig. 15 Solid-State MTO FCL during Fault State.

Fig. 15 illustrates after tertiary winding is removed from the circuit. Now there is no canceling flux and the MTO FCL makes high impedance.

And this model adopted 'entangled structure'. Entangled structure can be achieved by separating the tertiary winding into several branches and combining them with core across one another. By doing this, we can reduce the leakage flux more effectively.

4.2 Ongoing Experiment

In this paper we introduce a new Solid-State 220[V]/30[A] experimental MTO FCL model. Switching device is made out of normal switch controllers in order to observe current change step by step.

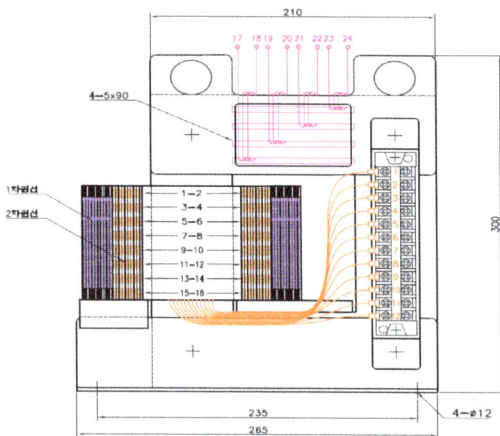

Fig. 16. Solid-State MTO FCL Experimental Model Overview.

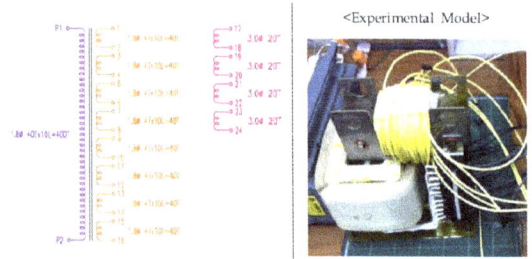

Fig. 17. Solid-State MTO FCL Experimental Model Windings.

Table 3. Overall Performance of the Solid-State MTO FCL Experimental Model

Voltage[V]	219.2	231.4	242.7
Before Current[A]	28.4	29.9	31.4
After Current[A]	0.034	0.036	0.038

Experimental result showed that we can reduce the fault current effectively by using MTO FCL. But the voltage is far lower than the transmission voltage, so we need to make more practical model in further research. More precise manufacturing technique is required, because higher voltage model will introduce much more flux leakage.

5. Conclusion

In response increasing electricity demand, South Korea has expanded its power capacity to meet the demand. But as 40% of the demand is concentrated around capital city, existing CBs became unable to afford the fault current. Upgrading CBs required a lot of cost, and the alternatives like network separation made the power system weaker than before. Series reactors or BTB HVDCs are discussed they are also not ultimate solutions in this situation[8].

In this paper, a new type of FCL is proposed. They control magnetic flux to increase impedance, while not using any superconducting materials. MTO FCL is basically based on the transformer structure and every step of current limiting process is controlled by thyristors.

By using MTO FCL, we don't need to separate substation buses or transmission lines any more. So the power system reliability will be greatly enhanced. And we can also suppress demand of upgrading CBs.

But the experiment and test model should be advanced further. Switching device should be implemented by thyristor controller to get more practical experiment results. A new model applying higher voltage is also needed in further research.

References

[1] Department Of Energy Homepage. Available: http://energy.gov/oe/technology-development/advanced-cables-and-conductors/fault-current-limiters

[2] Alexander Abramovitz, "Survey of Solid-State Fault Current Limiters", IEEE Transactions on Power Electronics, Vol. 27, No. 6, pp. 2770~2782, June 2012

[3] Jin-Seok Kim, "Study on Application Method of Superconducting Fault Current Limiter for Protection of Protective Devices in a Power Distribution System", IEEE Transactions on Applied Superconductivity, Vol. 22, No. 3, pp.1~4, June 2012

[4] A. R. Fereidouni, "Enhancement of Power System Transient Stability and Power Quality Using a Novel Solid-state Fault Current Limiter", Journal of Electrical Engineering & Technology Vol. 6, No. 4, pp. 474~483, 2011

[5] Gurjeet Singh Malhi, "Studies Of Fault Current Limiters for Power Systems Protection", Aug 2007.

[6] Ghanbari, "Development of an Efficient Solid-State Fault Current Limiter for Microgrid", IEEE Transactions on Power Delivery, Vol. 27, No. 4, pp. 1829~1834, Oct. 2012

[7] Geun-Yang Ji, "The research for a structure of current limiter using a phasic similitude of magnetic circuit", Journal of the Korean Institute of Electrical Engineers, Vol.58, No.11, pp.2128~2135, 2009

[8] Ryu Heeyoung, "The Influence of Current Limit Reactor installed in 345kV power systems on Transient Recovery Voltage", Korean Institute of Electrical Engineers Summer Conference, pp.222~223, Jul. 2011

31

Measurement of the Permeability in a Ferrite Core by Superimposing Bias Current

Kousuke Kikuchi*, Tomohiko Kanie** and Takashi Takeo†

Abstract – In this study, we investigate measurement of the magnetic permeability in a ferrite core at RF frequencies when bias current is superimposed on an RF signal with a view to adaptively controlling performance of RF transformers using ferrite cores. A measurement arrangement used comprises a short microstrip line (MSL) circuit including a coaxial conductor (CC) structure consisting of an electrically grounded metal pipe, a center conductor and a sample between them. A bias tee network is incorporated into this MSL-CC circuit in order to superimpose direct current on an RF signal. Using this arrangement, a dependence of permeability on an amplitude of superimposed bias current was measured at frequencies of 10 MHz to 500 MHz. Reliability of the measurement results is discussed based on several experimental data, implying that the method has an accuracy less than 10 % at most of the above frequencies.

Keywords: Bias current, Bias tee network, Ferrite, Permeability measurement

1. Introduction

Ferrite is used in many RF devices, such as a transformer, and EMC components [1]-[2]. For example, when designing an RF transformer, which is a typical RF device, engineers often select appropriate permeability or physical dimensions of a ferrite core to satisfy a specification designated for the device. Since this approach is costly and time-consuming with respect to manufacturing, the authors have proposed in a previous work [3] an adaptive method of controlling the permeability of a ferrite core of RF transformers by superimposing bias current on an RF signal. This proposed method provides us a means for a given ferrite core to vary its permeability and thus enables us to alleviate the above mentioned issue. In addition, this technique is beneficial when applied to compensation of temperature dependent nature of ferrite, for the permeability of a magnetic material is generally dependent on temperature [4].

However, how the permeability of a ferrite core changes when bias current is superimposed should be known or measured so as to actually implement the above mentioned adaptive control of RF device performance. In this study, we propose a relatively simple and practical measurement

method employing a combined microstrip line-coaxial conductor (MSL-CC) circuit equipped with a bias tee network and report experimental results obtained with this method. Furthermore, the accuracy of the measurement is also discussed through several experiments and an electromagnetic simulation technique.

2. Measurement Procedure

2.1 Measurement Circuit

The arrangement used here to measure permeability when a bias current is superimposed on an RF signal is illustrated in Fig. 1.

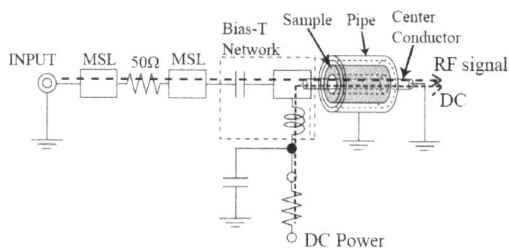

Fig. 1. Measurement circuit consisting of the combined microstrip line-coaxial conductor equipped with the bias tee network.

In this circuit, a coaxial conductor (CC) structure

---

† Corresponding Author: Graduate School of Engineering, Mie University, Japan (takeo@phen.mie-u.ac.jp)
* Graduate School of Engineering, Mie University, Japan (412m603@m.mie-u.ac.jp)
** Kanie Professional Engineer Office, Japan (kanie@aioros.ocn.ne.jp)

composed of a center conductor and an electrically grounded coaxial metal pipe is located at the end of microstrip lines each having a characteristic impedance of 50 ohms [5]. A measurement sample is housed between the center conductor and the pipe. In addition, a resistor (50 ohms) is connected to the MSLs for achieving better measurement accuracy. Furthermore, in front of the sample, a bias tee network is incorporated to superimpose bias current at the sample.

2.2 Equivalent Network

We utilize an equivalent network as shown in Fig. 2 for the above mentioned circuit so as to determine the permeability of a sample from the circuit impedance Z_M'. Here, the coaxial conductor structure is expressed as the L type network enclosed by the dotted rectangle, where circuit elements R_m, L_m, and C_m are given as a function of parameters (sizes and permeability) of the sample and the coaxial conductor as described in the following.

Namely, when electrical current flows through the center conductor and the metal pipe, inductance and resistance given by

$$L_m = \frac{l_s}{2\pi}\left\{\mu_0 \ln\left(\frac{R_2 R_4}{R_1 R_3}\right) + \mu' \ln(R_3 / R_2)\right\} \quad (1)$$

and

$$R_m = \frac{\omega \mu'' l_s}{2\pi} \ln(R_3 / R_2) \quad (2)$$

are caused. Here, $\mu = \mu' - j\mu''$ is the complex permeability of the sample, μ_0 is the permeability in vacuum, ω is the angular frequency, l_s is the length of the sample, R_1 is the outer radius of the center conductor, R_2 and R_3 are the inner and outer radii of the sample, respectively, and R_4 is the inner radius of the metal pipe. The equivalent network for the sample is expressed as a series connection of the resistance R_m and inductance L_m [6].

Furthermore, the two conductors (the center conductor and the electrically grounded pipe) generate a capacitive component C_m, which is given by

$$C_m = 2\pi l_s \frac{1}{\frac{1}{\varepsilon_0}\ln\left(\frac{R_2 R_4}{R_1 R_3}\right) + \frac{1}{\varepsilon}\ln(R_3/R_2)} \quad (3)$$

where ε and ε_0 are the permittivities of the sample and vacuum, respectively.

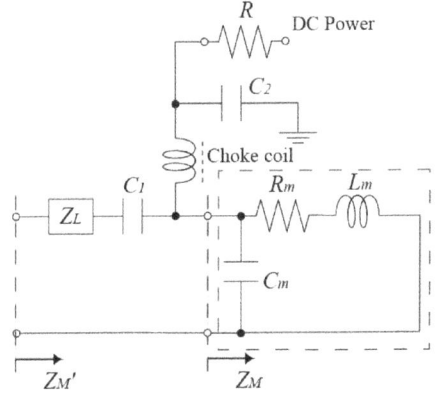

Fig. 2. Equivalent network for the measurement circuit.

If the impedance Z_M of the coaxial conductor structure expressed by the L type network in Fig. 2 is determined, one can calculate the permeability of the sample by a circuit analysis using eqs. (1) through (3) as will be described in the next subsection.

2.3 Permeability Determination Procedure

As mentioned in the previous subsection, the impedance Z_M of the measurement portion indicated by the dotted rectangle in Fig.2 should be known in order to obtain the permeability of the sample. One of methods to accomplish this is to measure the impedance Z_M' of the whole measurement circuit and then remove contributions of the circuit components other than Z_M, i.e. the load Z_L, the bias tee network, etc.

Following this procedure, the first step in the permeability determination procedure is subtraction of the impedance of the load Z_L and the capacitor C_1 from the whole circuit impedance Z_M' obtained through measurement. Then, the effect of the components such as the choke coil connected in parallel with the sample was canceled by an elementary circuit analysis. In this way, we obtain the impedance Z_M of the coaxial structure. This impedance value of the CC structure including the sample should be equal to the impedance enclosed by the dotted rectangle in Fig. 2. Namely,

$$Z_M = \frac{A\omega\mu''}{\alpha_1 \mu'^2 + \beta_1 \mu' + \alpha_1 \mu''^2 + \gamma_1} + j\frac{\alpha_2 \mu'^2 + \beta_2 \mu' + \alpha_2 \mu''^2 + \gamma_2}{\alpha_1 \mu'^2 + \beta_1 \mu' + \alpha_1 \mu''^2 + \gamma_1} \quad (4)$$

where

$$A = \frac{l_S}{2\pi} \ln\left(\frac{R_3}{R_2}\right) \tag{5}$$

$$\alpha_1 = \omega^4 C_m{}^2 A^2 \tag{6}$$

$$\beta_1 = 2\omega^2 C_m A(\omega^2 C_m B - 1) \tag{7}$$

$$B = \frac{\mu_0 l_S}{2\pi} \ln\left(\frac{R_2 R_4}{R_1 R_3}\right) \tag{8}$$

$$\gamma_1 = (\omega^2 C_m B - 1)^2 \tag{9}$$

$$\alpha_2 = -\omega^3 C_m A^2 \tag{10}$$

$$\beta_2 = \omega A(1 - 2\omega^2 C_m B) \tag{11}$$

$$\gamma_2 = \omega B(1 - \omega^2 C_m B) \tag{12}$$

Eq. (4) is a quadratic complex equation for unknown variables μ' and μ''. By solving this equation, we obtain the complex permeability.

3. Electromagnetic Simulation

We conducted an electromagnetic simulation based on a finite element method to check the validity of the procedure for determining the permeability of a ferrite sample described in the previous section. A measurement circuit model (referred to as C1 hereinafter) used in the simulation is shown in Fig. 3 and its major parameters are listed in Table 1. In the simulation, we employed results of impedance measurement for the choke coil in the bias tee network, which will be described in the next section. Namely, we used the measurement results for the impedance of the choke coil in the bias tee network. The electromagnetic simulation done in this way gives us the impedance Z_M' of the measurement circuit.

Table 1. Parameters of the measurement circuit and the sample

Initial core permeability	Core length	Core O.D.	Core I.D.	Metal pipe I.D.	Center conductor O.D.
2000	8.0mm	3.5mm	0.7mm	5.0 mm	0.65 mm

The value of Z_M, which was obtained by removing the effect of the load Z_L and the bias tee network was substituted into eq. (4) and the equation was solved with a Newton-Raphson method. The results are illustrated in Fig. 4. As can be seen, the values of the sample permeability obtained from the simulation are in agreement with the true values which are those used as the sample permeability in the simulation, indicating the validity of the present measurement procedure, though the error for μ' is slightly

large at higher frequencies.

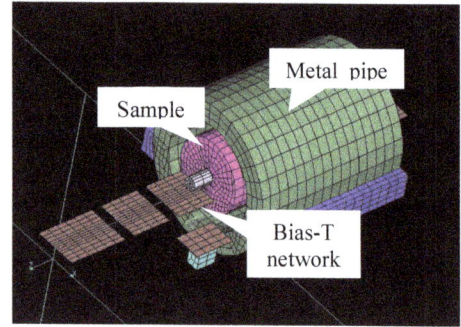

Fig. 3. Simulation model.

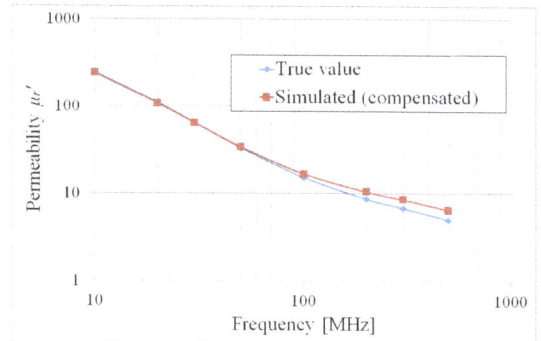

(a)

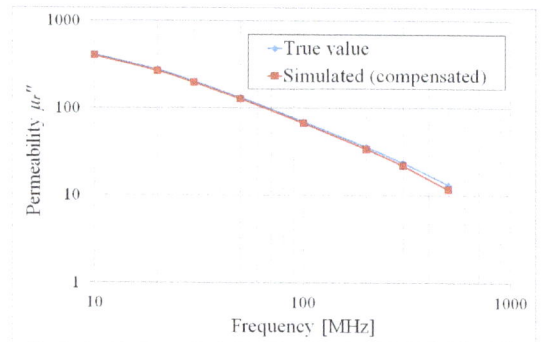

(b)

Fig. 4. Results of an electromagnetic simulation for the permeability measurement.

4. Results of Permeability Measurement

Before measuring the bias current dependence of the permeability in a ferrite core, we have experimentally checked the cancellation procedure of the bias tee network. For that purpose, in addition to the circuit C1, we prepared another measurement circuit (Circuit C1′) which is the same as C1 except that it does not have the bias tee network.

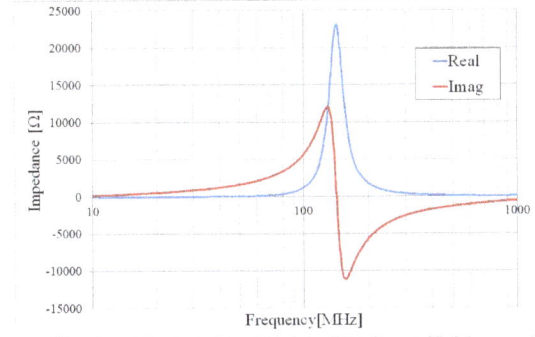

Fig. 5. Measurement results for the impedance of the choking coil.

(a)

(b)

Fig. 6. Results of the effect of the choking coil cancellation.

Then, the impedances of the two circuits were measured. In this measurement, the bias current I_b was set to be zero

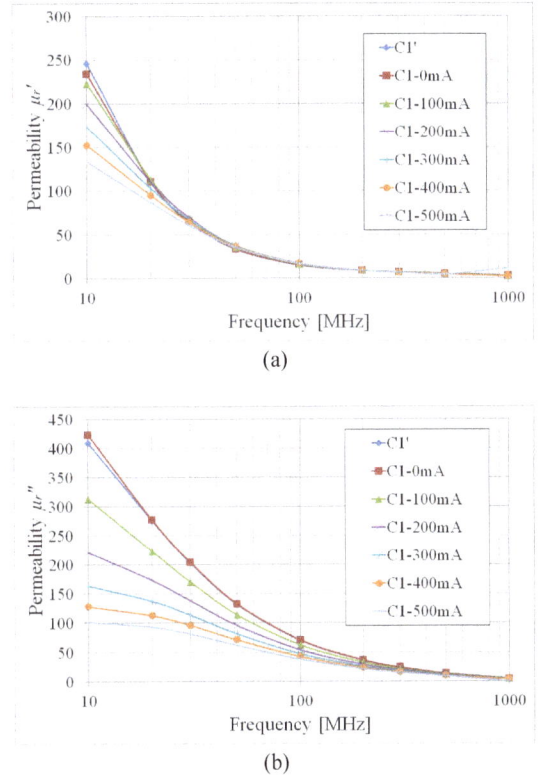

for the circuit C1. In addition, with regard to the circuit C1, we subtracted the influence of the bias tee network, especially of the choke coil, from the measured circuit impedance $Z_M′$. As illustrated in Fig. 5, the measured impedance of the choke coil has a resonant nature in the frequency range of interest. Nevertheless, the difference between the impedance of C1 when $I_b = 0$ and the effect of the bias tee network was canceled and that of C1′ was less than a few percents as shown in Fig. 6, validating our procedure for measuring the dependence of permeability on the bias current.

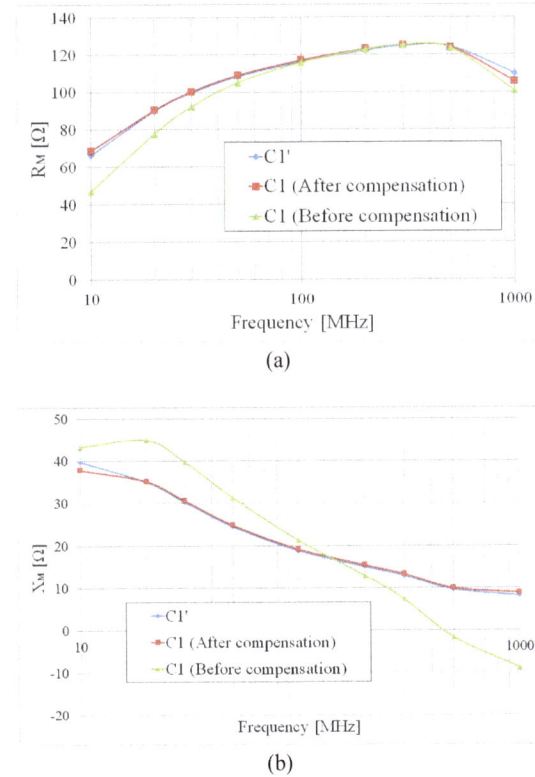

(a)

(b)

Fig. 7. Measurement results for the permeability by superimposing bias current: (a) real and (b) imaginary components.

After confirming the validity of our procedure experimentally in this way, we proceeded to permeability measurements by applying bias current. Ferrite samples having the parameters listed in Table 1 were purchased from Tomita Electric Co., Ltd. [7]. In the measurements, bias currents of up to 500 mA were superimposed on an RF signal from a network analyzer. Measurement results are illustrated in Fig. 7, indicating that both the real and imaginary components of the complex permeability vary significantly according to the bias current change especially at low frequencies. More specifically, when the bias current was varied from 0 mA to 500 mA, the real and imaginary components of permeability at a frequency of 10 MHz decrease from about 240 and 400 to about 135 and 100, respectively.

5. Accuracy Evaluation

In order to check the measurement results in the previous section or examine the accuracy of the measurement, we prepared another circuit C2 which has the same sample as the circuit C1 but the inner diameter of the metal pipe is different ($R_2 = 3.5$ mm) from C1.

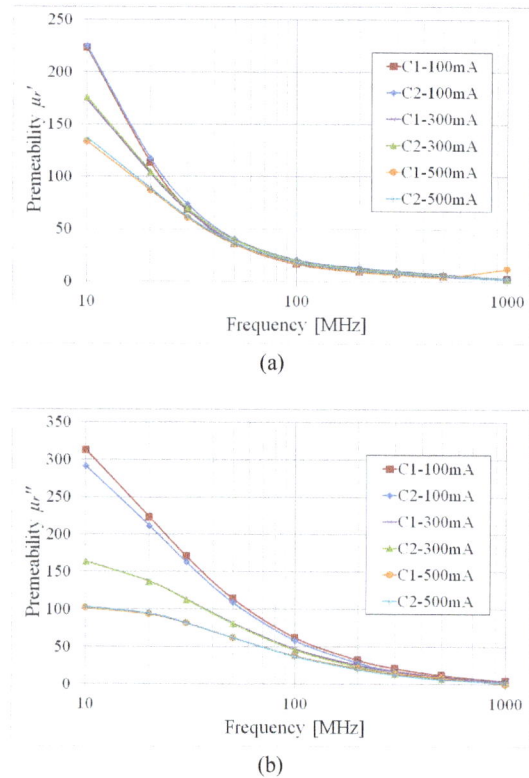

(a)

(b)

Fig. 8. Comparison of the measurement results for circuit C1 and C2: (a) real and (b) imaginary components.

Since the samples of the two circuits (C1 and C2) are the same, the permeability values measured for C1 and C2 should coincide. Permeability values obtained from C1 and C2 are given in Fig. 8 for bias currents of 100 mA, 300 mA

and 500 mA. Although the difference between the results for C1 and C2 is slightly large for the imaginary component in the case of $I_b = 100$ mA, it is relatively small for the other conditions, and implying that the accuracy in this measurement is comparable to that with conventional methods [8], [9].

As another way to examine the accuracy in the permeability measurement or check the validity of the results for the permeability measurements obtained in the previous section, we measured transmission characteristics of the ferrite sample using a circuit shown in Fig. 9 and compared them to those predicted by simulation. Microstrip lines and a coaxial conductor in this circuit were designed in the same way as described in Fig. 1 and Fig. 2, while another bias tee network is added after the sample so as to allow the bias current to flow to ground and the RF signal to flow to the output port, respectively.

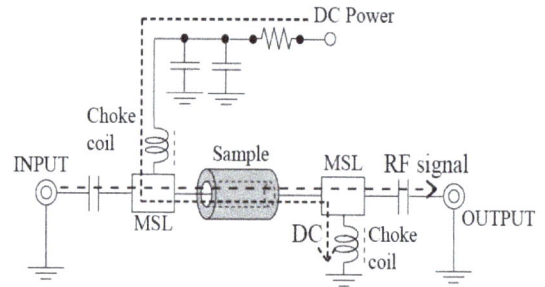

Fig. 9. Arrangement for measuring the transmission characteristics of the ferrite sample.

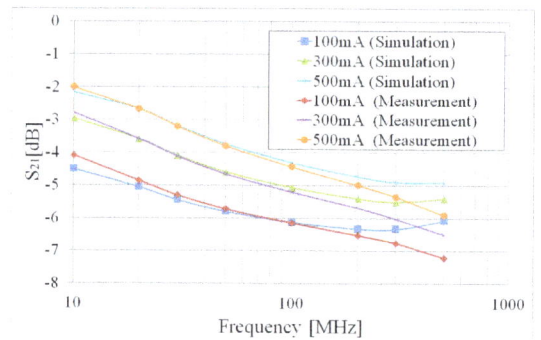

Fig. 10. Comparison between the results of the experiments and the electromagnetic simulation for the transmission characteristics.

Measured transmission S_{21} is plotted in Fig. 10 for the bias currents of 0 mA to 500 mA. As can be seen, it monotonically increases with the bias current. In Fig. 10, transmission characteristics predicted by an electromagnetic simulation using the permeability values of the sample obtained by the measurement in the previous section are also plotted. Although the difference between the measurement and the simulated results are slightly larger in the frequency range of 300 MHz to 500 MHz, it remains within 0.5 dB in terms of S_{21} at most frequencies. To examine how much this difference or error in S_{21} can be in terms of permeability, S_{21} was calculated by changing permeability values by 10 % to 30% from the measured one in the case of $I_b = 100$ mA. Results are shown in Fig. 11, which indicates that the difference between the measurement and the simulation corresponds to about 10 % in terms of permeability.

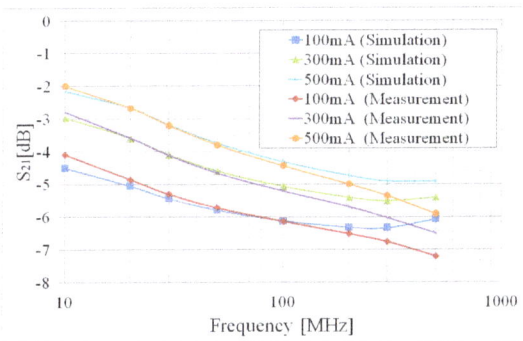

Fig. 11. Variation of the transmission characteristics S21 by changing the permeability value by 10% to 30%.

6. Conclusion

Permeability measurement with bias current superimposed has been investigated at RF frequencies (10 MHz to 500 MHz) using the combined microstrip line-coaxial conductor arrangement equipped with a bias tee network. The experimental results show that the permeability of a ferrite core can vary significantly, e.g. from 250 to 140 for μ'_r and from 400 to 100 for μ''_r at a frequency of 10 MHz by applying a bias current of 500 mA. In order to check the validity of the results, two kinds of investigations were further made. Firstly, permeability measurement was conducted for two different circuits having the same sample on it and the results from each were compared. In the other investigation, transmission characteristics of a transformer having a ferrite sample as a core material were compared between experiments and an electromagnetic simulation. Both investigations for checking the validity of the permeability measurement

indicate that the error remains within 10 % at most of the frequencies. The measurement technique presented in this study enables us to discuss performance of an RF transformer when bias current is superimposed on it. This is the subject of future study.

Acknowledgements

The authors would like to thank R. Goudy of Nissan Technical Center N.A. for his helpful comments.

References

[1] R. M. Bozorth, Ferromagnetism, IEEE Press, New York, 1993.MIT, 1981, p.75-94.

[2] A. Goldman, Handbook of Modern Ferromagnetic Materials, Kluwer Academic Publishers, Norwell, 1999.

[3] T. Aoyama, Y. Shibata, T. Kanie, and T. Takeo, "Active Control of RF Splitter Isolation by Superimposing Bias Current," IEICE Trans. Electron., vol. E95-C, no.7, pp.1297-1299, July 2012.

[4] T. Aoyama, Y. Shibata, T. Kanie, Y. Noro, and T. Takeo, "Adaptive Compensation Method for the Temperature Dependence of RF Transformer Isolation," Journal of ICEE, vol.2, no.4, pp.358-366, Oct., 2012.

[5] D. M. Pozar, Microwave Engineering, John Wiley & Sons, New Jersey, 2005.

[6] T. Aoyama, M. Katsuda, T. Kanie, and T. Takeo, "Alternative Method for Determining Permeability of a Ferrite Core by Using a Combined Microstrip Line-Coaxial Conductor," IEICE, Vol.E95-C, No.11 (2012)

[7] Tomita Electric Co., Ltd., Homepage, http://www.tomita-electric.com/en/

[8] J. B. Jarvis, M. D. Janezic, J. H. Grosvenor, Jr., and R. G. Geyer, "Transmission/Reflection and Sort-Circuit Line Methods for Measuring Permittivity and Permeability," Natl. Inst. Stand. Technol., Tech. Note 1355-R, December 1993.

[9] J. B. Jarvis, M. D. Janezic, B. F. Riddle, R. T. Johnk, P. Kabos, C. L. Holloway, R. G. Geyer, and C. A. Grosvenor, "Measuring the Permittivity and Permeability of Lossy Materials: Solids, Liquids, Metals, Building Materials, and Negative-Index Materials," Natl. Inst. Stand. Technol., Tech. Note 1356, February 2005.

32

Study on Voltage Regulation in Distribution System Using Electric Vehicles - Control Method Considering Dynamic Behavior

Yuki Mitsukuri[†], Ryoichi Hara, Hiroyuki Kita**, Keiichi Watanabe***,
Kenjiro Mori***, Yasuhiro Kataoka***, Eiji Kogure*** and Yuji Mishima***

Abstract – A surge of needs for the low carbon society promotes a spread of electric vehicle (EV). EVs could be charged at night simultaneously, as a result, severe voltage drop may happen. The authors have proposed the method which can compensate the voltage drop caused by EV charging by means of adjusting charging schedules and controlling reactive power. And, we have confirmed the effectiveness of the method by estimating steady state in order to figure out the limitation of the control capability. In this paper, from a practical viewpoint, we propose the method to consider dynamic behavior. In this method, the EV can not only finish charging effectively but also control minimal reactive power to keep admissible voltage with monitoring system voltage.

Keywords: Electric vehicle, Voltage drop, Distribution system, Reactive power control, Dispersed voltage control

1. Introduction

Electric vehicles (EV) have been attracting great attentions and expectations since they can run at good fuel efficiency without exhaust gases. Therefore, it is expected that the EVs will become more and more popular. Generally, the usage of EV strongly depends on our lifestyle; most EVs run in the daytime and are parked and charged at night. The current time-of-use tariff program offered by the utility company will enlarges the EV charging in nighttime. If most of EVs are charged from the distribution systems through electricity plug at home, emergence of EVs would enlarge the peak load at night in residential areas [2]. Therefore, it is necessary to consider the impact that the change of the load has an influence on voltage profile in distribution system. It is also significant to investigate some new methods to manage voltage if needed.

If EV consists of self-commutated inverter, it can control both active and reactive powers in principle. We have focused that point and proposed basic algorithm to keep voltage for avoiding voltage violation by simultaneous EV charging at nighttime. In this algorithm, EV adjusts

†   Corresponding Author : Hakodate National College of Technology, Japan (yuki.mitsukuri@gmail.com)
*   Hakodate National College of Technology, Japan
**   Graduate School of Information Science and Technology, Hokkaido University, Japan
***   The Tokyo Electric Power Company, Japan

charging schedule and control reactive power [3]. In [3], we have considered steady state and evaluated how the number of EV can be introduced in a distribution system without voltage violation by adopting the algorithm. In this paper, from a practical viewpoint, we propose the method to consider dynamic behavior. In this method, the EV can not only fulfill charging effectively but also control minimal reactive power to keep admissible voltage with monitoring system voltage.

2. Voltage control in distribution system

2.1 Voltage Control in two node system

In a simple two-node distribution system shown in Fig.1, the voltage drop along the feeder $V_s - V_r$ can be approximated as (1).

$$V_s - V_r \cong \frac{PR + QX}{V_r} \qquad (1)$$

where, Vs, Vr are the voltages at sending and receiving nodes, P, Q are the active power and the reactive power consumed at receiving node, R, X are the resistance and reactance of feeder. This equation implies that the voltage at receiving node can be controlled by P and Q.

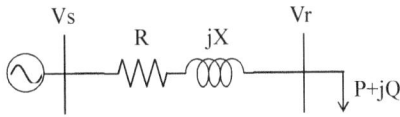

Fig. 1. Two nodes distribution system model.

2.2 Modeling of EV

In this paper, it is supposed that EV has self-commutated inverter. In principle, the self- commutated inverter can control active and reactive powers simultaneously. Therefore, an EV is modeled as a load which can control active and reactive power consumption freely within its capacity. Here, the discharge from the EV's battery is not considered in this paper.

2.3 Minimal reactive power for avoiding voltage violation

Fig. 2 expresses the relationship between active power (P) that EV battery charge and reactive power (Q) that EV battery consume. EV can control P and Q independently within inverter capacity. If typical EVs used at the present day is connected to a distribution system for charging through their inverter, EVs charge their battery by maximum inverter capacity as the point a in Fig. 2 expresses because EVs finish charging in the shortest possible time. The charging at the point a is called "normal charging" hereafter. If EVs are penetrated massively, voltage drop in the system by increase of charging power would be large. On the other hand, the point o in the Fig. 2 means operation point before start to charge. The voltage on line oc whose slope is $-(R/X)$ can be approximately regarded as same as the voltage at point o, depending on (1) which is approximation formula for voltage calculation. In other words, voltage drop by P equals voltage rise by Q on the line oc. As seen above, same voltage value can be expressed by the line which runs parallel to the line oc in the figure. If the operation point of the inverter is left below side, the operation point is included in "excessive compensation range" where the voltage will rise before charging. And, there is the line bd on which the voltage is the same as minimum voltage level if voltage violation occurs by normal charging as "voltage violation range" expresses in the figure.

Therefore, if the operation point is left below side of the line bd, voltage violation can be avoided. Among them, it is th e point b to avoid voltage violation with decrease suppression of charging power minimally.

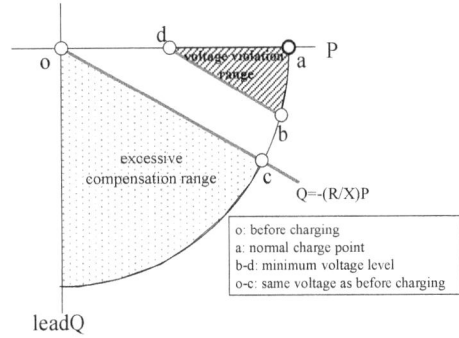

Fig. 2. The image of minimal reactive power for avoiding voltage violation.

3. Voltage control method by EV

3.1 Proposed method

Therefore, if the operation point is left below side of the line bd, voltage violation can be avoided. Among them, it is the point b to avoid voltage violation with decrease suppression of charging power minimally.

In order to realize to keep voltage with minimal suppression of charging power as described in previous section, the operation point should be point b in the Fig. 2. However, in fact, the accurate point b cannot be known from EV before start to charge because the point depend on loads around it and system parameter. Therefore, in order to search point b, control quantity is gradually adjusted with monitoring voltage of connected node in the proposed method.

In the Fig. 3, the line bd means minimum voltage level, same as the Fig. 2. Line ef is supposed to be reference voltage considering the margin for the minimum level, and node voltage has to be kept within the line ef level. At the beginning of charging, charging power is gradually increased from point b as Fig. 3. When the operation point comes at point f, that is, node voltage decreased at reference value, the operation point is moved across the line ef (Fig. 3(a)). In detail, the operation point is moved according to the rule shown in Fig. 3(b) in mode1 (the mode at start of charging). That is, if the node voltage is upper than the reference value $V_{low} + m$ (m is the margin for considering error), P is increased by minute quantity. And if the voltage is lower than the reference value, lead Q is increased by minute quantity. By continuing this every minute time unit, the point can reach at point e.

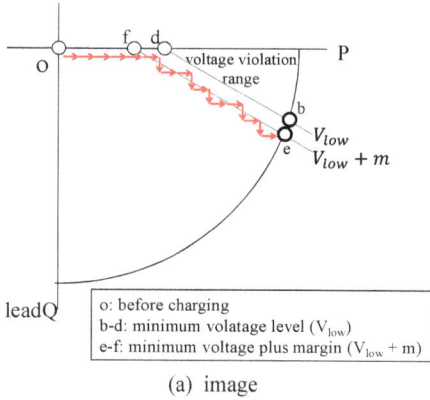

(a) image

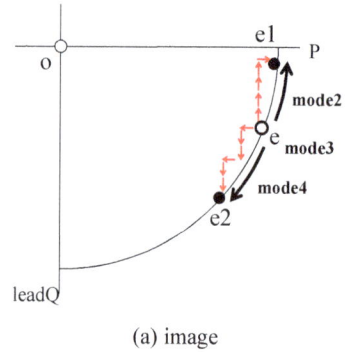

(b) P and Q adjustment corresponding to voltage

Fig. 3. PQ control image (mode1).

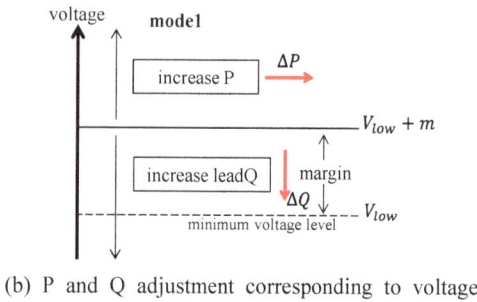

(a) image

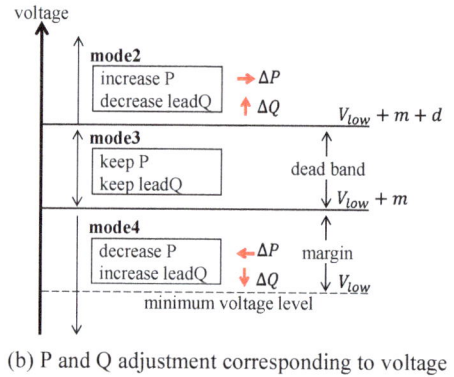

(b) P and Q adjustment corresponding to voltage

Fig. 4. PQ control image (mode2-4).

Next, let's consider the situation that when operation point is point e, the point e itself moves by other load change. At that time, operation point should move to new point e after load change because line ef itself moves. In Fig. 4(a), point e is the voltage before load change. Point e1 is new reference voltage point in the situaion that voltage rises with load decrease. Point e2 is new reference voltage point in the situation that voltage drops with load increase. If the voltage rises, the point is moved to right upper direction across the circle in order to finish charging by short time (it is called "mode2" hereafter). If the voltage drops, the point is moved to left lower direction across the circle in order to keep voltage within admissible range (it is called "mode4" hereafter). If the voltage change is small, the operation point is not moved (it is called "mode3" hereafter). In detail, the operation point is moved according to the rule that Fig. 4(b) shows. That is, if the node voltage is upper than dead band, P is increased and lead Q is decreased (mode2). If the voltage is within dead band, P and Q is not changed (mode3). If the voltage is lower than dead band, P is decreased and lead Q is increased (mode4). As described above, these sets of control method is called "PQ control" in this paper.

3.2 Possibility of hunting

As described in previous section, in this control method, the voltage is monitored and P or Q is adjusted every minute time unit, and next mode is decided. When P or Q is adjusted according to mode2 or mode4 and voltage after adjustment become within dead band, the operation mode is mode3 and adjustment is finished. However, in the case of narrow dead band, hunting occurs, that is, the voltage overshoots the dead band and mode2 and mode4 is repeated alternately, and the control can be unstable (Fig. 5). To avoid the instability, it is necessary to figure out amount of node voltage change by adjustment of P or Q and to set wider dead band than it. Though the amount of voltage change can be approximately calculated by (1) in theory, R and X in the system cannot be known from EV. Therefore, some countermeasures are needed. For example, it is necessary to set wide enough dead band with considering typical amount of R and X.

In addition to that, if there are some other EVs at same node or near node, control by an EV affects voltage at other nodes. As a result, though the dead band is enough for single EV control, hunting can occur as shown in Fig. 6.

Therefore, wide enough dead band is needed even if there are a number of EVs in the neighborhood.

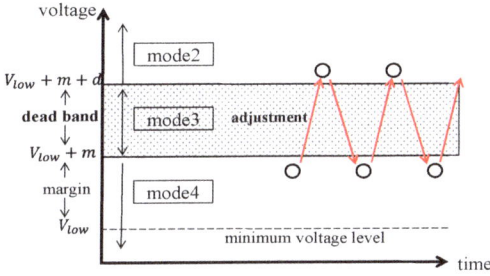

Fig. 5. Hunting image by single EV.

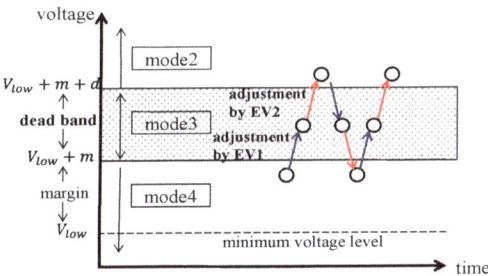

Fig. 6. Hunting image by two EVs.

4. Simulation

4.1 Simulation condition

The validity of the proposed methods was investigated through computational simulations using the distribution system model shown in Fig. 7. The applied model consists of 6 primary nodes (#S and #A~#E) as shown in Fig. 7(a). Node #S corresponds to the sending point in distribution substation. The sending voltage is set to be 6.6kV from 8AM to 10PM (heavy load period) and 6.35kV from 10PM to 8AM (light load period). Each primary node except #S has 30 pole transformers as shown in Fig. 7(b). Total length of secondary feeder line is all 80m. Each pole transformer feeds 12 residential loads as shown in Fig. 7(c). Each node which is the same position in terms of impedance is named from #a to #f. In addition, each load is named 1 or 2. The most severe voltage violation would occur at the secondary node #f in the primary node #E (written "#E_f" hereafter), which is the farthest load from the distribution substation (#S) in this system model.

Assumed load curve of each residential load is shown in Fig. 8 (inductive reactive power is defined as positive and active and reactive power of EV is not included). Fig. 9 illustrates the time-sequential voltage profile at secondary feeder nodes without EVs. In this case, the voltage at #A_a (closest to #S) is highest while the voltage at #E_f (farthest node) is lowest. In other words, the voltages at other nodes

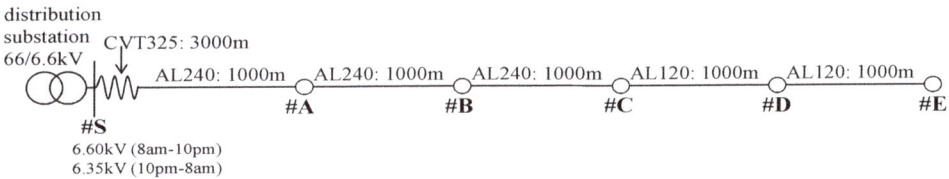

(a) Primary feeder model

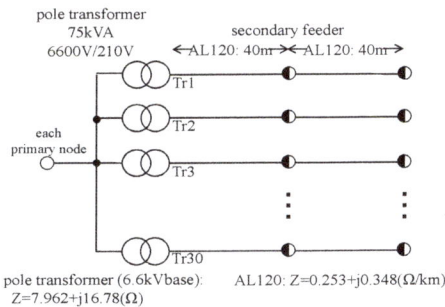

(b) High voltage node model

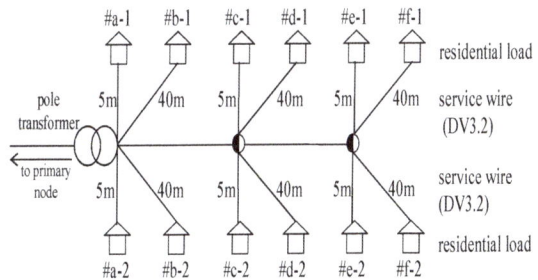

(c) Secondary feeder model

Fig. 7. Simulation system.

remain between them. Hereafter, discussions are mainly focused on the voltage at #E, the severest voltage drop.

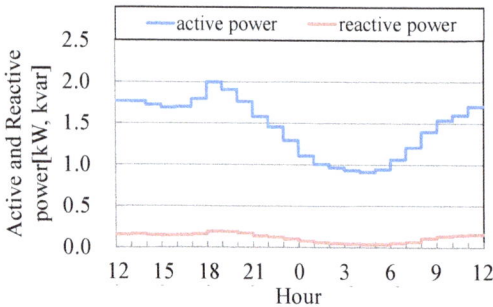

Fig. 8. Load curve of one residential load without EV.

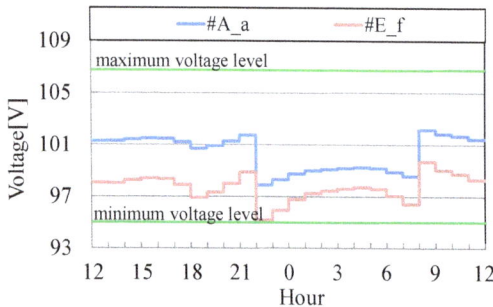

Fig. 9. Voltage at each node without EV.

Introduced EVs are supposed to have all same characteristics shown in table 1. Adjustment of P and Q in each mode of the proposed method is supposed to be done every one second. ΔP and ΔQ, the amount of adjustment of P and Q is 0.1kW and 0.1kvar respectively.

Table 1. Parameters of EV

Parameters		Value
Required energy to be charged	Ec	6kWh
Time to start charging	st	23
Time of departure	et	7
Capacity of inverter for EV	S	3kVA
Adjustment quantity of active power	ΔP	0.1kW
Adjustment quantity of reactive power	ΔQ	0.1kvar

4.2 Dynamic behavior of proposed control method

In this section, dynamic behaviors of voltage, active and reactive power in the situation that EVs are controlled by proposed method are investigated. It is supposed that EVs are intensively introduced to 12 residential loads which

exists in one pole transformer connected to #E. Fig.10 expresses dynamic behavior of each value described before in representative three nodes (#E_b, #E_d and #E_f) of secondary nodes in #E at immediate 23 o'clock and 0 o'clock. As seen in Fig. 10(a), all EVs are controlled by mode1 at 23 o'clock, start time to charge shown in Fig. 10(a). White point in the scatter diagram of active and reactive power means operation point before charging. Blue point means operation point every second. Red point means the last operation point. Because voltage at #E_b is higher than terminal side, EV doesn't control any reactive power and uses all inverter capacity for only charging. At node #E_f, lead reactive power continues to be incre ased from the time when charging power is 1.1kW. This track doesn't pass the line bh shown in Fig. 3(a). The reason is that not only charging power at its own node is increased, but also charging power at #E_b and #E_d is increased, and that made the voltage at #E_f lower as a result. Voltages at each node were kept within admissible range shown in the graphs below the scatter diagram of P and Q.

On the other hand, base residential load is simultaneously changed from the value at 23 o'clock to the value at 0 o'clock shown in Fig. 8. Fig. 10(b) illustrates dynamic behavior at that time. Because base load at 0 o'clock is smaller than base load at 23 o'clock, voltage goes up shown in Fig. 10(b). Therefore, EV at #E_d and EV at #E_f increase charging power according to the rule of mode2, and they can charge power more effectively. Charging power of the EV at #E_d is 3kW, and that of the EV at #E_f is increased by 2.8kW at the end of the control. Voltage at #E_f could also be stable near 95.4V, reference value.

If the charging power is compared, source side is larger than terminal side. Table 2 shows length of time for finishing charging. EV at source side finishes charging at first and EV at terminal side finishes later. The reason is that the proposed method is distributed autonomous voltage control as EV controls by monitoring only voltage of its own node.

Table 2. Time to finish charging

Node	#E_b	#E_d	#E_f
Time to finish charging [mm]	120	134	156

5. Investigation of dead band and adjustment quantity of output power

5.1 Simulation condition

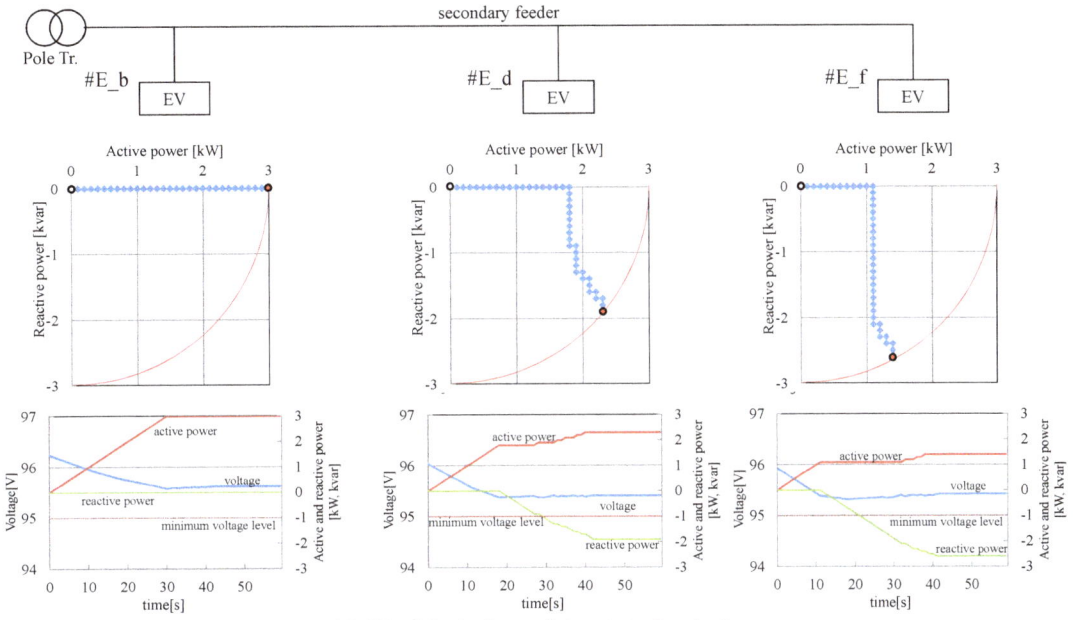

(a) 23 o'clock (immediate start charging)

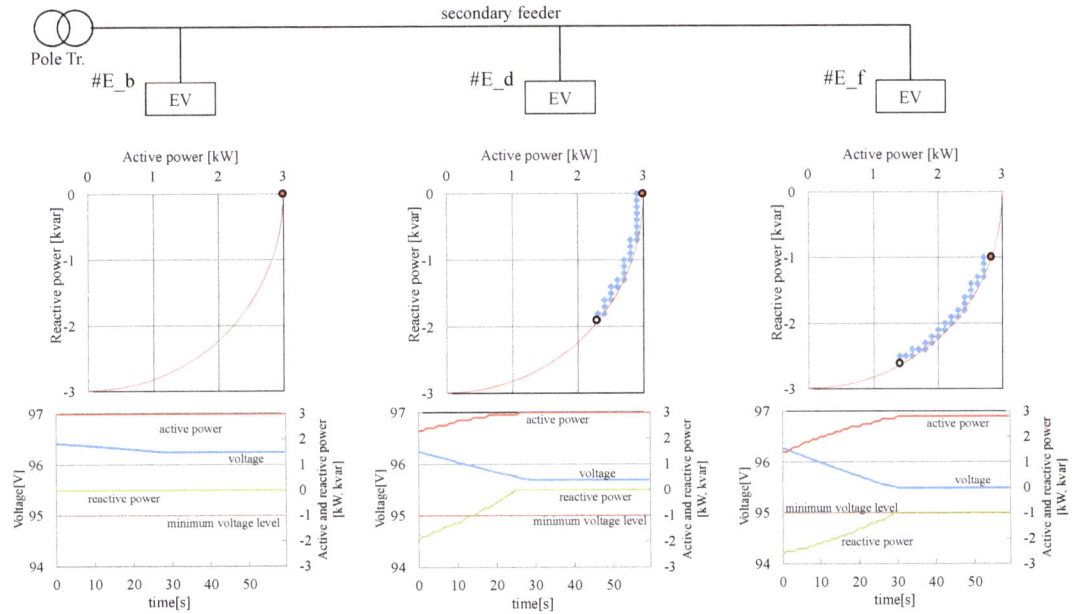

(b) 0 o'clock (immediate load change)

Fig. 10. Dynamic behavior by the proposed method.

The relationship between dead band and adjustment quantity of output power was investigated through computational simulations using the distribution system model used in chapter 4. It is noted that sending voltage at time period from 10PM to 8AM was set up as 6.30kV in order that hunting occur with only one EV introduced.

5.2 Simulation results

Voltage sensitivity to charging power at #E_f was 0.22[V/kW] if it is calculated by approximation formula shown in (1). It is noted that voltage sensitivity to reactive power was smaller than the one to charging power. Quantity of voltage change by one adjustment of charging power is 0.022[V] considering ΔP=0.1[kW]. So simulation was run with setting up the dead band for 0.010[V] and 0.024[V]. Fig. 11 shows voltage at #E_f when only one EV is introduced at #E_f and the dead band is 0.010[V]. The time of this graph means time after load change at 0 o'clock. Charging power had increased according to mode2 because voltage rose at t=0. Hunting occurred after t=52 because dead band is narrow compared to voltage change. On the other hand, Fig. 12 shows voltage at #E_f with setting up the dead band for 0.024[V]. In this case, hunting was avoided owing to wide dead band.

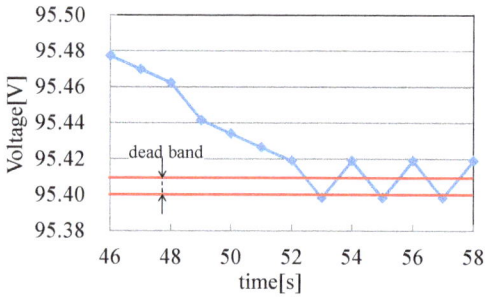

Fig.11. Dynamic behavior with dead band 0.010[V].

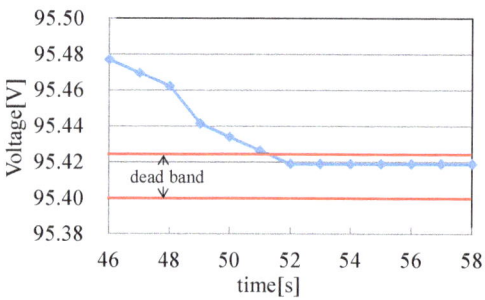

Fig.12. Dynamic behavior with dead band 0.024[V].

Such simulations as this were run with changing the number of introduced EVs and dead band and occurrence of hunting was found out. The results were shown in table 2. When the number of introduced EV was increased by 12 from 1, hunting occur even if the dead band was 0.024[V]

because a number of EV adjust P and Q simultaneously. When the dead band was set up wider for 0.030[V], hunting didn't occur.

Wide dead band is needed to avoid hunting. However, the wider dead band is, the less EV can increase charging power according to mode2. It could take excessive time as a result. Therefore, if some countermeasures that a function to lock mode change with detecting hunting for example are adopted, hunting can be avoided perfectly.

Table 3. Summary of simulation result

Number of EVs	Dead band	Hunting
1	0.010V	occur
1	0.024V	not occur
12	0.024V	occur
12	0.030V	not occur

6. Conclusions

In this paper, from a practical viewpoint, we propose the method that EV can not only fulfill charging effectively but also control minimal reactive power to keep admissible voltage with monitoring system voltage. In addition, it was found out that proper dead band can realize a stable behavior.

However, unfair situation cannot be avoided by distributed autonomous voltage control like the proposed method. Therefore, we will investigate some methods to resolve the unfair situation utilizing information communication in future.

This study was supported by research grant from Japan Power Academy.

References

[1] K. Qian, C. Zhou, M. Allan and Y. Yuan: "Modeling of Load Demand Due to EV Battery Charging in Distribution Systems", *IEEE Trans. Power Systems*, Vol.26, No.2 pp.802-810 (2011)

[2] Y. Mitsukuri, R. Hara, H. Kita, E. Kamiya, S. Taki, N. Hiraiwa and E. Kogure: "A Study on Compensating Voltage Drop in Distribution Systems due to Nighttime Simultaneous Charging of Electric Vehicles Utilizing Charging Power Adjustment and Reactive Power Injection", *IEEJ Trans. PE*, Vol.133, No.2 pp.157-166 (2013) (in Japanese)

33

Study of Credibility Evaluation of Dynamic Models and their Parameters

Guo Weimin†, Ye Xiaohui**, Zhong Wuzhi**, Su Yi***, Tang Yaohua*,
Song Xinli** and Liu Tao**

Abstract – The time-domain simulation plays an important role in planning, designing and operation of the power system, but mismatches between its results and the actual system response show that the credibility evaluation of dynamic models is critical in all the above aspects of power systems. This paper discusses the technologies of credibility evaluation, including hybrid dynamic simulation with PMU measurements, error analysis and the calculated indices. Moreover, a phase shift method is presented to implement hybrid dynamic simulation in PSD-FDS. Then the paper proposes a credibility evaluation scheme to provide solution for the evaluation. At last, one sample is used to validate the implementation of hybrid dynamic simulation, while the other sample illustrates the effectiveness of the credibility evaluation approach.

Keywords: Credibility Evaluation, Hybrid Dynamic Simulation, Phase Shift method, Error Analysis, Phasor Measurement Unit (PMU)

1. Introduction

The time-domain simulation plays an important role in planning, designing, and operation of the power system. And its results are used not only in making decision about the operation strategies, but also in parameters' setting of control equipments [1]. Therefore, pessimistic or optimistic results will give operators a false sense of system security. The simulation result of WSCC outage in August 10, 1996 found that it cannot repeat the collapse as the real power system using existing simulation models [2]. In March 2004, artificial three-phase short-circuit test was carried out in Northeast power grid of China [3], and the simulations are optimistic. Those mismatches warn people of the crediblity of power system dynamic simulation, so more attention should be paid onto the credibility evaluation of the dynamic models and their parameters.

In large power system, the credibility evaluation of dynamic models is a difficult task, because all the models of components will influent the simulation and it is impossible to find out the model with bad parameters. The traditional method uses equivalent models for the external

system. Since the equivalent model is always an approximate of the physical system, error is brought into the simulation unavoidably.

Recently, the widely install of PMU provides a good opportunity to credibility evaluation. PMU can record synchronized phasors, which accurately measures the real-time states of the power system, including the voltage, angle, active and reactive power. And hybrid dynamic simulation [4, 5, 6 ,7] employs PMU measurements as the equivalent of the external system to make the external system having the same behaviors.

This paper proposes a credibility evaluation scheme which integrates hybrid dynamic simulation, error analysis of signals. The remainders of this paper are organized as follows. In section II, the hybrid algorithms are introduced, and the details of implementation are presented in PSD-FDS. And then, it researches the characteristic of dynamic signals from PMU measurements and simulation results, and then the error analysis methods are discussed to build an indices system. In section IV, the above methods are combined to credibility evaluation of dynamic models. In section V, two samples are present to illustrate the effective of the proposed method. Finally, the conclusions are drawn in Section VI.

2. Hybrid Dynamic Simulation

Time-domain dynamic simulation is one of the most important tools for the planning and operation of electrical power system, and the appearance of PMU increases the accuracy of the simulation. So the hybrid dynamic

†   Corresponding Author: Henan Electric Power Research Institute, Henan, China
*   Henan Electric Power Research Institute, Henan, China
**   Power System Department of China Electric Power Research Institute, Beijing, China
*** Fujian Electric Power Dispatch & Telecommunication Center, Fujian, China

simulation combining PMU measurement technology and traditional simulation have the advantages of these two technologies [5]. This section firstly discusses the time-domain method for dynamic simulation. Then, traditional methods of the hybrid dynamic simulation are introduced and compared with each other. At last, the details of implementation in PSD-FDS are present.

2.1 Time-Domain Simulation

The time-domain simulation uses the numerical integration algorithm to deal with a set of high order differential algebraic equations (DAEs) formed from dynamic models and the network of power system [1, 8]. The DAEs used in the time-domain simulation of power systems can be expressed as:

$$\begin{cases} \dfrac{dx}{dt} = f\left(x, \dot{V}\right) & (1) \\ I\left(x, \dot{V}\right) = Y\dot{V} & (2) \end{cases}$$

where (1) represents nonlinear first-order differential equations that models the dynamic components and (2) represents algebraic equations formed form transmission network, with x, $\dot{V}$, $\dot{I}$, and t representing state variables, bus voltage vectors, injection current vectors and time, respectively.

The Y in formula (2) is admittance matrix formed from network. And this matrix is symmetric and constant in almost the whole simulation period.

In a modern large-scale interconnected power system, the calculation is complex and time-consuming. And the accuracy of simulation results depends on the credibility of dynamic models.

2.2 Introduction of Hybrid Dynamic Simulation

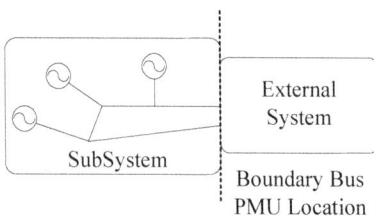

Fig. 1. Location of PMU.

Hybrid dynamic simulation has the advantages of PMU measurement technology and traditional simulation. And applications of such technology include system event

playback, model validation, software validation [9], and locating the disturbance source for the low-frequency oscillation [10].

Hybrid dynamic simulation injects external signals to simulation process and opens traditional dynamic simulation loops for interaction with external signals. As in Fig. 1, the boundary point can divide the whole system into subsystem and external system. With PMU measurements injected at its boundary, the hybrid dynamic method allows for the dynamic simulation of a subsystem without introducing errors caused by an external system. The PMU measurements record the system behavior at the boundary accurately, so it deals only with subsystem and there is no need to model the external system.

The PMU measurements include the voltage, angle, real power and reactive power. Theoretically hybrid dynamic simulation can solve with any two of them, so different methods are proposed. The phase shift method [9] uses an artificial generator and ideal phase shifter to obtain equivalent circuit of external system as shown in Fig. 2. By adjusting the turn ratio k and phase shift α of ideal phase shifter, both the voltage V and the angle θ at the boundary bus are ensured the same as the PMU measurements during the simulation. In Fig. 3, Equivalent impedance method [6] replaces the external system by a dynamic load. And its impedance $R + jX$ changes at each simulation step according to the Norton equivalence method. Another method uses a large synchronous machine with fast-responding exciters and governors (Fig. 4) to make sure that simulated bus voltage and frequency are forced to follow very closely the recorded voltage V^* and frequency f^* [11].

In the meantime of adding new equipments, the simulation error is generated. So [7] use PMU measurements directly when solving the DAEs. In this method, there is no equivalent model, so the voltage and angle of the boundary bus is equal to the measurements exactly. However, this method must modify the program, which is not possible especially when commercial program is used.

As discussed above, the phase shift method is easy to implement, therefore, it is adopted in this paper.

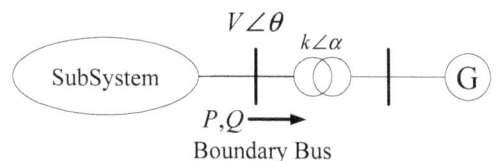

Fig. 2. Diagram of phase shift method.

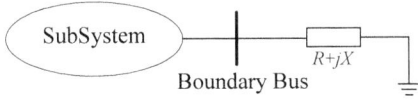

Fig. 3. Diagram of equivalent impedance method.

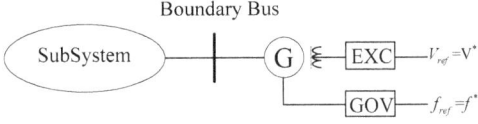

Fig. 4. Diagram of fast-responding generator method.

2.3 Implementation with PSD-FDS

PSD-FDS (Power System Department - Full Dynamic Simulation) [8, 12] is selected for the implementation of the proposed hybrid simulation for its commercial availability and its popularity in power industry, especially in China.

PSD-FDS has been in development at the China Electric Power Research Institute (CEPRI) since 1998. This program employs the modified Gear algorithm, and is capable of electric-mechanical transient, medium-term and long-term dynamic courses.

In this method, the voltage and angle are used as input of hybrid dynamic simulation, while active and reactive powers are used to compare with those got from hybrid dynamic simulation. As illustrated in Fig. 2, a generator and a phase shifter is added, they should meet the following requirements:

A. The generator is modeled as a classical model and provides an ideal voltage source with constant amplitude and angle.

B. The phase shifter has a near-zero impedance (0.0001 pu. in this program). Meanwhile, its turn ratio k and phase shift α change at every simulation step to make sure the voltage and angle of boundary bus as same as PMU measurements.

To satisfy the second requirement, the turn ratio k and phase shift α of the phase shifter is,

$$\begin{cases} k_t = \dfrac{V_t}{V_0} \\ \alpha_t = \theta_t - \theta_0 \end{cases} \quad (3)$$

where, 't' denotes the simulating time. And the formula of the phase shifter in admittance matrix is calculated as follow,

$$\begin{cases} Y_{ii} = Y_T \\ Y_{ij} = \dfrac{-Y_T}{a_t + jb_t} \\ Y_{ji} = \dfrac{-Y_T}{a_t - jb_t} \\ Y_{jj} = \dfrac{Y_T}{a_t^2 + b_t^2} \end{cases} \quad (4)$$

and,

$$a_t + jb_t = k_t \left(\cos\theta_t + \sin\theta_t \right)$$
$$Y_T = \dfrac{1}{(R + jX)}$$

where, R, X are resistance and reactance of the phase shifter respectively, and is 0, 0.0001 in the program.

In formula (4), the addition of the phase shifter makes the admittance matrix asymmetric and changed every simulate step. To overcome this advantage, the elements of phase shifter in admittance matrix are calculated as (5), which is symmetric and constant. And then, the proposed method injects currents in its two terminal buses as (6) to simulate the changing turn ratio and phase shift.

$$\begin{cases} \overline{Y}_{ii} = Y_T \\ \overline{Y}_{ij} = \overline{Y}_{ji} = \dfrac{-Y_T}{k_t} \\ \overline{Y}_{jj} = \dfrac{Y_T}{k_t^2} \end{cases} \quad (5)$$

$$\begin{cases} \dot{I}_i = \left(Y_{ii} - \overline{Y}_{ii} \right)\dot{V}_i + \left(Y_{ij} - \overline{Y}_{ij} \right)\dot{V}_j \\ \dot{I}_j = \left(Y_{ji} - \overline{Y}_{ji} \right)\dot{V}_i + \left(Y_{jj} - \overline{Y}_{jj} \right)\dot{V}_j \end{cases} \quad (6)$$

where, $\dot{V}_i$, $\dot{V}_j$ are voltage of phase shifter's two terminal buses.

3. Error Analysis of Simulation

After hybrid dynamic simulation, we get two signals, one from measurements, and the other from hybrid dynamic simulation. So the key of the credibility evaluation is error analysis between them. Indices from error analysis can be used in credibility evaluation of dynamic models and their parameters [13]. The error analysis can be categorized into qualitative one and quantitative one.

The main method of qualitative analysis is visual

observation, which observes the differences between signals and then adjudges whether they are similar or not. So this method takes the advantage of human experience, but it is not time-consuming in case of a great quantity of curves. Meanwhile this method cannot be used in choosing the better ones among thousands of signals.

On the other side, the quantitative analysis overcomes the limitation of qualitative one, and gives the indices for credibility evaluation and parameter optimization. The quantitative methods can be divided into two categories: residual analysis and eigenvalue analysis.

3.1 Residual Analysis

Residual analysis uses $\left\{ y_i, i = 1, 2, \cdots, N \right\}$ from real measurements and $\left\{ \hat{y}_i, i = 1, 2, \cdots, N \right\}$ from simulation to get the difference sequence $\left\{ \Delta y_i = \left| y_i - \hat{y}_i \right|, i = 1, 2, \cdots, N \right\}$, and then the indices are calculated as [5]:

$$\varphi = \sum_{i=1}^{N} w_i \Delta y_i \tag{7}$$

or,

$$\varphi = \sum_{i=1}^{N} w_i \left(\Delta y_i \right)^2 \tag{8}$$

where, $\left\{ w_i, i = 1, 2, \cdots, N \right\}$ is the weight sequence. And the residual analysis methods are different in formula of weight sequence. In this paper, it is defined as follow [14]:

$$w_i = \frac{1}{\sum_{i=1}^{N} \left(y_i - y_s \right)^2} \tag{9}$$

so, we get

$$\varphi_r = \frac{\sum_{i=1}^{N} \left(y_i - \hat{y}_i \right)^2}{\sum_{i=1}^{N} \left(y_i - y_s \right)^2} \tag{10}$$

The indices from residual analysis can reflect the whole relative error of the simulation results, but it cannot show the fluctuation of the sequences. So there is more character information needed to study.

3.2 Eigen-value Analysis

Eigen-value Analysis methods extract the eigenvalue of the signals, and then use it to measure the error level.

Fourier, wavelet and prony algorithms are the effective methods for the analysis of transient signals. And prony method extracts valuable information from the signal and builds a series of damped exponentials of sine function, which include frequency, amplitude, damp, and phase angle. So it is widely used in low frequency oscillation analysis, design of controller, parameter identification, and so on [15].

However, the orders of the prony results are usually very high, and in most case, the results are in different orders so as to be non-comparable. So in this paper, only the main mode is used to calculate the credibility indices as follows:

$$\varphi_{ef} = \frac{\left| f_m - \hat{f}_m \right|}{\left| f_m \right|} \tag{11}$$

$$\varphi_{e\zeta} = \frac{\left| \zeta_m - \hat{\zeta}_m \right|}{\left| \zeta_m \right|} \tag{12}$$

where, 'm' denotes that value is got from main mode, f, and ζ stand for frequency and damp getting from measured signals respectively, while $\hat{f}$ and $\hat{\zeta}$ have the same meaning calculated from simulated signals.

3.3 Error Indices Calculation

Beyond the above indices, the first swing is very important, so we get another index:

$$\varphi_{max1} = \frac{\left| M_1 - \hat{M}_1 \right|}{\left| M_1 \right|} \tag{13}$$

$$\varphi_{min1} = \frac{\left| N_1 - \hat{N}_1 \right|}{\left| N_1 \right|} \tag{14}$$

where, M_1 and N_1 are the maximum and minimum of first swing of measured signals, and $\hat{M}_1$ and $\hat{N}_1$ have the same meaning from simulated signals.

In short, the formulas (10)-(14) are used as indices in this paper.

4. Credibility Evaluation

In power system, there are many nonlinear factors, like eddy current, saturation of the generator. Therefore, the dynamic models and their parameters from off-line testing may be not in accordance with on-line condition. It is very valuable of credibility evaluation using on-line signals. And the install of PMU enables such evaluation.

The signals from PMU include voltage, angle, active and reactive powers on bus. Considering the possibility of PMU fault, the measurements should be pre-processed before use. Pre-process includes the signals checking (bad data repairing, missing data making), and signals rebuilding because of the mismatch between PMU sampling time and simulation step.

The model credibility evaluation process with hybrid dynamic simulation is summarized as follows (see Fig. 5):

Step1: acquire the measurements from PMU, including voltage, angle, real power and reactive power;

Step2: pre-process the measurements, such as checking and fixing the missed points in sequence, re-generating the sequence suiting for the simulation step;

Step3: prepare network and models of subsystem, and then do the hybrid dynamic simulation using the voltage and angle from the measurement;

Step4: after getting the simulated P, Q and measured P, Q, analyze the difference between these two results, and calculate the error indices.

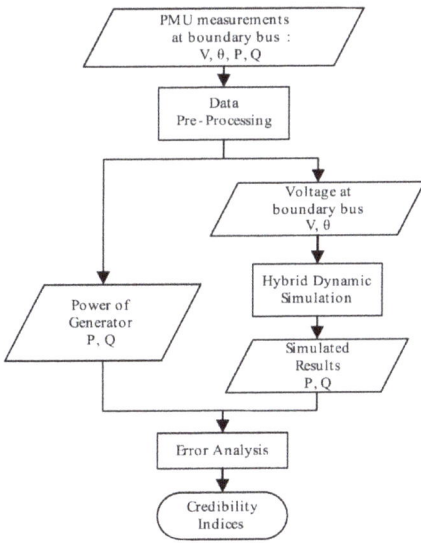

Fig. 5. Flow chart of credibility evaluation.

5. Case Study

5.1 Validation of the Hybrid Dynamic Simulation

The method to validate the program is to compare time-domain simulation results of whole system and hybrid dynamic simulation results of subsystem. A three-generator,

9-bus power system, as shown in Fig. 6, was used to validate the proposed method for hybrid simulation. And BusA and BusC are used as boundary buses to separate the whole system into two parts.

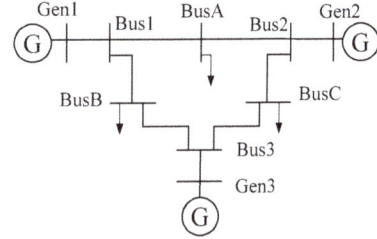

Fig. 6. Diagram of IEEE 9 bus system.

Firstly, the time-domain simulation is done with the system in Fig. 6 as benchmark simulation, and voltage, angle, active power and reactive power of BusA and BusC are recorded. Then hybrid dynamic simulation is processed in subsystem as Fig. 7, there are two generators and two phase shifters are added to BusA and BusC respectively.

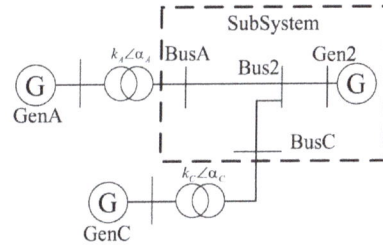

Fig. 7. Diagram of subsystem.

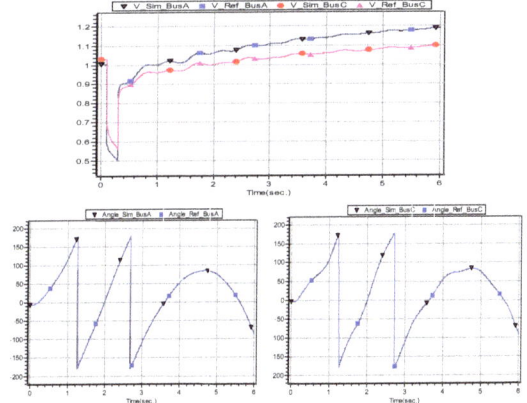

Fig. 8. Voltage and angle of BusA and BusC.

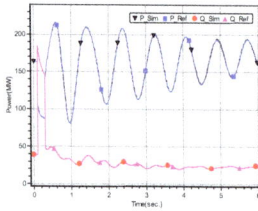

Fig. 9. Active power and reactive power of Gen2.

The results of the above two simulations are illustrated in Fig. 8 and Fig. 9. Each plot contains two traces, which are identical so as not to be distinguished. From Fig. 9, we see that the phase shift method has succeeded in keeping the voltage and angle of BusA and BusC in hybrid dynamic simulation the same as in time-domain simulation. Then the power curves are same between these two simulations. So it is easy to get the conclusion that the subsystem in hybrid dynamic simulation has accurately repeat the dynamic behavior.

5.2 Credibility Evaluation Example

This section presents the results of simulation studies for evaluating and verifying the proposed identification scheme. In this experimental case, one generator in the power system of China is studied, and the generator bus is chosen as boundary bus. The time-domain dynamic simulation of this system is used as Base Case, from which the voltage, angle, active, and reactive powers are recorded to be used to evaluate the credibility of the model.

The models include the generator, governor, turbine, exciter and power system stabilizer (PSS). There are there cases studied:

Case A: generator with correct models and parameters;

Case B: the limiter of PSS is more strict, that means, the parameters of PSS are wrong.

Case C: kinetic energy of generator is wrong.

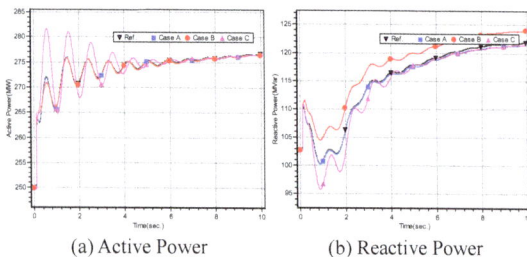

(a) Active Power   (b) Reactive Power

Fig. 10. The results of hybrid simulation.

Table 1. Results of prony analysis

		amplitude	damp	frequency
P	Base Case	0.8204	-0.4734	1.1477
	Case A	0.8536	-0.4711	1.1474
	Case B	0.8651	-0.4776	1.1479
	Case C	1.8268	-0.7712	1.1050
Q	Base Case	0.4346	-0.5941	1.1470
	Case A	0.4372	-0.5905	1.1510
	Case B	0.4600	-0.6180	1.1470
	Case C	0.9254	-0.7108	1.1060

Table 2. Indices of credibility evaluation

		φ_r	φ_{ef}	$\varphi_{e\zeta}$	φ_{max1}	φ_{min1}
P	Case A	0.012	0.005	0.000	0.000	0.000
	Case B	0.017	0.009	0.000	0.004	0.002
	Case C	0.423	0.629	0.037	0.035	0.003
Q	Case A	0.003	0.006	0.003	0.011	0.001
	Case B	0.162	0.040	0.000	0.017	0.042
	Case C	0.048	0.196	0.036	0.011	0.045

The simulation curves of all the three cases and the base case are shown in Fig. 10. And from the figure, the results of Case A and Base Case are identical so as not to be distinguished. Case B has the same active power outputs and different reactive power ones. In Case C, active and reactive powers are different from the Base Case.

After visual observation, the prony analysis is processed, and the results are shown in Tab. 1. According to the formulas in section 3.3, the indices are calculated as in Tab. 2. From the Tab. 2, we draw the same conclusion as visual observation. So it proves that the phase shift method can reliably present the system dynamic response for model variation purposes, and the proposed credibility evaluation and its indices are valuable.

6. Conclusion

A new credibility evaluation scheme is proposed in this paper to evaluate the accuracy of dynamic models from PMU-based measurements recorded during disturbances. The scheme integrates the hybrid dynamic simulation and error analysis technologies. And the implementation of the phase shifter method is presented in PSD-FDS, and this method is easy realization and programming implementation. Moreover, the results from two simulation cases show the effective of proposed credibility evaluation.

Credibility evaluation with hybrid dynamic simulation could be used not only in small-scale system, but also in large-scale system. But the accuracy of estimation result still has some room for improvement. If possible, multiple disturbance events should be use to provide a more reliable indices. Meanwhile, this method cannot replace the traditional power system equipment testing, but only provide an auxiliary means to evaluate the accuracy of simulation. And more application in real system will be done in further work.

References

[1] P. Kundur, Power System Stability and Control, McGraw Hill, 1994.

[2] Dmitry N. Kosterev, Carson W. Taylor, Model Validation for August 10, 1996 WSCC System Outage, Power System, IEEE Trans. on,1999, 14(3).

[3] Wang Gang, Tao Jiaqi, Xu Xingwei, et al, 500kV Man-Made Three-Phase Earthing Short Circuit Experiment in Northeast Power Grid. Power System Technology, 2007, 31(4) 42-48 (in Chinese).

[4] Huang Z, Kosterev D, Gutteromson R, et al, Model Validation with Hybrid Dynamic Simulation, Proceedings of the 2006 IEEE Power Engineering Society General Meeting: Vol 2, June 18-22, 2006, Montreal, Canada.

[5] Han Dong, He Renmu, Power System Dynamic Simulation Validation Based on Similarity Theory and Analytical Hierarchy Process, International Conference on Power System Technology, 2006: 1-7.

[6] Ma Jin, Han Dong, Shen W. J., et al., Wide area measurements-based model validation and its application, Generation, Transmission & Distribution, IET, 2008, 2(6): 906-915.

[7] Wu Shuangxi, Wu Wenchuan, Zhang Boming, et al, A Hybrid Dynamic Simulation Validation Strategy by Setting V-θ Buses with PMU Data, Automation of Electric Power Systems, 2010, 34(17):12-16 (in Chinese).

[8] Song Xinli, Tang Yong, Zhong Wuzhi, Liu Wenzhuo, Wu Guoyang, Liu Tao, New Mixed Integral Algorithm for Unified Dynamic Power System Simulations of Transient, Medium-term and Long-term Stabilities,IEEE PES 2011 Power Systems Conference & Exposition, March 20-23, 2011, Phoenix, AZ US.

[9] Huang Z., Guttromson R. T., Hauer J.F., Large-scale Hybrid Dynamic Simulation Employing Field Measurements, the IEEE Power Engineering Society General Meeting, Denver, USA, June 2004.

[10] Ma Jin, Zhang P. F., Fu Hongjun,et al, Application of Phasor Measurement Unit on Locating Disturbance Source for Low-Frequency Oscillation, Smart Grid, IEEE Trans on, 2010, 1(3): 340-346.

[11] Kosterev D. N., Hydro turbine-governor model validation in Pacific Northwest, Power Systems, IEEE Trans. On, 2004, 19(2) : 1144- 1149.

[12] Song Xinli, Tang Yong, Bu Guangquan, et al, Full dynamic simulation for the stability analysis of large power system, Power System Technology, 2008, 32(22): 23-28 (in Chinese).

[13] Shestakov A. L., Dynamic Error Correction Method, Instrumentation and Measurement, IEEE Trans. on, 1996, 45(1): 250 – 255.

[14] Li Wei, Yang Hongxia, Xiong Peihua, et al, Generator Model Parameter Adjustment Based on PMU Measured Data, Power System Technology, 2009, 33(2): 89-93 (in Chinese).

[15] Hauer J. F., The Use of Prony Analysis to Determine Modal Content and Equivalent Models for Measured Power System Response, In Eigenanalysis and Frequency Domain Methods for System Dynamic Per formance, IEEE Publicatio n 90TH0292- 3- PWR, 1989: 105- 115.

34

Evaluation of Improvement Effect of Voltage Quality by Reactive Power Control with Available Capacity of Residential FC Inverter

Shoichi Koinuma[†], Yu Fujimoto*, Yasuhiro Hayashi**, Takao Shinji***, Yosuke Watanabe*** and Masayuki Tadokoro***

Abstract – The installation of residential fuel cell systems (FCs) with photovoltaic generation systems (PVs) in residences has begun in Japan. Because of the reverse power flow from PVs, the line voltage may exceed the proper range during daytime operation. In addition, because residential PVs are connected to a single-phase distribution system, the voltage imbalance rate may increase. Here, we focus on an FC with a large available capacity during the daytime, so as to adjust the voltage by reactive power control with the available capacity of the FC. We propose a method of reactive power control that uses the available capacity of FCs to improve the voltage magnitude and voltage imbalance rate in a distribution network.

Keywords: Reactive power control, Fuel cell, Photovoltaic generation, Distribution system

1. Introduction

The need for renewable energy resources is rapidly being recognized throughout the world from the standpoint of the global warming. In Japan, large numbers of photovoltaic generation systems (PVs) have been installed in standard residences for the purpose of preventing global warming. Residential fuel cell systems (FCs), which have also been installed in residences in Japan, have high energy efficiency because they can generate thermal power whenever they produce electrical power. However, distributed generations (DGs) such as PVs and FCs are generally connected to a single-phase distribution network. Consequently, DGs may raise the line voltage magnitude and adversely affect the voltage imbalance rate in the distribution network as the number of DGs increases.

Reactive power control by using a grid-connected inverter is one method for improving voltage quality. However, the inverter of a PV runs out of capacity in the daytime owing to the increase in generation. In contrast, the inverter of an FC has sufficient capacity for reactive power control because FC generation is not performed in the daytime. Therefore, using the FC inverter for reactive power control leads to effective use of PV generation.

Thus far, several control methods have been proposed to improve either the voltage magnitude or voltage imbalance rate [1]-[6]. However, a control method to improve both the voltage magnitude and imbalance, which vary depending on the output of DGs, has yet to be proposed and evaluated. Because reactive power control in a single-phase distribution system affects the three-phase current, the voltage imbalance rate might be exacerbated by using the conventional method despite improving the voltage magnitude.

In this paper, we propose a method of reactive power control that uses the available capacity of residential FCs. The amount of reactive power output of each FC is determined centrally by using the voltage-reactive power sensitivity. By using the relationship between voltage and reactive power in the distribution system, both the voltage magnitude and voltage imbalance is improved. The validity of the proposed method is confirmed by evaluating the voltage magnitude and imbalance in numerical simulations.

2. Reactive Power Control with Available Capacity of Fuel Cell System

2.1 Reactive Power Control with FC

Usually, a grid-connected inverter is operated at a power factor of unity. When the voltage at the point of common coupling increases, changing the power factor causes lagging of the reactive power, and the voltage can be reduced. However, a PV inverter has no capacity for

---

[†] Corresponding Author: Dept. of Electrical Engineering and Bioscience, Waseda University, Japan (sic.koinuma@gmail.com)
*   Inst. for Nanoscience & Nanotechnology, Waseda University, Japan
** Dept. of Electrical Engineering and Bioscience, Waseda University, Japan
*** Tokyo Gas Co., Ltd., Japan

reactive power control during the daytime.

The profiles of residential FCs are generally decided by each residential thermal demand. Because most of the residential thermal demand is concentrated in the morning and evening, the inverters of FCs have large available capacities in the daytime. The voltage is controlled by outputting lagging reactive power in the daytime by using FCs when PVs cause voltage increases and voltage imbalances. The available capacity is shown in the equation below.

$$Q_{available} = \sqrt{S_{inv}^2 - P_{FC}^2} \qquad (1)$$

where $Q_{available}$ is the available capacity of the inverter, S_{inv} is the inverter capacity of the FC, and P_{FC} is the electrical generation of the FC.

2.2 Voltage-reactive Power Sensitivity

The reactive power output at a certain point affects the voltage at all points. Fig. 1 shows the concept of sensitivity.

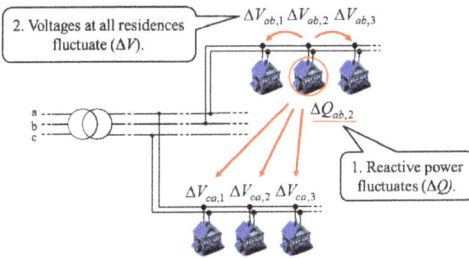

Fig. 1. Influence of reactive power fluctuation.

In our method, the voltage-reactive power sensitivity is used to balance the three-phase voltage. The sensitivity represents a change in the voltage vector of all points per unit output of reactive power:

$$
\begin{bmatrix} \Delta V_{ab,1} \\ \Delta V_{ab,2} \\ \vdots \\ \Delta V_{ab,N} \\ \Delta V_{bc,1} \\ \vdots \\ \Delta V_{ca,N} \end{bmatrix} = \begin{bmatrix} \sigma_{ab1,ab1} & \sigma_{ab1,ab2} & \cdots & \sigma_{ab1,caN} \\ \sigma_{ab2,ab1} & & & \\ \vdots & & & \\ \sigma_{abN,ab1} & & & \\ \sigma_{bc1,ab1} & & & \\ \vdots & & & \\ \sigma_{caN,ab1} & & \cdots & \sigma_{caN,caN} \end{bmatrix} \begin{bmatrix} \Delta Q_{ab,1} \\ \Delta Q_{ab,2} \\ \vdots \\ \Delta Q_{ab,N} \\ \Delta Q_{bc,1} \\ \vdots \\ \Delta Q_{ca,N} \end{bmatrix} \qquad (2)
$$

where $\sigma_{Xn,Ym}$ ($X,Y \in \{ab,bc,ca\}$, $n,m=1,...,N$) is the vector

of voltage change at node n of line X by the unit output of lagging reactive power at node m of line Y, ΔV_{Xn} is the vector of voltage change at node n of line X, ΔQ_{Ym} is the amount of the reactive power at node m of line Y, and N is the number of nodes of each line.

2.3 Method of Selecting the Amount of Reactive Power Output

Fig. 2 shows the concept of the method to determine the amount of reactive power output from each FC. The voltage vectors bc and ca are rotated. Minimizing the distance between each pair of line voltage vectors leads to improving the rate of voltage unbalance.

The amount of reactive power output from each FC is determined by solving the following optimization problem on the basis of the voltage-reactive power sensitivity.

Minimize:

$$
\sum_{n=1}^{N} \left\{ \begin{array}{l} \left| \dot{V}_{ab,n}(\Delta \boldsymbol{Q}) - a \times \dot{V}_{bc,n}(\Delta \boldsymbol{Q}) \right|^2 \\ + \left| a \times \dot{V}_{bc,n}(\Delta \boldsymbol{Q}) - a^2 \times \dot{V}_{ca,n}(\Delta \boldsymbol{Q}) \right|^2 \\ + \left| a^2 \times \dot{V}_{ca,n}(\Delta \boldsymbol{Q}) - \dot{V}_{ab,n}(\Delta \boldsymbol{Q}) \right|^2 \end{array} \right\}
$$
$$
+ \alpha \times \sum_{n=1}^{N} \left\{ \left| Q_{ab,n,new} \right| + \left| Q_{bc,n,new} \right| + \left| Q_{ca,n,new} \right| \right\} \qquad (3)
$$

Subject to:

$$ 95[V] \le \left| V_{X,n} \right| = \left| \Delta V_{X,n} + V_{X,n,old} \right| \le 107[V] \qquad (4) $$

$$ \left| Q_{Y,m} \right| = \left| \Delta Q_{Y,m} + Q_{Y,m,old} \right| \le Q_{max}. \qquad (5) $$

In eq. 3, the line voltages are calculated by eq. 2; the variables of this optimized calculation are ΔQ_{Ym}. The first term of eq. 3 consists of the sum of the distances between each pair of vectors; this is an index of voltage unbalance. Further, the second term of eq. 3 consists of the sum of the amounts of reactive power at each node. Here, a is the vector operator that rotates the vectors 120° counter-clockwise, and Q_{max} is the capacity of the FC inverter. The subscript *old* indicates a non-updated variable, and *new* indicates the updated variable.

We solve this optimization problem by using the particle swarm optimization (PSO) method [7]. This method optimizes a problem by iteratively improving a candidate solution. PSO achieves optimization by having a population of candidate solutions (agents), and moving the agents around in the search-space according to the following equations:

$$x(i,new) = x(i,old) + v(i,new),\qquad(6)$$

$$v(i,new) = w \times v(i,old) + c_1 \times r_1 \times (pbest(i) - x(i,old))$$
$$+ c_2 \times r_2 \times (gbest - x(i,old))\qquad(7)$$

where $x(i, new)$ is an updated point of agent i, $v(i, new)$ indicates a new direction for this update calculated on the basis of old v, $pbest$, and $gbest$. Here, $pbest(i)$ is the point of the best objective value evaluated by agent i and $gbest$ indicates the point of the global best objective value among all agents searched in the past steps. Fig. 3 shows a concept of PSO.

In this paper, the amount of reactive power for each line is determined, and the others are given linearly as in Fig. 4. Therefore, an agent has three variables.

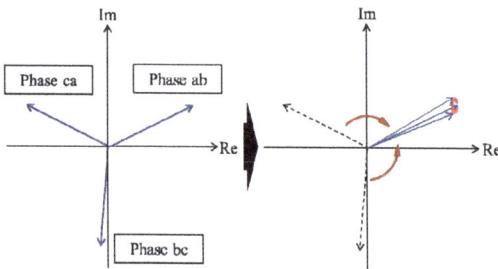

Fig. 2. Concept of the first term of eq. 3.

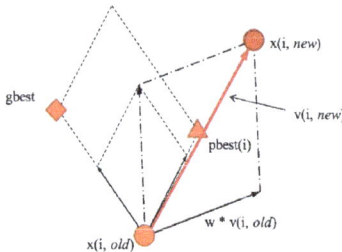

Fig. 3. Concept of PSO.

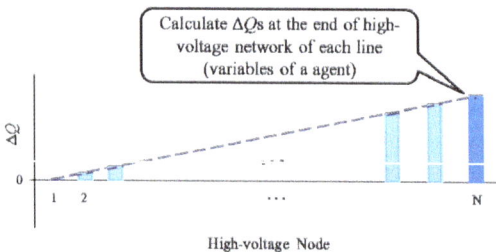

Fig. 4. Determination of reactive power at high-voltage nodes of one line.

3. Numerical Simulation

To affirm the validity of the proposed method, the voltage magnitude and voltage imbalance rate are evaluated in a numerical simulation of a distribution network model by comparing performance with and without the proposed control.

3.1 Simulation Conditions

The distribution network model is shown in Fig. 5. The model consists of 25 high-voltage nodes. One node has 2 pole transformers in each line. Low-voltage distribution networks, which consist of 10 electrical loads, are connected to the pole transformer. All electrical loads are given a typical electrical load profile in the residential area. The power factor of all electrical loads is set as 1.0. The proper range of line voltage is from 95 V to 107 V in a low-voltage distribution network. For simplicity, it is assumed that the sending voltage and line impedance are balanced on a three-phase circuit. Additionally, 20 nodes under one high-voltage node are reduced to one node. The following is an explanation of each type of load in the distribution network.

(i)   Electrical load: We used real-world residential load profiles as electrical loads; the 100 types of profiles used in this simulation are distributed uniformly and randomly to all electrical loads. Fig. 6 shows the example of daily load curves.

(ii) PV: We assume that residential PV systems have a rated output of 3.0 kW and were installed in residences. Real-world data of PV generation on a sunny day is used as a PV profile. For simplicity, we used a single profile for all the PVs. In addition, PV generation is not suppressed by power conditioning system (PCS) when problems of voltage elevation occur. Fig. 7 shows the daily PV curves.

(iii) FC: We assume that a residential FC system has a rated output of 1.2 kW and is installed for the customer who also installed the PV. The profile of FC generation is determined by thermal demand and load demand. In Japan, because reverse power flow from a FC is not allowed, FC generation cannot exceed the load demand. Additionally, because the primary purpose of FC installation is supplying thermal energy, the profile of the FC is determined in order to supply thermal energy. Therefore, the profile of an FC depends on thermal demand and load demand. In this paper, to confirm the effect of FC, FC power generation is not performed, and all of the inverter capacity is used for reactive power control.

The ratio of PV and FC installation is set for each line as follows: line ab, 10%; line bc, 10%; and line ca, 50%. DGs are introduced from the end of the low-voltage network. Note that the number of DGs introduced implicitly indicates the voltage unbalance.

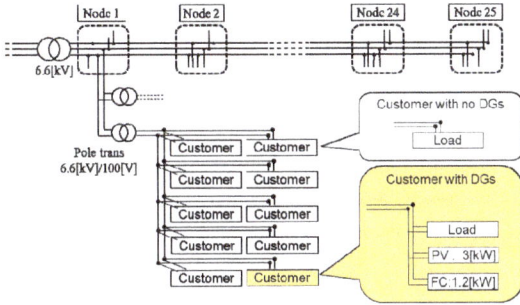

Fig. 5. Distribution network model.

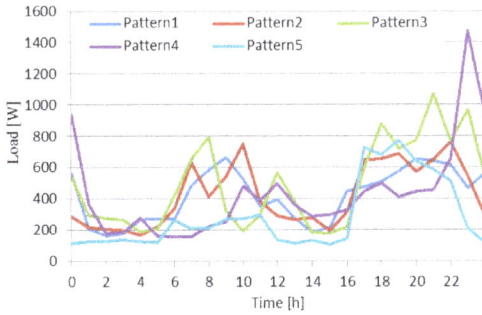

Fig. 6. Examples of load profile.

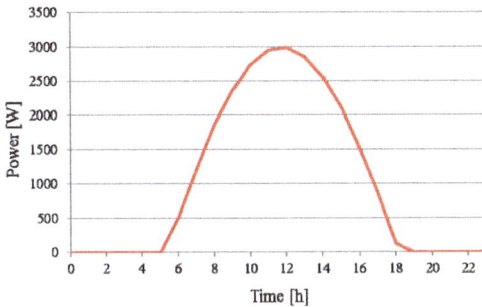

Fig. 7. PV curve.

3.2 Results of Numerical Simulation

Fig. 8 shows the voltage magnitude result at the end of the network. In Fig. 8, the curves with and without the proposed control method are compared. With the proposed control method, each voltage magnitude at the end of the distribution network is maintained within the proper range in Japan (95–107 [V]). That the voltage magnitudes at line ab and line ca are decreased by the proposed method is confirmed. Conversely, the voltage magnitude at line bc is increased by the proposed method. This suggests that three-phase voltages are balanced by the reactive power control method. In fact, the voltage imbalance rate is improved at all time points in Fig. 9. The voltage imbalance rate is maintained below the proper range in Japan (3%) by the proposed method.

The amount of reactive power of each line at the end of the high-voltage network is shown in Fig. 10. At lines ab and ca, lagging reactive power is injected, and the line voltages are reduced. Conversely, at line bc, leading reactive power is injected so that the line voltage is increased. Balancing reactive power output controls three-phase voltages and improves the voltage imbalance rate.

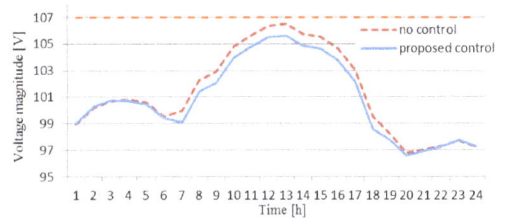

(a) Line ab

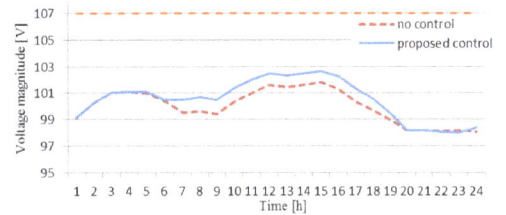

(b) Line bc

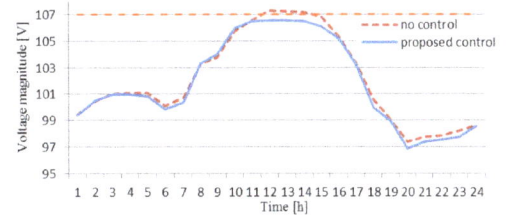

(c) Line ca

Fig. 8. Voltage magnitude at the end of distribution network.

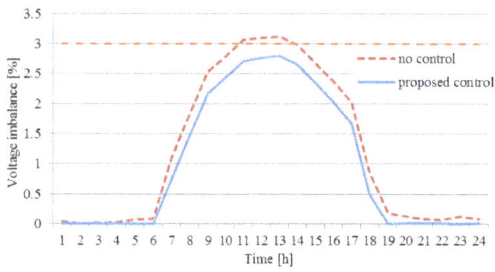

Fig. 9. Voltage imbalance rate at the end of high-voltage network.

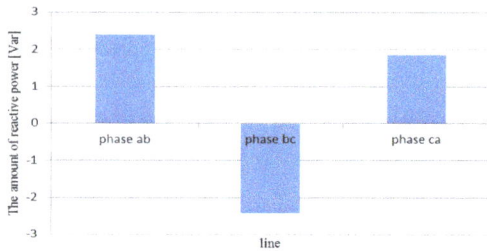

Fig. 10. Amount of reactive power at the end of high-voltage network.

4. Conclusions

In this paper, for the purpose of improving both the line voltage magnitude and the line voltage imbalance rate when large numbers of DGs are installed in the distribution system, we proposed a method of reactive power control that uses the voltage-reactive power sensitivity and available inverter capacity of residential FCs.

To confirm the validity of the method, the voltage magnitude and voltage imbalance rate in a numerical simulation were evaluated by using a distribution network model. Through numerical simulations, we showed that the proposed method is capable of maintaining voltage magnitudes in the distribution network within the proper range, and improving the voltage imbalance rate by balancing the outputs of reactive power from residential FCs, using the voltage-reactive power sensitivity.

References

[1] M. Sano, Y. Hanai, Y. Hayashi, T. Shinji and S. Tsujita, "Autonomous Decentralized Voltage Control of Residential Fuel Cell to Improve Voltage Imbalance and Margin of Distribution System", *2011 ANNUAL MEETING RECORD I.E.E. JAPAN,* IEEJ, 2011: vol.6, pp.240-241, 2011.

[2] S. Tsujita and T. Shinji, "Study on Coordinated Voltage Control for Distribution Line with High Penetration of Photovoltaic Generations using Fuel cells", *The papers of the International Conference on Electrical Engineering 2009*, ICEE, 2009.

[3] M. Oshiro, T. Senju, A. Yona, N. Urasaki and T. Funabashi, "Voltage Control in Distribution Systems Considered Reactive Power Output Sharing", *IEEJ Trans. PE*, vol. 130, no.11, pp.972-980, 2010.

[4] T. Tsuji, T. Oyama, T. Hashiguchi, T. Goda, T. Shinji and S. Tsujita, "A Study of Autonomous Decentralized Voltage Profile Control Method considering Power Loss Reduction in Distribution Network", *IEEJ Trans. PE*, vol.130, no.11, pp.941-954, 2010.

[5] Y. Hayashi, J. Matsuki, M. Ohashi and Y. Tada, "Computation Method to Improve Three-phase Voltage Imbalance by Exchange of Single-phase Load Connection", *IEEJ Trans. PE*, vol.125, no.4, pp.365-372, 2005.

[6] S. Tsujita, T. Shinji, T. Oyama and T. Tsuji, "A Study on Influence under High Penetration of Photovoltaic Generation Conditions and Coordinated Voltage Control by Reactive Power using Distributed Generator", *The Papers of Joint Technical Meeting on Power Engineering and Power Systems Engineering, IEEJ*, 2009.

[7] J. Kennedy and R. Eberhart, "Particle swarm optimization", *IEEE International Conference on Neural Networks*, vol.4, pp.1942-1948, 1995.

35

Power Supply Assessment Model of Renewable Energy Generators - Focused on Wind Turbine Generator and Solar Cell Generator

Sunghun Lee*, Yeonchan Lee*, Jaeseok Choi[†], Seunggu Han** and Jinsu Kim**

Abstract – This paper proposes power supply assessment model of renewable energy generators(REG). The REGs have uncertainty of resource supply. Therefore, the REGs have probabilistic power characteristics. The analytical probabilistic power assessment model of the REGS is developed in this paper. This paper is focused on Wind Turbine Generator(WTG) and Solar Cell Generator(SCG). The assessment model based on probabilistic model considering uncertainty of resources(wind speed and solar radiation) is developed. The power is assessed from expected value based the proposed model. The proposed model is applied two simple systems, which have a WTG or a SCG in this paper.

Keywords: Power Supply Assessment Model, Renewable Energy, Wind Turbine Generator, Solar Cell Generator.

1. Introduction

The renewable energy generators(REGs) with uncertainty of resource supply are dramatically increased in world recently. While the power supply of conventional central controllable load dispatching generator(CCG), which can be estimated and controlled, is used for supply reserve power or load dispatch of Independent System Operator(ISO), the REGs are operated independently with power supply control center(PSCC). Therefore, the REGs are categorized as central uncontrollable or no controllable load dispatching generator (NCCG).

Additionally, because of the uncontrollable circumstances of the REGs, some situations such as intermittent and dramatic power change of the REGs may yield controller or operators of the PSCC to make more difficult operation of power system. The difficulty comes mainly from the reasons that the power output uncertainty of a most of NCCG depends on weather condition according to REG's place and resources supplied time. It makes the PSCC to predict the power output very difficulty. If the REGs are not in-cooperated with ESS during the period of source supply, it is certain that the generators using renewable energy source supply the uncertain and various power output.

If the power supply of the NCCG, which are recently enlarged and small scale, is estimated more precisely, the operation reserve power can be estimated in valuable.

Relatively higher power reliability may be certified. The system operator, further more, can handle as like as new load dispatchable generator. If the power supply of the NCCG is estimated, also, it is possible to evaluate operating reserve power in valuable. Therefore, it is important to develop power supply assessment system of wind turbine generator, solar cell generator and tidal generator, which characterize uncertainty of supply of resources. Especially, the accuracy of the non-central load dispatching generator power supply assessment system is determined by the methodology. It is closely related to the predict precision of the weather. Contemporary system largely contributes to develop the model but the uncertainty of the machine or the flexibility of the power output curve according to the weather condition are necessary. Therefore, visible and objective non-central load dispatching generator estimation tool which enables system operator at load dispatching center is desperately required to certify more reliable power system operation.

This paper proposes an algorithm and model which can assess the power supply of wind turbine and solar cell generators. The proposed power supply prediction model is based on analytical method combining the power characteristics function and probabilistic resource supply function of the REGs. Eventually, the proposed model will be extended to power estimation system of REG's in near future.

† Corresponding Author: Dept. of Electrical Engineering, Gyeongsang National Univeristy, Korea (jschoi@gnu.ac.kr)

* Dept. of Electrical Engineering, Gyeongsang National Univeristy, Korea (huniiya@nate.com, kkng1914@gnu.ac.kr)

** Korea Power Exchange, Korea ({hansgoo, jincuboy}@kpx.or.kr)

2. Power Supply Assessment Methods of the REGs

The conventional four types of domestic and abroad prediction model of the REGs are:

1. Physical Approach: As shown at beneath, the predicting method using wind turbine generator and solar cell generator output curves.
2. Statistical Approach: The predicting methods using Time Series Models such as ARMA model from output performance data.
3. Learning Approach: Including predictive value method from connection between input data and output data.
4. Mixed Approach: Predicting method mixing upper three ways.

Looking into developing state of prediction system abroad based on former methods follows:

- Denmark: Considerable contribution to decrease system operating costs owing to the practice of wind power prediction technology. In 1994, practical use of wind power prediction tool WPPT.
- Germany: In 2001, practical use of wind power prediction tool.
- U.S: Application of power system operation at major electric power company.
- Japan: Under developing Japanese model suitable to the characteristics of geography and meteorology in Japan based on European, Wind Power Prediction tool.

Fig. 1 shows that German, ISET wind turbine operational control system among commercial wind power prediction tools which are well known to the other countries.

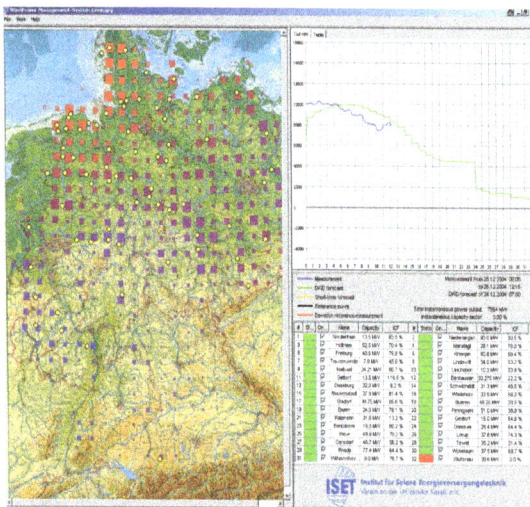

Fig. 1. Snap shot screen of German operating ISET wind turbine operational control system.

3. Supply Capacity Assessment Model of Wind Turbine and Solar Cell Generators.

3.1 Probabilistic output prediction model of wind turbine generator[1,2]

Fig. 2 shows the output curve of wind turbine generator which has broad using range contemporarily. The theoretical output of wind turbine generator is explained at equation (1). As you see in theoretical output equation of wind turbine generator, the output of wind turbine generator depends greatly of wind speed. However the actual output of wind turbine can't produce the electric power infinitely according to equation (1).

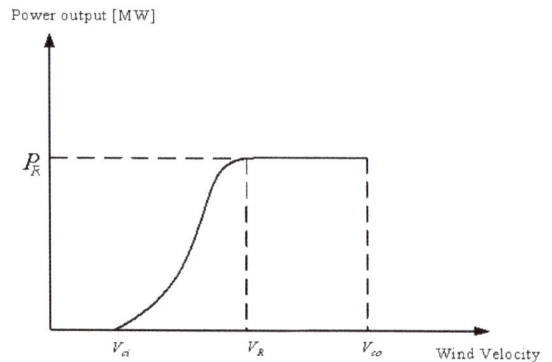

Fig. 2. Typical output curve of wind turbine generator.

$$\begin{aligned}
P_{SWbi} &= 0, & 0 \leq SW_{bi} < V_{ci} \\
&= P_R(A + B \times SW_{bi} + C \times SW_{bi}^2), & V_{ci} \leq SW_{bi} < V_R \\
&= P_R, & V_R \leq SW_{bi} < V_{co} \\
&= 0
\end{aligned} \quad (1)$$

Where,
Vci: The cut-in speed [m/sec]
VR: The rated speed [m/sec]
Vco: The cut-out speed [m/sec]
PR: Rated power [kW]
PSWbi: Output of wind turbine generator corresponding to average wind speed SWbi of wind speed band #i [kW]
Nb: Total number of bands.

At the equation (1) coefficients of the curve A, B and C between the cut-in speed an the rated speed are calculated speed as equation (2), (3) and (4).

$$A = \frac{1}{(V_{ci} - V_R)^2} [V_{ci}(V_{ci} + V_R) - 4(V_{ci}V_R)(\frac{V_{ci} + V_R}{2V_R})^3] \quad (2)$$

$$B = \frac{1}{(V_{ci} - V_R)^2} [4(V_{ci} + V_R)(\frac{V_{ci} + V_R}{2V_R})^3 - (3V_{ci} + V_R)] \qquad (3)$$

$$C = \frac{1}{(V_{ci} - V_R)^2} [2 - 4(\frac{V_{ci} + V_R}{2V_R})^3] \qquad (4)$$

Generally, the wind speed is changed greatly according to space and time. Until now, the distribution of wind speed is known to be presented as Weibull probability distribution similar to normal probability distribution and the model is already developed. After getting the output probability distribution function of multi-state wind turbine generator which combines wind power model considering the uncertainty of wind power source supply, and wind turbine generator output characteristic model, to calculate the expectation output using previous value is encouraged. Fig. 3 explains that the state of acquiring multi-state. Available capacity probability distribution function of wind turbine generator is followed by combing wind speed probability density function and wind turbine generator's output curve. In this figure (PSWbi, PBSWbi) mean output and wind speed of wind turbine generator when the wind speed is SWbi, wind speed of i wind speed band. The equation (5) formulates predicted output expectation EPkt at t time zone of k wind farm according to numerical analytical solution suggested by this research, ρki is set of wind speed on k wind farm in certain time period t.

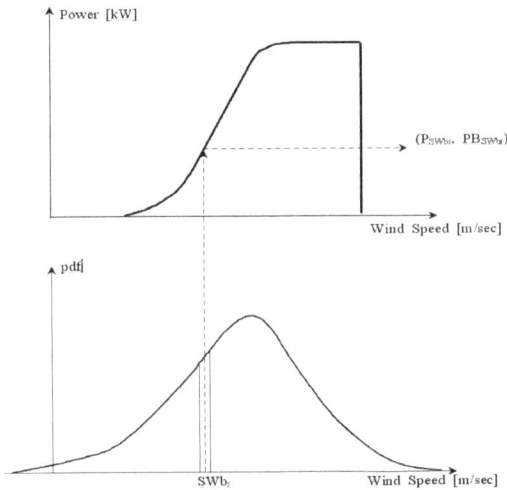

Fig. 3. Probabilistic output prediction model of wind turbine generator.

$$EP_{WTG_{kt}} = \sum_{i \in \rho ki} (P_{SWi} \times PB_{SWbi}) \qquad (5)$$

3.2 Probabilistic output prediction model of solar cell generator[3,4]

The input and output curve of the solar cell generator isn't well known. In this paper approximated output-equation according to the change of solar radiation is used as equation (6)[3,4]. The approximated output curve of solar cell generator efficiency according to the changes of the temperature and solar radiation affecting the out put of solar cell generator can be made as shown as Fig. 4.

$$\begin{aligned} P_{bi}(G_{bi}) &= \frac{\eta_c}{R_c}(G_{bi}^2), \, 0 \le G_{bi} < R_c \\ &= \eta_c G_{bi}, R < G_{bi} \le G_{std} \\ &= P_{sn}, G > G_{std} \end{aligned} \qquad (6)$$

Where,
i : Number of solar radiation band(=1,2,..,Nb)
Nb : Total number of band
Pbi : Solar cell generator(SCG) output of #i band [MW]
η_c: Efficiency of solar radiation (Psn/Gstd)
Gbi : #i Solar radiation of band [W/m2]
Gstd : Solar radiation under standard condition (Normally 1,000[W/m2])
Rc : Point of inflection of output curve (Normally 150[W/m2])
Psn : Rated power of SCG [MW]

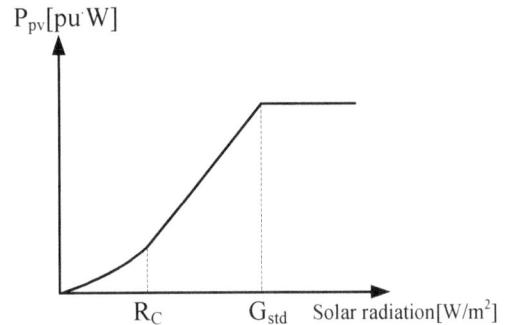

Fig. 4. Output curve of solar cell generator.

Fig. 5 demonstrates that the state of getting multi-state available capacity probability distribution function of solar cell generator after combing solar cell generator output curve and solar radiation probability distribution function. Pi and PBi means that the input and solar radiation of solar cell generator when solar radiation of i solar radiation band is SRbi. Available capacity probability distribution function of solar cell generator can be induced a the equation (7).

$$f_{oi} = f(P_{oi}, PB_{oi}) \tag{7}$$

Where,

foi : #i Available capacity probability distribution function of solar cell generator

Poi : #i Available capacity variables of solar cell generator(=Psni-Pik)[MW]

Psni : #i Rated capacity of solar radiation [MW] Rated capacity of solar cell generator

PBoi : #i Available capacity probability variables of solar cell generator(=PBi)

EPkt, t time zone predicted output expectations of k solar farm according to numerical analytical solution, therefore, is formulated as Eq(8). Where, ρki is the set of solar radiation of k solar farm at a certain t time zone.

$$EP_{SCGkt} = \sum_{i \in \rho ki} (P_{SWi} \times PB_{SWbi}) \tag{8}$$

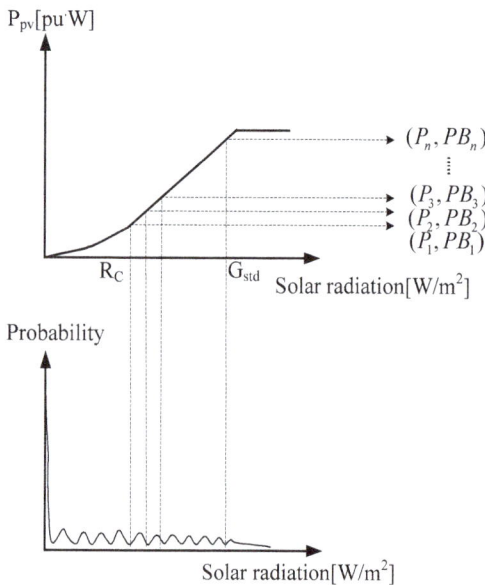

Fig. 5. Combining of solar radiation distribution model and output curve in order to make the available capacity probability distribution function of solar cell generator.

4. Study Case

In this study used generator output and weather data(wind speed and solar radiation) in April, 2014.and weather forecast data is April.25 to 26,2014

4.1 Probability predicted output of solar cell generator

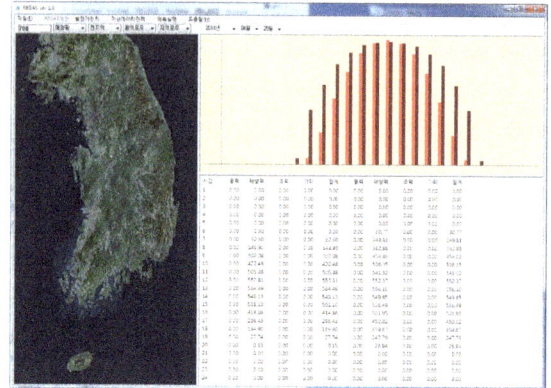

Fig. 5. Solar generator output (2014.04.25).

Table 1. Solar output data [MWh] (2014.04.25)

	Actual solar generator power	Predicted solar generator power	Error rate[%]
5	0	0	
6	0.3	30.77	10157
7	148.9	249.81	68
8	302.08	362.88	20
9	422.48	454.03	7
10	505.88	506.35	0
11	550.81	541.32	2
12	564.49	552.37	2
13	548.13	556.1	1
14	501.1	549.65	10
15	414.86	536.49	29
16	286.43	501.95	75
17	134.9	450.02	234
18	27.74	359.67	1197
19	0.15	247.79	165093
20	0	23.84	23.84
21	0	0	

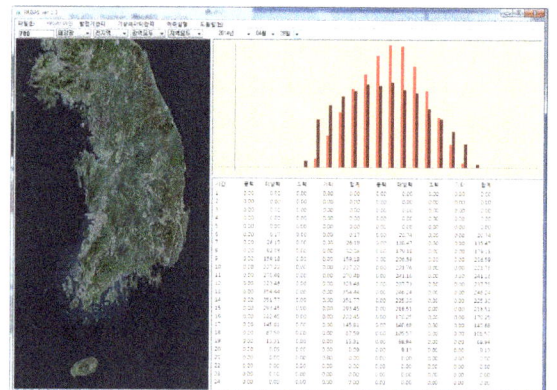

Fig. 6. Solar generator output (2014.04.26).

Table 2. Solar output data [MWh] (2014.04.26)

	Actual solar generator power	Predicted solar generator power	Error rate[%]
5	0	0	
6	0.17	20.74	12100
7	26.19	138.47	429
8	92.03	179.18	95
9	159.18	206.59	30
10	227.22	221.76	2
11	270.48	241.16	11
12	323.46	237.73	27
13	354.44	246.24	31
14	351.77	225.3	36
15	293.45	216.51	26
16	222.45	170.25	23
17	145.81	140.68	4
18	67.59	105.57	56
19	13.31	168.94	1169
20	0.09	9.13	10044
21	0	0	

In the case of solar generator, an error ranging from around 60MWh to 100MWh is visible in reference to overall per-time output by using the per-time table.

4.1 Probabilistic output prediction of wind power generator

In the case of wind generator, an error ranging from around 30MWh is visible in reference to overall per-time output by using the per-time table.

Table 3. Wind output data [MWh] (2014.04.25)

	Actual solar generator power	Predicted Wind generator power	Error rate[%]
1	49.12	60.52	23
2	48	59.57	24
3	48.82	58.26	19
4	40.25	57.78	44
5	41.59	56.91	37
6	46.36	56.03	21
7	37.68	59.54	58
8	22.57	63.45	181
9	14.01	65.97	371
10	8.29	67.66	716
11	6.62	69.7	953
12	7.4	71.14	861
13	9.75	75	669
14	16.89	79.94	373
15	27.57	83.67	203
16	31.65	84.23	166
17	34.64	85.02	145
18	30.39	85.72	182
19	35.69	80.61	126
20	33.32	74.99	125
21	30.65	67.03	119
22	29.69	60.42	104
23	41.8	51.72	24
24	46.15	48	4

Table 4. Wind output data [MWh] (2014.04.26)

	Actual solar generator power	Predicted Wind generator power	Error rate[%]
1	51.98	60.63	17
2	51.36	61.32	19
3	56.73	61.73	9
4	55.57	61.52	11
5	53.71	61.3	14
6	34.93	61.17	75
7	32.52	65.27	101
8	33	69.97	112
9	51.17	73.11	43
10	44.21	75.58	71
11	64.69	78.63	22
12	95.07	80.88	15
13	116.99	85.14	27
14	113.7	90.31	21
15	111.23	94.6	15
16	112.25	91.76	18
17	121.31	88.4	27
18	111.46	86.22	23
19	85.29	84.04	1
20	86.14	80.95	6
21	106.05	78.39	26
22	123.85	76.98	38
23	140.82	74.84	47
24	161.98	72.97	55

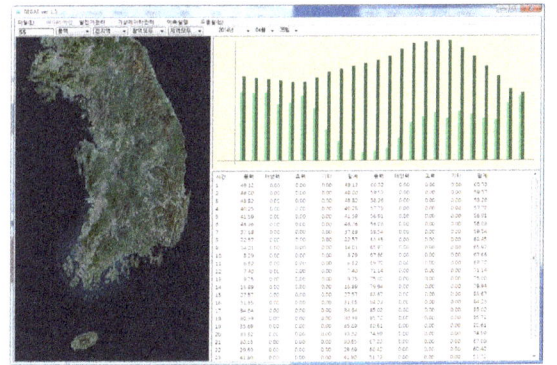

Fig. 6. Wind generator output (2014.04.25).

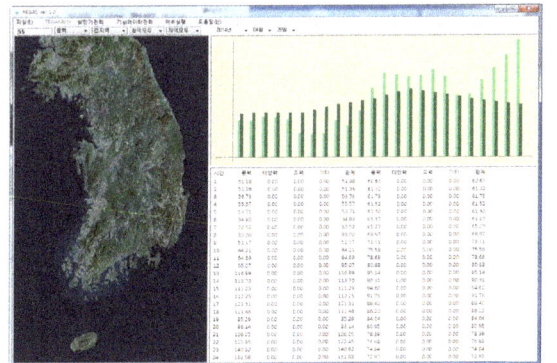

Fig. 7. Wind generator output (2014.04.26).

5. Conclusion

In this paper, the method of estimating supply capacity of wind and solar generator was introduced through the use of output forecasting method by considering the Weibull probability distribution function. Also, this paper introduced the development of a system that can forecast the non-scheduled generators output of the next day by inputting the previous day's weather forecast data. In future researches, if errors in wind forecasts can be reduced and output errors can be mitigated through studies in rainfall probability conversion method of solar generator, REGASII will enable the acquisition of estimation standards of appropriate power supply capability for real operations.

Acknowledgements

This work was supported by the Korean National Research Foundation (NRF) (No,#2012R1A2A2A01012803) and Korea Power Exchange(KPX).

References

[1] Jeongje Park, Wu Liang, Jaeseok Choi, Seungil Moon, "A Study on Probabilistic Reliability Evaluation of Power System Considering Wind Turbine Generators", The Transactions of the Korean Institute of Electrical Engineers, Vol.57, No.9, pp.1491-1499, September 2008. ISSN :1975-8359.

[2] Jeongje Park, Wu Liang, Jaeseok Choi, Junmin Cha, "Probabilistic Production Cost Credit Evaluation of Wind Turbine Generators" The Transactions of the Korean Institute of Electrical Engineers, Vol.57, No.9, pp.2153-2160, December 2008. ISSN :1975-8359

[3] Jeongje Park, Wu Liang, Jaeseok Choi, and Junmin Cha, "A Study on Probabilistic Reliability Evaluation of Power System Considering Solar Cell Generators", J. of KIEE, Vol.58, No.3, pp.486-495, March 2009.

[4] Jeongje Park, Jaeseok Choi, " A Study on Probabilistic Production Costing for Solar Cell Generators", J. of KIEE, Vol.58, No.4, pp.700-707, April 2009. ISSN : 1975-8359.

[5] Wu Liang, Jeongje Park, Jaeseok Choi, Junmin Cha, Kwangyeon Lee, "A Study on Wind Speed Using the ANN at Cheju Island", KIEE, May 22-23, 2009.

[6] Jaeseok Choi, Jeongje Park, Kyeonghee Cho, Taegon Oh and Mohammad Shahidehpour, "Probabilistic Reliability Evaluation of Composite Power Systems

36

Study on the Influence of UHVDC Dynamic Response on Recovery Characteristic of Receiving End Power Grid

Zheng Chao[†], Zhang Kai, Sheng Canhui*, Lin Junjie*, Xue Jinying***,
Chen Dezhi*, Zhang Zhiqiang* and Luo Bangyun***

Abstract – Aiming at the three UHVDC controller model in the electromechanical simulation software BPA which widely used in China, the structural features were analyzed. By the time-domain simulation, the influence of different simulation models and the key controller parameters on the recovery characteristics of the receiving end power system were studied. Simulation results show that the commutation failure simulated method impacts the recovery characteristic significantly for weak power grids, increasing the voltage measurement time constant of constant dc power control module and enhancing the start voltage threshold of voltage dependent current order limit (VDCOL) can decrease the reactive power consumption of inverter during the disturbance, and can help to improve the recovery characteristic of receiving end power grid.

Keywords: UHVDC, Controller simulation model, Receiving end power grid, Recovery characteristic, Commutation failure, Voltage measurement time constant, VDCOL

1. Introduction

Ultra high voltage direct current (UHVDC) transmission plays an important role in the trans-regional large-capacity transmission and is an effective transmission technology to optimize the allocation of resources[1,2]. In china, the primary energy and load centers have the reverse distribution characteristic. In order to transmit the large-capacity power generated by hydro, thermo and renewable resources in the western and northern regions to load density areas located at center and east regions, more than thirty UHVDC transmission lines will be put into operation until 2020. Majority of these UHVDC transmission lines are with dc voltage at ±800kV or higher.

After the UHVDC infeed, the coupling degree between UHVDC and receiving end power grid will be aggravated, the dynamic characteristic and stability level will be changed significantly[3]. For the stability and security analysis of UHVDC/AC hybrid system, the selected electromechanical simulation model of UHVDC will affect the coupling characteristic and the stability level of receiving end power grid[4,5]. With the rapid development of UHVDC, the east china grid and center china grid will be

formed mulit-infeed UHVDC receiving end power grid. Under this background, it is necessary to study and evaluate the influence of dynamic characteristic of UHVDC during disturbance on the recovery characteristic of ac system, identify the critical parameters of dc controller and propose the optimized tuning strategy.

In this paper, the structure characteristics of three different UHVDC controllers simulated by BPA software which widely used in the power system electromechanical transient analysis in china are studied. Aimed to the receiving power gird with ±800kV/8000MW UHVDC infeed, the influences of controller's electromechanical transient simulation model and critical parameter's settings on recovery characteristic under large disturbance are revealed. Simulation results show it is necessary to enhance the UHVDC simulation modeling accuracy in order to master power system dynamic behavior characteristic accurately, guide receiving power planning and formulate the control measures effectively.

2. Electromechanical Transient Simulation model of UHVDC

2.1 quasi steady-state model of converter

UHVDC converter consists by semi-controlled thyristors and is a typical time-varying circuit. In the power system electromechanical transient simulation, only the fundamental

[†]   Corresponding Author: China Electric Power Research Institute, China (zhengch@epri.sgcc.com.cn)

*   China Electric Power Research Institute, China

**   North China Electric Power University, China

***   Xinjiang Hami Electric Power Bureau, Hami 839000, China

component in ac side and the direct component in dc side are considered, the quasi steady-state model of UHVDC converter can be simulated by the followed equations:

$$U_{d0} = 3\sqrt{2}KBU_{ac} / \pi \qquad (1)$$

$$U_d = U_{d0} \cos\alpha - 3BX_c I_d / \pi \qquad (2)$$

$$P_d = U_d I_d \qquad (3)$$

$$Q_d = P_d \tan\varphi \qquad (4)$$

$$\mu = -\alpha + \cos^{-1}(\cos\alpha - 2X_c I_d / \sqrt{3}E_m) \qquad (5)$$

In the equations, U_{ac} and E_m are the phase-to-phase RMS voltage and the peak voltage of commutating bus respectively. U_d and U_{d0} are the dc operation voltage and no-load voltage. K is the converter transformer ratio. B is the number of commutation bridge of each pole. α is the firing delay angle. μ is the overlap angel. X_c is the commutation reactance. $3BX_c I_d/\pi$ is the commutation voltage drop. I_d is the dc current. P_d is dc active power. Q_d is the reactive power consumed by converter. φ is the converter's power factor.

2.2 HVDC controller's simulation model

2.2.1 Pacific intertie HVDC controller's model

Earlier developed pacific intertie HVDC controller has been used ever since. This model is corresponding to the D card model in BPA software. The control modes which can be simulated by this model include the constant power and constant dc current control, constant α_{min} control and constant γ_{min} control.

The features of pacific intertie HVDC simulation model including:

1) The input of constant power control modular and the voltage dependent current order limit (VDCOL) modular use the same dc voltage measurement channel. This means the voltage measuring time constant cannot be set respectively.

2) The adjustable parameter of VDCOL modular is only the value of voltage starting threshold. This makes it hard to precisely simulate the characteristic of VDCOL as used in modern HVDC system.

3) This model cannot simulate the commutation failure process except for equipped with the DF card which use the commutation voltage as the characteristic quantity to distinguish the commutation failure. When the commutation voltage less than threshold settings or the changing rate exceed threshold settings, the

commutation failure will be assumed occurring. Then the active power and the reactive power of the inverter will be set to zero, which have differences from the actual power characteristics when commutation failure happened.

4) Less regulation means to adjust dc controll's dynamic response characteristics. It is difficult to precisely simulate the actual dc controller's characteristic just by adjusting the parameters of VDCOL and the proportion differential current regulation controller.

5) Equipped with additional controller model, the function of emergency power control, active power modulation and bilateral frequency modulation can be simulated.

2.2.2 CIGRE HVDC controller's model

The CIGRE (international council on large electric systems) HVDC benchmark model is the standard testing model for HVDC controller's characteristic research. The controller is belonging to the pole control layer. The model had been developed in BPA corresponding to the DM/DZ card model[6].

The basic control modes of the rectifier include constant power control, constant current control and α_{min} control. The control modes of inverter include constant extinction angle γ control and constant current control.

The features of CIGRE HVDC simulation model including:

1) The input of constant power control modular and the VDCOL modular use the independent dc voltage measurement channel. This means the voltage measuring time constant can be adjusted separately according to the needing of the receiving end power gird.

2) Three segment polyline can be used to simulate the characteristic of VDCOL. The dc current order varying rate can be limited to ensure the current changing smoothly and help to reduce the reactive power demand caused by the rapid dc current increasing after fault disturbed.

3) Uses the inverter extinction angel γ_{min} as characteristic quantity to distinguish the commutation failure automatically during simulation.

4) More regulation means to adjust dc controller dynamic response characteristics such as varying the VDCOL polyline curve or parameters of PI controller can be used to optimize the dynamic response of receiving end power grid.

5) Equipped with additional controller model, the function of emergency power control, active power modulation

and bilateral frequency modulation can be simulated.

2.2.3 SEIMES HVDC controller's model

To accurately simulate the dynamic characteristic of Gui-Guang and Tian-Guang HVDC transmission project, the SEIMES HVDC controller's model had been developed in BPA corresponding to the DN/DZ card model[6].

According to the principle of reserving main control modular which has strong correlation with the electromechanical transient response characteristic, the complex control system was simplified.

Compared with CIGRE HVDC model, the SEIMES HVDC model is closer to the actual characteristics and more perfect in dynamically limiting link.

3. UHVDC Receiving End Power Grid and ESCR evaluation

The local power gird of a provincial power gird designed according to "Twelfth Five-Year Plan" is showed as Fig. 1. The HVDC connecting with rectifier A and inverter A is a back-to-back project with small capacity. The HVDC connecting with rectifier B and inverter B is UHVDC with ±800kV dc voltage and 8000MW transmission capacity. The local power grid is connected with main grid by three ac channels. The load level of the local gird is 13330MW and the total output power by ac channels is 2730MW.

The UHVDC control mode is set as constant power and the dc voltage measuring time constant is set as 0.09s. The value of voltage starting threshold of VDCOL is set as 0.7p.u. and the γ_{min} is set as 7°.

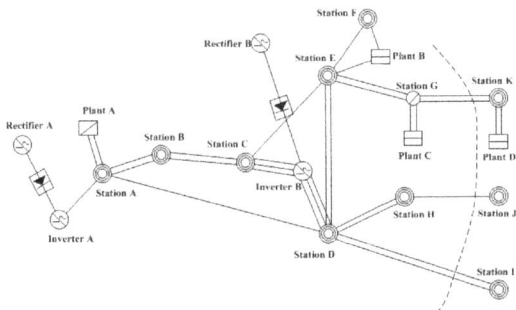

Fig. 1. The receiving end power system of UHVDC.

Under different structures of the receiving power grid, the calculation result of effective short circuit ratio (ESCR) for the UHVDC inverter is showed as table.1. From the static point of view, the ESCR can be maintained above 3.0. The influence of the ac channel connected station D with station I breaking off on ESCR is greater relatively. It shows this the ac channel breaking off will impact the security and stability more serious.

Table 1. ESCR under different grid structures

Operation Mode	ESCR	△ESCR
Normal	4.59	-
Substation G-K breaking off	4.09	-0.50
Substation H-J breaking off	4.31	-0.28
Substation D-I breaking off	3.81	-0.78

4. Influence of UHVDC dynamic response on recovery characteristic of receiving end power grid

4.1 Fault response under different UHVDC simulation models

When suffered by single fault disturbance, the receiving end power grid can maintain security and stability. Under the ac channel connected station D and I suffered three-phase permanent short-circuit fault and breaking off, the voltage recovery characteristic using different UHVDC electromechanical transient simulation model is showed as Fig. 2. It can be seen from the simulation results got form pacific intertie HVDC model that the converter bus voltage drop rapidly and loss voltage stability before losing enerator's rotor angle stability. By using CIGRE and SEIMES UHVDC models, the voltage can maintain stability and the recovery characteristic are roughly the same.

The pacific intertie HVDC model cannot simulate the commutation failure. Due to the inverter's bus still maintain small amplitude voltage during disturbance, the reactive power remains a certain amount of consumption according to the quasi steady-state model of converter. Furthermore, along with the active power fast recovery after fault clearing, the quality of reactive power consumption increases rapidly. These reactive consumption characteristics are unfavorable to the voltage stability of receiving end power grid.

By using the CIGRE and SEIMES HVDC model, the extinction angle of inverter will less than the γ_{min} during disturbance, the transient simulation program discriminate the commutation failure occurred and set the inverter's active and reactive power to zero automatically. After fault cleared, the active power increases slowly and the growth rate of reactive power consumed by inverter is relatively small. These reactive consumption characteristics are favorable to the voltage stability of receiving end power grid.

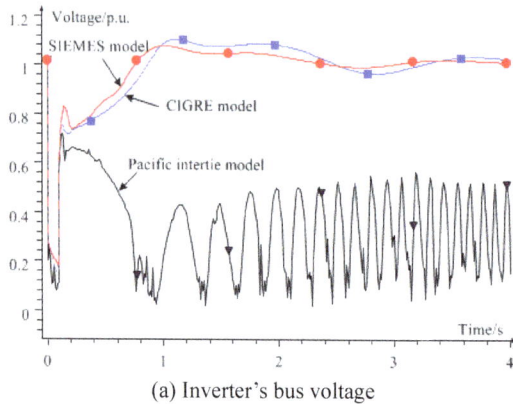

(a) Inverter's bus voltage

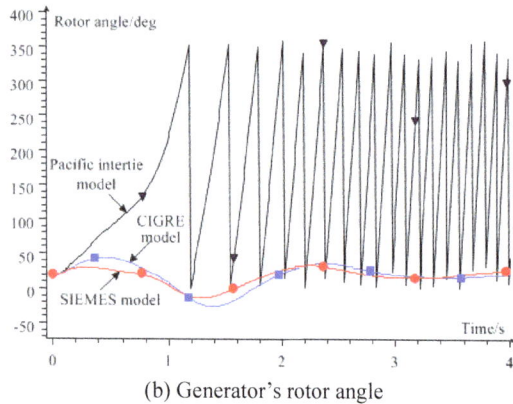

(b) Generator's rotor angle

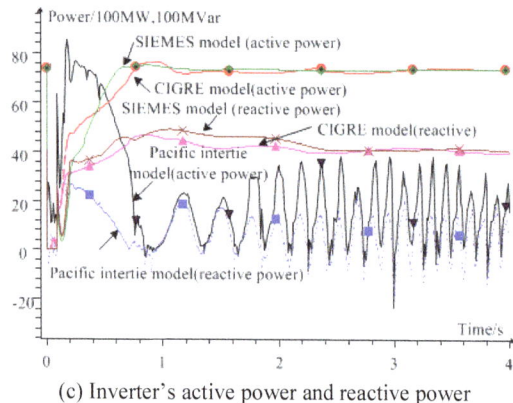

(c) Inverter's active power and reactive power

Fig. 2. Transient response with different UHVDC model.

In summary, the UHVDC transient response will affect the voltage recovery characteristics of receiving end power

grid significantly.

4.2 Influence of commutation failure simulation method on fault recovery characteristic

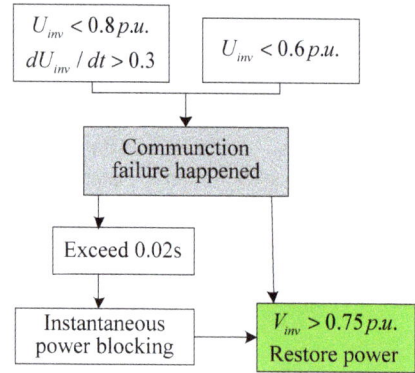

Fig. 3. Commutation failure criteria.

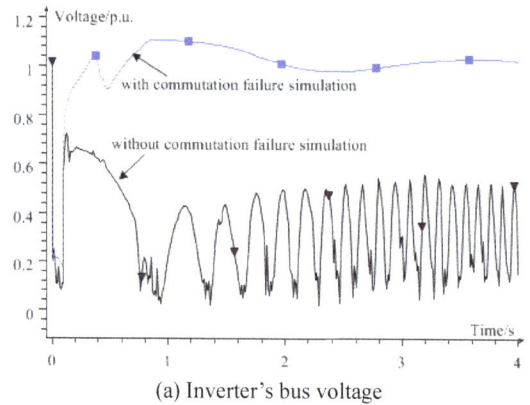

(a) Inverter's bus voltage

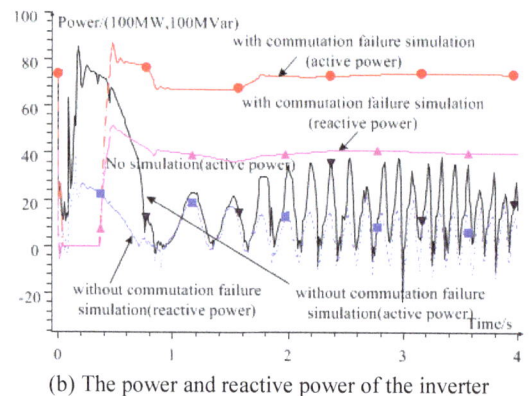

(b) The power and reactive power of the inverter

Fig. 4. Recovery characteristic with or without failure commutation simulation.

Using the inverter's bus voltage as the characteristic quantity and combined with corresponding voltage threshold, the pacific intertie HVDC model can simulate the commutation failure approximately. However, due to the voltage threshold settings are closely related with the grid characteristics, so the simulation method lacks adaptability and versatility.

In accordance with the discriminant logic of commutation failure and voltage threshold settings as showed in Fig. 3, the compared of voltage recovery characteristic curve with or without commutation failure simulation after the ac channel connected station D and I suffered three-phase permanent short-circuit fault and breaking off, is showed as Fig. 4.

It can be seen from the simulation results that the active power and the reactive power will be set to zero when commutation failure discriminated. Then the reactive power outputted from the filter will inject into the receiving end power grid and the voltage recovery characteristic will be improved significantly.

So it should make a detailed simulation for commutation failure characteristics to avoid getting optimistic conclusions on voltage stability, especially for the weak receiving end power grid.

2.4 Influence of controller parameters on fault recovery characteristic

2.4.1 DC voltage measuring time constant

Under constant power control mode, the time constant of dc voltage measuring module has great influence on the recovery characteristic of receiving end power grid. During the voltage dropping process after fault disturbed, the dc current order will increase faster when take smaller time constant. This will cause the reactive power consumption increasing substantially and threat the voltage stability of the receiving end power grid.

The simulation results corresponding to the dc voltage measuring time constant T_d set as 0.09s and 9.0s are showed as Fig. 5. The VDCOL limit the dc current only when the dc voltage less than the voltage starting threshold U_{lim}. Therefore, the dc current order will research the maximum current corresponding to the overload current of converter valves before dc voltage greater than the U_{lim}. So during fault disturbance and dc voltage still greater than U_{lim}, the smaller T_d will cause the dc current order increase more rapidly. The increased reactive power consumption will make the receiving end power grid lose voltage stability. Increasing T_d, the dc current increasing rate can be limited and the inverter's reactive power consumption can

be significantly reduced.

So increasing the dc voltage measuring time constant can improve the voltage stability just as showed in Fig. 8.

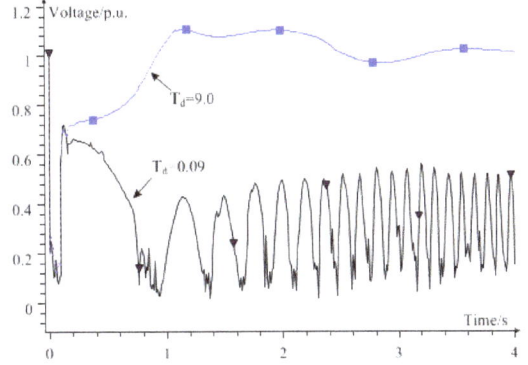

(a) Inverter's bus voltage

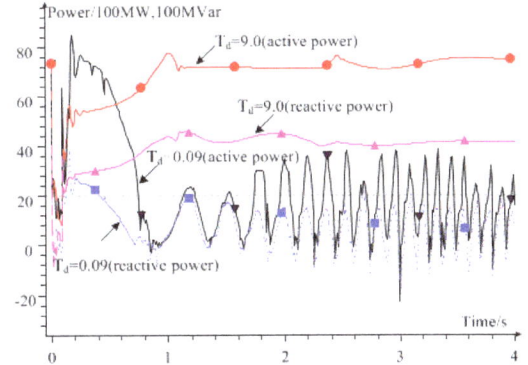

(b) Active and reactive power of the inverter

Fig. 5. Influence of T_d on transient response with the Pacific intertie HVDC model.

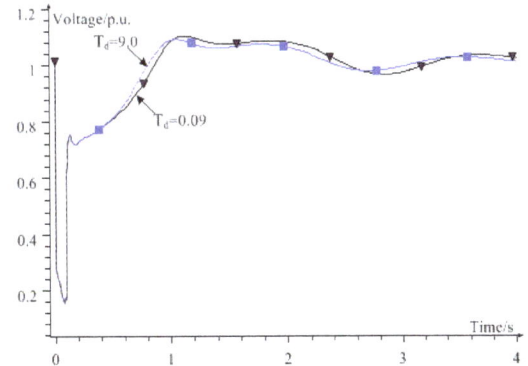

Fig. 6. Influence of the time constant on transient response with the CIGRE HVDC model.

The CIGRE and SEIMES HVDC model use three segment polyline to simulate the characteristic of VDCOL and can flexibly set according to the actual dc controller's characteristics. The influence of T_d on recovery characteristic using CIGRE HVDC model is showed as Fig. 6. It can be seen from the simulation results that the dc current order is effectively limited by VDCOL and the influence of T_d on recovery characteristic is not obvious.

Furthermore, due to pacific intertie HVDC model using the same dc voltage measuring channel, therefore increasing T_d is not conducive to limit the dc current order increasing of VDCOL.

2.4.2 Voltage starting threshold of VDCOL

The value of voltage starting threshold U_{lim} of VDCOL module is an important parameter to the recovery characteristic of receiving end power grid. Enhancing the U_{lim} can play VDCOL function faster and help to maintain the voltage stability. The compared recovery characteristic curves with U_{lim} setting at 0.7p.u. and 0.9p.u., are showed as Fig. 7(a) and Fig. 7(b) respectively .

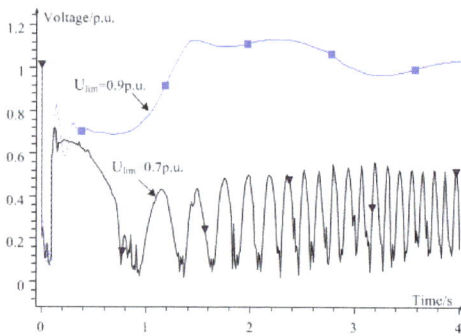

(a) Using the pacific intertie HVDC model

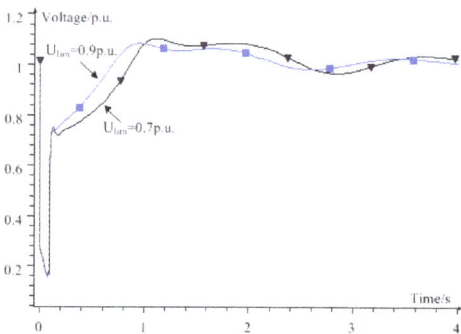

(b) Using the CIGRE model

Fig. 7. Influence of U_{lim} on recovery characteristics.

From the simulation results, it can be seen that the value of U_{lim} has significantly affect on the voltage recovery characteristic for the different HVDC model. It is an effective measure to enhance voltage stability level by increasing U_{lim}.

5. Conclusion

(1) The UHVDC inverter has the characteristics such as complex dynamic response and reactive power consumed in large quantity, and so on. The difference of UHVDC dynamic response will affect the voltage stability level of receiving end power grid significantly, especially for the weak ac system. In order to provide the better guidance for power system planning and operating, it should be modeled the HVDC precisely according to the actual project.

(2) Reasonable grid structure is the material basis for the security and stability of the power grid. Along with the UHVDC vigorously developed, it should expedite UHVAC power grid construction to ensure the dc power consumed safely and reliably.

(3) Combined with the UHVDC/AC hybrid system characteristic, it is proposed to optimize the dc voltage measuring time constant in constant power control module and the value of voltage starting threshold U_{lim} in VDCOL module in order to enhance the stability level of receiving end power grid.

Acknowledgements

Supported by State Grid Corporation of China, Major Project on Planning and Operation Control of Large Scale Grid(SGCC-MPLG001-2012).

References

[1] SUI Guoping. Power safe system analysis of ±800 kV Yun-Guang HVDC when bipole blocking[J]. High Voltage Engineering, 2012, 38(2): 421-426.

[2] MA Weiming, NIE Dingzhen, CAO Yanming. Optimizing design of UHVDC converter stations[J]. High Voltage Engineering, 2012, 38(12): 3109-3112.

[3] CHENG Limin, LI Xiyuan, LIU Jian. Optimal nonlinear coordinated control to enhance transient stability of AC/DC power system[J]. High Voltage Engineering, 2011, 37(4): 1029-1034.

[4] WU Hongbin, DING Ming, LIU Bo. Analysis on commutation process of converters in transient simulation of hybrid AC/DC power systems[J]. Power System Technology, 2004, 28(17): 11-14.

[5] CHEN Shuiming, YU Zhanqing, XIE Haibin, et al. Transient interaction of HVAC and HVDC in converter station in HVDC, III: effects of AC sides on DC sides[J]. High Voltage Engineering, 2011, 37(5): 1082-1092.

[6] User manual of PSD-BPA transient stability program[M]. Beijing: China electric power research institute, 2008, 187-213.

37

Study of MPPT Control Method for Large-scale Power Conditioning System in Hokuto Mega-solar System

Hiroo Konishi[†]

Abstract – Feed-in Tariff system was started from July 2012 in Japan. According to the act, the electricity energy generated from large-scale Potovoltaic (LSPV) generating stations is bought for 42 yen per kWh including tax for 20 years. This price is about 3 times as large as a normal electricity rate. However it is still important to improve efficiency of the LSPV system in order to return high investment as soon as possible. The New Energy and Industrial Technology Development Organization (NEDO) advertised for consignment research business called "Verification of Grid Stabilization with Large-scale Photovoltaic Power Generation Systems" in 2006. 2MW PV system was constructed for this research in Hokuto City and five-year tests were finished in 2011. In the research, we clarified a MPPT control method for large-scale power conditioning systems (LS-PCS) to get more PV energy efficiently. Results were confirmed by ATP (Alternative Transients Program) simulations and also the field. The verification test results of grid stabilization systems by the developed LS-PCS and PV systems were also described and discussed.

Keywords: Power conditioning system, MPPT, ATP simulation, PV system evaluation

1. Introduction

A Solar generation system is one of the promising measures for reducing global warming and is important for future energy resources, especially in Japan. An installed capacity target of solar generation systems in our country is planned to 28GW in 2020, but the cumulative installed capacity is still 4.80GW in 2011. About 80% of its capacity is mainly residential use such as roof-top types. They are 3 kW to 5 kW systems and very small size. Some large-scale_ solar generation systems should be constructed to attain the target. Therefore, feed-in tariff system (FIT) was started from July 2012 in Japan. According to the act, the electricity energy generated from large-scale photovoltaic (LSPV) generating stations is bought for 42 yen per kWh including tax for 20 years. This price is about 3 times as large as a normal electricity rate. However it is still important to improve efficiency of the LSPV system in order to return high investment as soon as possible.

The New Energy and Industrial Technology Development Organization (NEDO) advertised for consignment research business called "Verification of Grid Stabilization with Large-scale Photovoltaic Power Generation Systems" in 2006. Two sites were selected for the verification tests, which were Hokuto City in Yamanashi Prefecture and

Wakkanai City in Hokkaido. Mega-solar systems (PV systems) of about 2MW and about 5MW were constructed, respectively. Five-year's evaluation and verification test were finished [1-5]. In the research, we clarified a MPPT control method to get more PV energy efficiently in this paper first, and the verification test results of grid stabilization systems by the developed LS-PCS and PV systems were also described and discussed.

2. Hokuto mega-solar system

Fig. 1 shows configuration of Hokuto mega-solar system. In first stage, we constructed 600kW PV systems consisting of 24 kinds of advanced type PV modules in 2006. They have seven different PV technologies; they are single crystalline silicon (sc-Si), multi-crystalline silicon (mc-Si), amorphous silicon (a-Si), a-Si/μ-Si multi-junction, sc-Si backside contact, a-Si/sc-Si with hetero-junction and cupper indium (gallium) di-selenide modules (CI(G)S). Sixty 10kW PV arrays with the same 30-degree tilt angle and a due south orientation were constructed for evaluating the generation characteristics. They were connected to a 6.6kV distributed grid through sixty conventional 10kW PCSs (power conditioning systems) and a 6.6kV/210V transformer.

In the second stage started from 2008, we constructed about 1200kW PV systems consisting of 4 kinds of PV

---

† Corresponding Author: *NTT Facilities Inc., Tokyo, Japan*
(konish36@nttf.co.jp)

modules which were selected from the first stage evaluation. For connecting the PV systems to an existing power grid, we developed a novel 400kW large-scale PCS (LS-PCS) to prevent the power grid from being affected undesirably by the PV system behaviours. Developed grid stabilization controls and function of the LS-PCS were voltage fluctuation suppression control due to the PV generation fluctuations, harmonics current suppression control and fault-ride-through function in the grid faults. Specifications and developed targets are shown in table 1. At first, we were connected the LS-PCS to the 6.6kV distribution grid through a 6.6kV/420V transformer, then a 66kV/6.6kV transformer was equipped and the connection was changed from the 6.6kV distributed grid to a 66kV extra-high voltage transmission grid for mega-solar generation in 2009. The PV modules were selected for future mega-solar systems in consideration of cost [yen/W], generation efficiency [Wh/W], compactness [m^2/W], gentleness for the environment [J/w], productivity ability of the makers, and so on. As a result, two 200kW PV systems consisting of the sc-Si/a-Si with hetero-junction, 200kW sc-Si PV system, 200kW CI(G)S PV system and two 200kW mc-Si PV systems were connected to the three 400kW LS-PCS.

In the third stage, other 4 kinds of 10kW PV systems with newly developed advanced types were constructed. Finally, total installed capacity in Hokuto becomes 1,840kW in 2010. And the mega-solar system has been put into test operation.

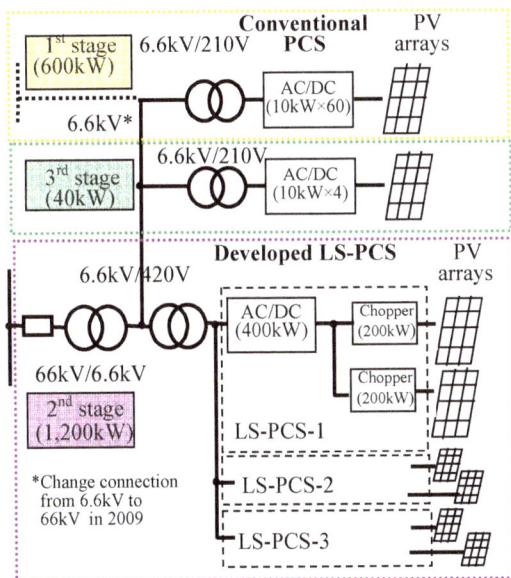

Table 1. Specification and developed target of LS-PCS

Items	Contents & specifications
AC voltage	420Vac $\pm$ 10%
Converter transformer	Transformer-less
DC voltage	600Vdc
Input DC voltage	230-600Vdc
Switching freq.	4kHz
Conversion efficiency	> 95% from 30% - 100%
Control / protection functions	/ MPPT (by choppers)
	/ Suppression of Δ Vac
	/ Suppression of low-order harmonics
	/ Continuous operation

3. MPPT control method for LS-PCS

At first, the LS-PCSs were controlled by (DC) current-control method in order to operate continuous operation easily when the AC grid faults occur, because the protection from over-currents and over-voltages in the faults can perform surely. For this reason, a maximum power point tracking (MPPT) of the PV arrays were also carried out by current-control method, which finds the maximum power point (MPP) by scanning DC currents. However, we found that the method did not get PV generation outputs efficiently.

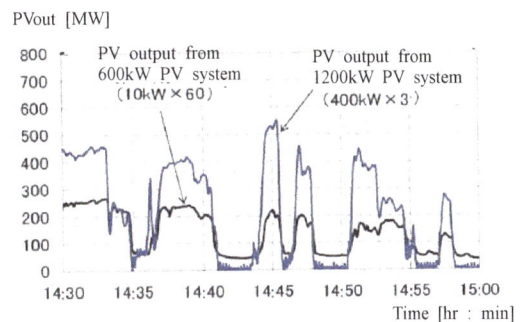

Fig. 2. PV outputs due to the difference in MPPT control method.

Fig. 2 shows PV generation outputs at the time from 14:30 to 15:00 of one cloudy day. The blue line shows 1,200kW PV system outputs from the developed three 400kW LS-PCSs with current-controlled MPPT, while the black line shows 600kW PV system outputs from conventional 10kW PCSs with voltage-controlled MPPT, When PV modules become shade, PV outputs of current-controlled MPPT often become zero compared to those of voltage-controlled MPPT.

Fig. 1. Hokuto mega-solar system.

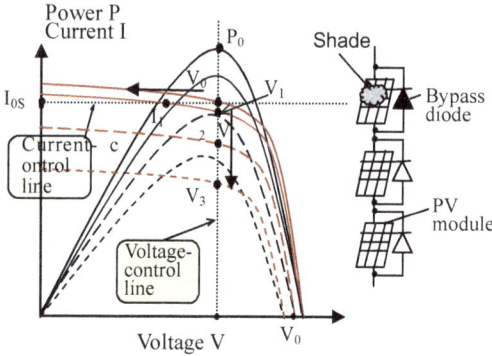

Fig. 3. Power and current versus voltage characteristics of arrays.

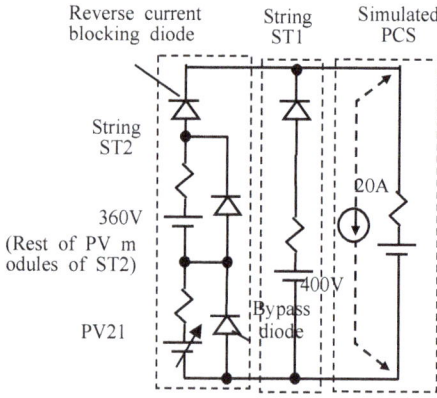

Fig. 4. Simulation circuit.

The reasons can be explained as follows.

In the current-controlled MPPT, when some of the PV modules become shade, as the currents are kept on the current setting value controlled by MPPT before the shade occurring. The bypass diodes parallel-connected to the shaded modules turn on and the current flow moves from the modules to the bypass diodes. Then the PV system outputs decrease and the MPPT loses the MPP. Once the bypass diodes turn-on, MPPT control needs more time to move to the new MPP when sunshine resumes. Because currents are not cut and the bypass diodes continue to flow currents, the MPPT control can not move to the new MPP.

Fig. 3 shows power and DC current characteristics of an array versus DC voltages. The figure indicates mentioned above. In the figure, Po, Vo and Io are MPP in normal condition. When shades occur, the operating points go along a current-control line to a direction of I_{os} through I_1 by bypassing the shaded PV modules in the current-controlled MPPT.

On the other hand, in the voltage-controlled MPPT, when a shade occurs, the operating points move to V_1, V_2, V_3, $\cdots$ along a voltage-control line shown in the same figure. As the voltage-controlled MPPT keeps on DC voltages to the controlled setting voltage value instantly, bypass diodes do not turn on. The strings which can not output the MPPT controlled setting voltage value, are separated from under the MPPT control by a reverse current blocking diode. Therefore, it is easy to move to the next MPP when sunshine resumes.

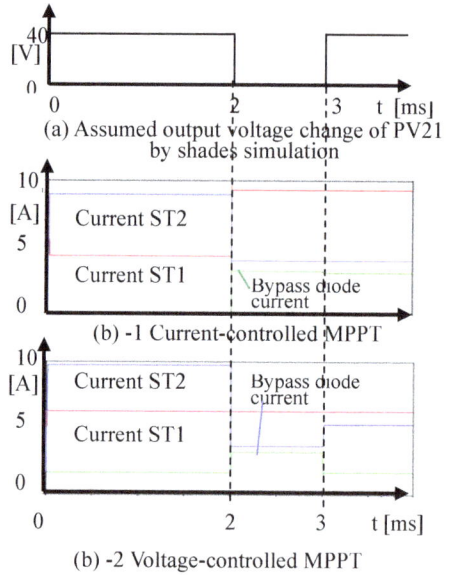

Fig. 5. Analysis results of flowing currents in strings and in bypass diode.

Fig. 4 shows ATP simulation circuit of a PV array controlled by LS-PCS. The array consists of 2 strings ST1 and ST2 and ST1 is normal condition, while ST2 is supposed that the output voltages of one module PV21 of the string vary as shown in Fig. 5(a) due to a shade. We connect a current source to the strings in the current controlled MPPT simulation, while we connect a voltage source in the voltage-controlled MPPT simulation. Analysis results of flowing currents in the strings are shown in Fig. 5(b). In the voltage-controlled simulation in Fig. 5(b)-2, The currents flowing in the bypass diode removes to the PV module when a sunshine resumes, while in the current-controlld simulation in Fig. 5(b)-1, the currents keeps on flowing in the bypass diode after resuming a sunshine. After this, we modified the control algorithm of the LS-

PCSs from the current-controlled MPPT to the voltage-controlled MPPT in normal condition. However, when AC grid faults occur, the control algorithm is changed to the original current-control for protecting the PV system surely.

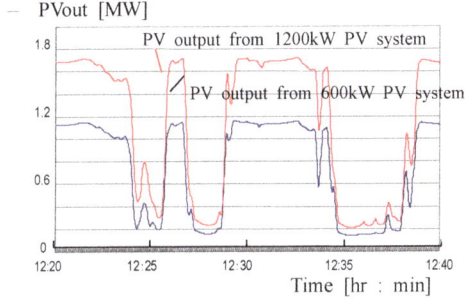

Fig. 6. PV output by voltage-controlled MPPT.

Fig. 6 shows the 600kW PCSs and 1,200kW LS-PCSs PV system outputs. Both are controlled by voltage-controlled MPPT. Zero PV outputs could not be seen in the 1200kW PV system outputs.

4. Grid stabilization test results

As for the grid stabilization, the voltage fluctuation suppression (VFS) control, the harmonics current suppression control and the fault-ride-through (FRT) function were installed in the developed LS-PCSs and tested.

Fig. 7 shows an example of test results with VFS control and without VFS control. In the figure, V, P and Q show distribution line voltages, PV generations and reactive powers of the LS-PCSs for suppressing voltage fluctuations, respectively. The voltage fluctuations due to the PV generation fluctuations were suppressed within 0.2% against 68% PV fluctuations. The VFS control satisfies the target value within 2%.

Fig. 8 shows measured harmonics. In the figure, In, Vn and n show the current harmonics against the fundamental current I_1, the voltage harmonics against the fundamental voltage V_1 and harmonics order, respectively. Where n= 1 shows a fundamental. The harmonics are small enough compared to the target of 80% of the guideline. The harmonics current suppression control was also confirmed.

However, the FRT function has not been confirmed yet because no fault occurs in the grid.

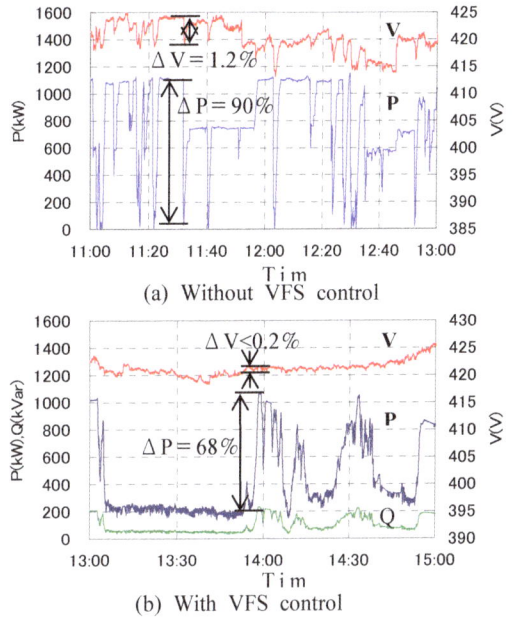

Fig. 7. Measured results of VFS control.

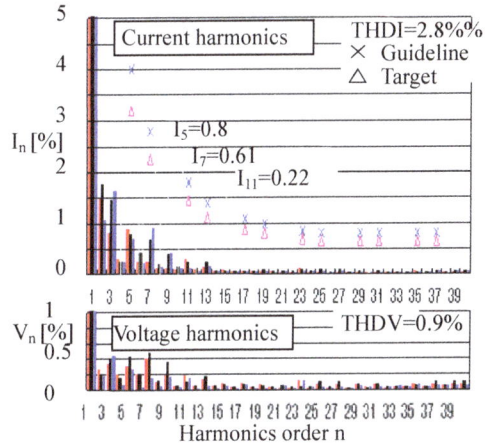

Fig. 8. Measured voltage and current harmonics.

5. PV evaluation test

In the PV evaluation tests, gain/loss evaluation and annual performance ratio (PR) for modules and systems have been measured and evaluated. Fig. 9 shows an example of the evaluation results of PRs, which are based on the power of the modules in STC measured by AIST, the power from the manufacture's test results and the power from the module specified power (nominal value). The generation power was measured in 2010. PRs are different by the based value. The tendency of different based value does not have regularity and is different by each PV technologies. It is necessary to standardize the definition of the PRs. PRs of installed modules in the site are high enough and were from 80% to 95%.

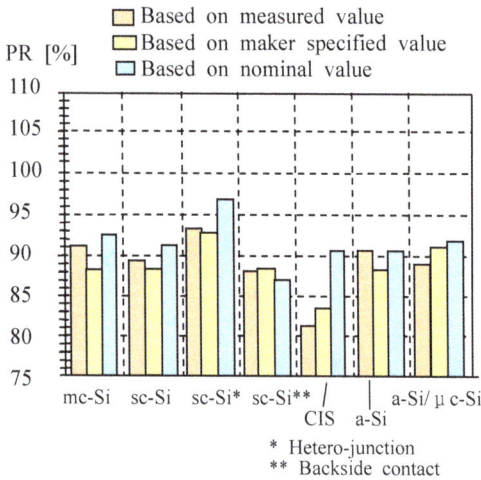

Fig. 9. Performance ratio calculated by different base.

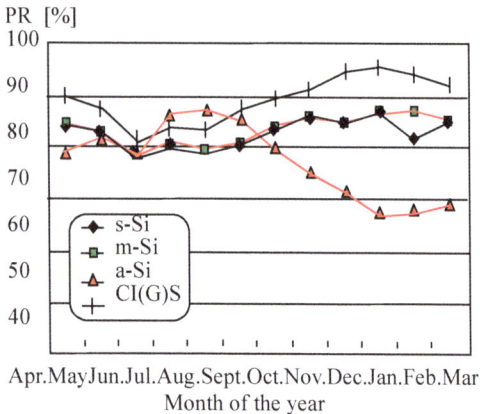

Fig. 10. Monthly performance ratio.

Fig. 10 shows monthly PRs in 2010. The PRs were based on the nominal value of the modules. The PRs of crystalline modules is low in summer and high in winter but that of amorphous type became opposite. The CI(G)S showed high PRs because of the light soaking effect. The annual degradation can be seen in amorphous type. The

degradation rate of the sc-Si was around 1.25%/yr in average and the thin film technologies showed larger degradation rate than the sc-Si.

6. Conclusions

We clarified a MPPT control method for LS-PCSs (large-scale power conditioning systems) to get more PV energy efficiently for Hokuto PV system. The results were confirmed by ATP (Alternative Transients Program) simulations and also in the PV system. The verification test results of grid stabilization systems and the evaluation of PV systems were satisfactory.

Although the verification tests were finished in March, 2011, we keep on collecting and evaluating the data.

References

[1] H.Konishi, R.Tanaka, T.Shiraki, "Introduction of Hokuto Mega-solar Project", PVSEC-**17**, 50-C6-5, Fukuoka (2007)

[2] H.Konishi, R.Tanaka, T.Shiraki, "Study on control and operation system of PCS in Hokuto Mega-solar project", **23**[rd] EU-PVSEC, No. 5BV2.16, Sept. 1-5, 2008, Valencia, Spain

[3] H.Konishi, R.Tanaka, T.Shiraki, "Operation and Moni-toring System of Hokuto Mega-solar Project", INTELEC 2008 (**30**[th] International Tele-communications Energy Conference), No.7.3, San Diego, CA, USA, Sept. 14-18, 2008

[4] H.Konishi, T.Iwato, M.Kudou, R.Tanaka, "Outlines and some results in the first stage of Hokuto mega-solar project", 24th EU-PVSEC, Sept. 21-25, 2009 Hamburg Germany

[5] H.Konishi, T.Iwato, and M.Kudou, "Development of large-scale power conditioner in Hokuto Mega-solar system", 25th EU-PVSEC/ WCPEC-5, No. 5AO7.2, Sept. 6-10, 2010, Valencia Spain

Permissions

All chapters in this book were first published in Sensors, by MDPI; hereby published with permission under the Creative Commons Attribution License or equivalent. Every chapter published in this book has been scrutinized by our experts. Their significance has been extensively debated. The topics covered herein carry significant findings which will fuel the growth of the discipline. They may even be implemented as practical applications or may be referred to as a beginning point for another development.

The contributors of this book come from diverse backgrounds, making this book a truly international effort. This book will bring forth new frontiers with its revolutionizing research information and detailed analysis of the nascent developments around the world.

We would like to thank all the contributing authors for lending their expertise to make the book truly unique. They have played a crucial role in the development of this book. Without their invaluable contributions this book wouldn't have been possible. They have made vital efforts to compile up to date information on the varied aspects of this subject to make this book a valuable addition to the collection of many professionals and students.

This book was conceptualized with the vision of imparting up-to-date information and advanced data in this field. To ensure the same, a matchless editorial board was set up. Every individual on the board went through rigorous rounds of assessment to prove their worth. After which they invested a large part of their time researching and compiling the most relevant data for our readers.

The editorial board has been involved in producing this book since its inception. They have spent rigorous hours researching and exploring the diverse topics which have resulted in the successful publishing of this book. They have passed on their knowledge of decades through this book. To expedite this challenging task, the publisher supported the team at every step. A small team of assistant editors was also appointed to further simplify the editing procedure and attain best results for the readers.

Apart from the editorial board, the designing team has also invested a significant amount of their time in understanding the subject and creating the most relevant covers. They scrutinized every image to scout for the most suitable representation of the subject and create an appropriate cover for the book.

The publishing team has been an ardent support to the editorial, designing and production team. Their endless efforts to recruit the best for this project, has resulted in the accomplishment of this book. They are a veteran in the field of academics and their pool of knowledge is as vast as their experience in printing. Their expertise and guidance has proved useful at every step. Their uncompromising quality standards have made this book an exceptional effort. Their encouragement from time to time has been an inspiration for everyone.

The publisher and the editorial board hope that this book will prove to be a valuable piece of knowledge for researchers, students, practitioners and scholars across the globe.

List of Contributors

Hui-Myoung Oh
Corresponding Author: Korea Electrotechnology Research Institute, Korea

Sungsoo Choi, Jimyung Kang and Won-Tae Lee
Korea Electrotechnology Research Institute, Korea

Taisuke Masuta
National Institute of Advanced Industrial Science and Technology, Tsukuba, Japan

Akinobu Murata
National Institute of Advanced Industrial Science and Technology, Tsukuba, Japan

Eiichi Endo
National Institute of Advanced Industrial Science and Technology, Tsukuba, Japan

Shunke Sui
Harbin Institute of Technology, China

Rongfeng Yang and Dianguo Xu
Harbin Institute of Technology, China

Toru Takahashi
National Institute of Advanced Industrial Science and Technology, Japan

Takayasu Fujino and Motoo Ishikawa
Dept. of Engineering Mechanics and Energy, University of Tsukuba, Japan

Baek-Ju Sung
Dept. of System Reliability, Korea Institute of Machinery & Materials, Korea

Jong-Hyun Kim
Hunter technology Co.,LDT, Korea

Soon-Ho Kim
Hunter technology Co.,LTD, Korea

Kyu-Hyoung Choi
Graduate School of Railroad, Seoul National University of Science and Technology, Korea

Zhao Yuan
Electric Power Research Institute, Jibei Electric Power Company Limited, Beijing 100045, China

Li Yu, Deng Chun and Yuan Yi-chao
Electric Power Research Institute, Jibei Electric Power Company Limited, Beijing 100045, China

He Jin-liang and Wang Xi
Department of Electrical Engineering, Tsinghua University, Beijing 10084, China

Christopher H. T. Lee
Dept. of Electrical and Electronic Engineering, The University of Hong Kong, Hong Kong SAR, China

K. T. Chau, Chunhua Liu and Fei Lin
Dept. of Electrical and Electronic Engineering, The University of Hong Kong, Hong Kong SAR, China

Kosuke Ito and Motoo Ishikawa
University of Tsukuba, Japan

Toru Takahashi
National Institute of Advanced Industrial Science and Technology, Japan

Takayasu Fujino
University of Tsukuba

Bai Dan-dan
Dept. of Electrical Engineering, Beijing Jiaotong University, China

He Jing-han, Tian Wen-qi, Wang Xiaojun and Tony Yip
Dept. of Electrical Engineering, Beijing Jiaotong University, China

Zhang Ji
Technology&Economic Research Institute of Hubei Electric Power Company, China

Shao Hanqlao, Peng Changyong and Du Zhi
Technology&Economic Research Institute of Hubei Electric Power Company, China

Deng Yong
Corresponding Author: Fujian Electric Power Dispatching and Controling Center, China

Mi Weimin and Xu Dandan
State Grid Electric Power Research Institute, China

Ren Xiaohui and Chen Zhengping
Fujian Electric Power Dispatching and Controling Center, China

Du Ning
Corresponding Author: China Electric Power Research Institute, China

Song Rui-Hua, Liu Chuan-Wen, Chen Zhen-Zhen, Xiang Zu-Tao and Ban Lian-Geng
China Electric Power Research Institute, China

Hyun-Chul Lee and Ki-Seok Jeong
Dept. of Electrical Engineering, Kyungpook National University, Korea

Ji-Ho Park
Dept. of Electrical Engineering, Koje Colleage, Korea

Young-Sik Baek
Corresponding Author: Dept. of Electrical Engineering, Kyungpook National University, Korea

Yang Xiao
Hebei Electric Power Research Institute, Shijiazhuang 050021, Hebei Province, China

Gao Zhiqiang and Fan Hui
Hebei Electric Power Research Institute, Shijiazhuang 050021, Hebei Province, China

Jialu Cheng
Corresponding Author: received his BS degree in electrical engineering in 2009 from the Southeast University, Nanjing, China

Diego robalino
Megger USA

Peter werelius and Matz ohlen
Megger Sweden

Ryohei Kawagishi
Dept. of Energy Science and Technology, Kyoto University, Japan

Daisuke Yamanaka
Dept. of Energy Science and Technology, Kyoto University, Japan

Yasuyuki Shirai
Dept. of Energy Science and Technology, Kyoto University, Japan

Seung-Hyun Sohn
Corresponding Author: College of Information and Communication Engineering, Sungkyunkwan University, Korea

Gyu-Jung Cho, Ji-Kyung Park, Yun-Sik Oh and Chul-Hwan Kim
College of Information and Communication Engineering, Sungkyunkwan University, Korea

Wan-Jong Kim, Hwa-Jin Oh and Junh-Jae Yang
Korea Power eXchange, KPX, Korea

Tomonobu Senjyu
Dept. of Electrical and Electronics Engineering, University of the Ryukyus, Japan

Toshihisa Funabashi
Meidensha Corporation, Japan

Chen Qiurong, Xu Gang, Liu Shu, Liu Zhichao and Shi Shan
Beijing Sifang automation co.,ltd

Shi Yu
Corresponding Author: Beijing Sifang automation co.,ltd, China

Jun-Min Cha
Dept. of Electrical and Electronic Engineering, Daejin University, Korea

Bon-Hui Ku
Dept. of Electrical and Electronic Engineering, Daejin University, Korea

Shao Yao
China Electric Power Research Institute, Haidian District, Beijing 100192, China

Tang Yong, Zhang Jian and Li Baiqing
China Electric Power Research Institute, Haidian District, Beijing 100192, China

Yang Xiu
Shanghai University of Electric Power, China

Li Cheng
Corresponding Author: Shanghai University of Electric Power, China

Liu Chunyan
Shenyang University of Chemical Technology, China

GuoJu Zhang
ABB (China) Corporate Research Center, Beijing, China

Yao Chen and Rongrong Yu
ABB (China) Corporate Research Center, Beijing, China

Lisa Qi and Jiuping Pan
ABB (US) Corporate Research Center, Raleigh, USA

Qiang Li
Graduate School of Information, Production and Systems, Waseda University, Japan

Jing Wang and Yasuaki Inoue
Graduate School of Information, Production and Systems, Waseda University, Japan

Dan Niu
Key Laboratory of Measurement and Control of CSE, Southeast University, Nanjing, China

Youngsun Kim
Corresponding Author: Power Telecommunication Research Center, Korea Electrotechnology Research Institute (KERI), Korea

Soon-Woo Rhee, Jae-Jo Lee and Sang Ki Oh
Power Telecommunication Research Center, Korea Electrotechnology Research Institute (KERI), Korea

Puming Li
Guangdong Power Dispatch Center of Guangdong Power Grid Co., China

Jianing Liu and Bo Li
Guangdong Power Dispatch Center of Guangdong Power Grid Co., China

Yuqian Song and Jin Zhong
Department of Electrical and Electronic Engineering, the University of Hong Kong, Hong Kong

Guanjun Ding
Beijing Information Technology Institute, Beijing, China

Bangkui Fan, Haibin Lan and Jing Wang
Beijing Information Technology Institute, Beijing, China

Teng Long
Beijing Institute of Technology, Beijing, China

Zhiyong Chen
National University of Defense Technology, Changsha, China

Kyeong-hee Cho and Seul-ki Kim
Smart Distribution Research Center, Korea Electrotechnology Research Institute, Korea

Eung-sang Kim
Smart Distribution Research Center, Korea Electrotechnology Research Institute, Korea

Kin-Wah Yeung
Corresponding Author: Technical Services Department, Transmission & Distribution Division, The Hongkong Electric Co. Ltd., Hong Kong

Jonathan
Technical Services Department, Transmission & Distribution Division, The Hongkong Electric Co. Ltd., Hong Kong

Ji-Seong Kang
Power Grid Protection Team, Korea Power Exchange, Korea

Young-Hyun Moon
Dept. of Electrical and Electronic Engineering, Yonsei University, Korea

Kousuke Kikuchi
Graduate School of Engineering, Mie University, Japan

Tomohiko Kanie
Kanie Professional Engineer Office, Japan

Takashi Takeo
Graduate School of Engineering, Mie University, Japan

Yuki Mitsukuri
Hakodate National College of Technology, Japan

Yuji Mishima
Hakodate National College of Technology, Japan

Ryoichi Hara and Hiroyuki Kita
Graduate School of Information Science and Technology, Hokkaido University, Japan

Keiichi Watanabe, Kenjiro Mori, Yasuhiro Kataoka and Eiji Kogure
The Tokyo Electric Power Company, Japan

Guo Weimin
Henan Electric Power Research Institute, Henan, China

Tang Yaohua
Henan Electric Power Research Institute, Henan, China

Ye Xiaohui, Zhong Wuzhi, Song Xinli and Liu Tao
Power System Department of China Electric Power Research Institute, Beijing, China
Su Yi

Fujian Electric Power Dispatch & Telecommunication Center, Fujian, China
Shoichi Koinuma
Dept. of Electrical Engineering and Bioscience, Waseda University, Japan

Yu Fujimoto
Inst. for Nanoscience & Nanotechnology, Waseda University, Japan

Yasuhiro Hayashi
Dept. of Electrical Engineering and Bioscience, Waseda University, Japan

Takao Shinji, Yosuke Watanabe and Masayuki Tadokoro
Tokyo Gas Co., Ltd., Japan

Sunghun Lee and Yeonchan Lee
Dept. of Electrical Engineering, Gyeongsang National Univeristy, Korea

Jaeseok Choi
Dept. of Electrical Engineering, Gyeongsang National Univeristy, Korea

Seunggu Han and Jinsu Kim
Korea Power Exchange, Korea

Zheng Chao
China Electric Power Research Institute, China

Sheng Canhui, Lin Junjie, Chen Dezhi and Zhang Zhiqiang
China Electric Power Research Institute, China

Zhang Kai
North China Electric Power University, China

Xue Jinying and Luo Bangyun
Xinjiang Hami Electric Power Bureau, Hami
839000, China

Hiroo Konishi
NTT Facilities Inc., Tokyo, Japan

Index